BACULOVIRUS EXPRESSION VECTORS

A Laboratory Manual

BACULOVIRUS EXPRESSION VECTORS

A Laboratory Manual

David R. O'Reilly
Imperial College

Lois K. Miller
University of Georgia

Verne A. Luckow
Monsanto Corporation

W. H. Freeman and Company
New York

Cover image by Jean Adams. Reprinted with permission from *Atlas of Invertebrate Viruses*. Copyright © CRC Press, Inc., Boca Raton, FL.

Library of Congress Cataloging-in-Publication Data

O'Reilly, David R.
Baculovirus Expression Vectors : a laboratory manual / David R. O'Reilly, Lois K. Miller, Verne A. Luckow.
p. cm.
Includes bibliographical references and index.

ISBN 0-7167-7017-2

1. Baculoviruses—Genetics. 2. Genetic vectors. 3. Gene expression.
I. Miller, Lois. II. Luckow, Verne. III. Title.
QR398.5.O74 1992
576′. 6484—dc20 92–9215
CIP

Printed in the United States of America

1 2 3 4 5 6 7 8 9 0 KP 9 9 8 7 6 5 4 3 2

Contents

Acknowledgments

Many people helped in a variety of ways during the preparation of this manual. David L. Clemm, a research scientist working with Ligand Pharmaceuticals, Inc., San Diego, California, wrote Chapter 16 and much of the introduction to Part IV. He also contributed the lipofection protocol and assisted with the PCR protocol in Chapter 13. Cheryl Isaac Murphy, a research scientist with Cambridge BioScience, Worcester, Massachusetts, wrote Chapter 17. Dr. Susumu Maeda (Department of Entomology, University of California at Davis) contributed the protocols for rearing and infection of *Bombyx mori* in Chapter 18. Credit for the electron micrographs used in this work goes to Jean Adams (USDA, Beltsville, Maryland; front cover, Figs. 1-1 and 1-2) and Greg Williams (Queen's University, London, Ontario; Fig. 1-4).

We would like to acknowledge the researchers who sent us information about their transfer vectors and who took the time to review the maps and sequence information we prepared. These workers are Robin Clark, John Kuzio, Susumu Maeda, Bob Possee, Chris Richardson, Just Vlak, Thierry Vernet, and Douwe Zuidema.

Many people read all or part of this manual during its preparation and provided many helpful comments and suggestions. We are particularly grateful to Robin Clark, David Clemm, and Paul Friesen, who reviewed the final version of the manuscript in its entirety. We would also like to thank Mark Birnbaum, Rollie Clem, Norman Crook, Russ Eldridge, Peter Faulkner, Glenn Godwin, Linda Hanley-Bowdoin, Don Jarvis, Susumu Maeda, David Murhammer, Cheryl Isaac Murphy, and William Whitford, all of whom read individual sections of the work. We also wish to acknowledge Ingrid Krohn for her encouragement and support.

Finally, we would like to express our appreciation to our families—Helen and Colm O'Reilly, Karl and Erin Espelie, and Ellen, Sara, and Evan Luckow—for their patience, support, and understanding throughout the preparation of this manual.

David R. O'Reilly, Lois K. Miller, and Verne A. Luckow

How to Use This Manual

This manual is designed to be a comprehensive guide to the generation, characterization, and use of a recombinant baculovirus vector overexpressing the heterologous gene or genes of your choice. It is divided into four parts: Part I, An Overview of Baculoviruses; Part II, Choosing a Transfer Plasmid and Parent Virus; Part III, Methods for Vector Construction and Gene Expression; and Part IV, Methods for Scale-up of Protein Production and Use of Insect Larvae.

Part I presents an intellectual framework for understanding how the baculovirus expression system works by providing basic information on the mechanism of the baculovirus infection process at the organismal, cellular, and molecular levels. The background information selected is particularly relevant to expression vector work. Part I concludes with a summary of the features that make baculoviruses particularly useful gene expression vectors and notes some of the features that the vector system cannot provide. To get a general understanding of the baculovirus system, you may want to read Part I. Alternatively, you may want to skim this part to find out what information is provided and then return to it as questions arise.

Part II provides information on how to design a baculovirus expression vector, what types of expression systems are available, and how to tailor your gene for optimal expression. You will want to read and consult this part if you are planning to construct your own expression vector or wish to optimize expression.

Part III and Part IV provide actual methods for baculovirus expression vector work. The protocols provided cover all stages in the generation and use of a recombinant baculovirus. However, methods are not provided for the insertion of a heterologous gene into the transfer plasmid, or for the preparation of transfer plasmid DNA. These are common molecular biology procedures and may be found in any good cloning manual such as Sambrook et al. (1989) or Berger and Kimmel (1987).

Part III includes methods for culturing insect cells, handling virus stocks, constructing and identifying a baculovirus expression vector, and characterizing the nature and level of gene expression. Part IV provides several approaches to the large-scale production of the expressed gene product and includes methods for the rearing and infection of insect larvae.

The appendixes provide additional information, which will be of use to all researchers using the baculovirus expression vector system. This information includes the nucleotide sequences of the AcMNPV *polyhedrin* and *p10* regions; a comprehensive, up-to-date list of genes that have been expressed using recombinant baculoviruses; the compositions of some commonly used stock solutions; the addresses of suppliers for the reagents mentioned in the manual; and a sample spreadsheet to facilitate the calculation of $TCID_{50}$ values.

I

AN OVERVIEW OF BACULOVIRUSES

This part of the book is divided into four chapters: Chapter 1 provides information on the nature and function of the budded and occluded forms of baculoviruses while focusing on the mechanism of virus infection at the organismal, cellular, and subcellular levels; Chapter 2 describes the organization of the *Autographa californica* nuclear polyhedrosis virus (AcMNPV) genome, the functions of known baculoviral genes, and the regulation of viral gene expression; Chapter 3 considers the effects of the virus on the host and host barriers to virus replication and gene expression; Chapter 4 summarizes the advantages of the baculovirus expression system based on the features described in Chapters 1–3.

The information presented is meant to provide an overview of the baculovirus infection process that will be useful in understanding how the expression system works. It introduces you to viral nomenclature and cites key references, which will allow you to find additional literature on this subject.

1

VIRUS STRUCTURE AND THE INFECTION PROCESS

BACULOVIRUS STRUCTURE

Virion Structure

Baculoviruses are a diverse group of viruses found mostly in insects. They are not known to have any nonarthropod hosts. The *baculo* portion of the name refers to the rod-shaped capsids of the virus particles. Baculovirus capsids are usually 40–50 nm in diameter and 200–400 nm in length (Harrap, 1972b). The length of the capsids can extend to accommodate larger DNA genomes such as those of recombinant viruses carrying large inserts (Fraser, 1986). The ends of the cylindrical capsids are structurally different (Fraser, 1986), giving a polarity to the rods.

Within the capsid, the DNA is condensed into a nucleoprotein structure known as the *core* (Tweeten et al., 1980). The proteins associated with this core structure include a predominant protaminelike protein (Tweeten et al., 1980; Wilson et al., 1987) known as p6.9 (also known as basic protein, the core protein, or VP12). The capsid plus the core are collectively referred to as the nucleocapsid. (See Table 1-1 for baculovirus structural nomenclature and commonly used abbreviations.)

The DNA genome of a baculovirus is double-stranded, covalently closed, and circular (Summers and Anderson, 1972). The length of baculoviral DNA is between 80–200 kilobasepairs (kbp) (Burgess, 1977). The DNAs of the two baculoviruses commonly used for expression vector work, *Autographa californica* nuclear polyhedrosis virus (AcMNPV) and *Bombyx mori* nuclear polyhedrosis virus (BmNPV), are both approximately 130 kbp.

Nucleocapsids are made in the nucleus of infected cells and are subsequently enveloped (i.e., they acquire a membrane) by one of two processes. Nucleocapsids can bud through the plasma membrane of the infected cell; such budded virus particles are released into the extracellular fluid with a loosely fitting membrane envelope. Figure 1-1 shows an electron micrograph of a budded baculovirus. The envelope near one end has surface projections or *peplomers*. Nucleocapsids may also acquire an envelope within the nucleus where they are produced. This membrane is often referred to in baculovirus literature as a "de novo" membrane (Stoltz et al., 1973). The origin of the intranuclear membrane segments that contribute to this process is not clear, but the membrane is a phospholipid bilayer membrane, which may be elaborated from the inner nuclear membrane. The envelope of virions embedded in occlusion bodies fits snugly around the nucleocapsids, and additional nuclear material may be compressed between the membrane and the nucleocapsids during the occlusion process.

TABLE 1-1 Baculovirus structural nomenclature

Budded virus (BV)	A form of virus released by budding from plasma membrane
Calyx	A carbohydrate-rich layer on the surface of OV (also known as the outer or polyhedron envelope)
Capsid	Rod-shaped protein shell surrounding the viral nucleoprotein core
Core	Viral DNA plus p6.9 (a basic protaminelike protein encoded by the *cor* gene) and other tightly associated proteins
Extracellular virus	Another name for budded virus (BV is preferred nomenclature)
Granulosis virus (GV)	A type of baculovirus forming small occlusion bodies comprising a single embedded virion
Multiply embedded NPV (MNPV)	An NPV having multiple nucleocapsids per envelope in the occluded form
Nonoccluded virus (NOV)	A term referring either to BV or to baculoviruses that lack an occluded form
Nuclear polyhedrosis virus (NPV)	A type of baculovirus forming large occlusion bodies with many embedded virions
Nucleocapsid	Virus capsid containing nucleoprotein core
Occluded virus (OV)	The occluded form of an NPV or GV
Polyhedral inclusion body (PIB)	The occluded form of an NPV; also known as a polyhedron (plural polyhedra) or occlusion body
Polyhedrin	The major matrix protein of polyhedra
Singly embedded NPV (SNPV)	An NPV having only one nucleocapsid per envelope in the occluded form
Virion	A virus particle consisting of at least one nucleocapsid and associated membrane envelope

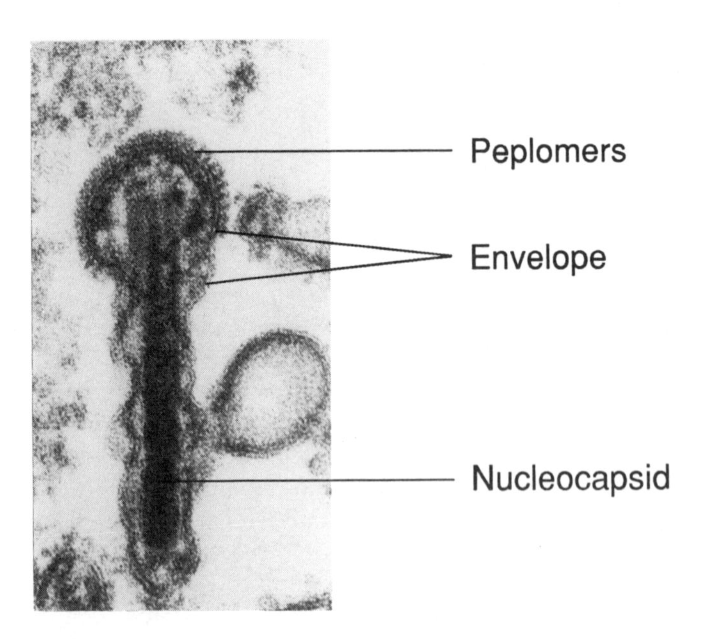

FIGURE 1-1 *Budded form of a nuclear polyhedrosis virus.* The rod-shaped nucleocapsid of the budded virus (BV) is surrounded by a baggy lipid bilayer membrane, which is referred to as the envelope. At one end of the nucleocapsid, the envelope contains peplomerlike structures that are thought to comprise the major virus-encoded glycoprotein, gp64. The electron micrograph shows the budded form of an MNPV of *Lymantria dispar* at a magnification of 159,999X. Reprinted with permission from Adams et al. (1977). (Courtesy of Dr. Jean R. Adams, USDA, Beltsville, MD.)

Membrane-enveloped nucleocapsids are referred to as virus particles or virions. The plasma membrane-budded form of the virus is referred to by different workers as *extracellular* virus (EV), nonoccluded virus (NOV) or budded virus (BV). The preferred terminology is BV. Virions that obtain their envelopes within the nucleus are usually also occluded within a crystalline protein matrix within the nucleus. Such virions may thus be referred to as *occluded virions,* but this term must be distinguished from the term *occluded virus,* which refers to the entire occluded virus particle including the crystalline protein matrix.

Occlusion Body Structure

Viral occlusion bodies are formed in the nucleus and comprise enveloped nucleocapsid(s) embedded in a crystalline protein matrix (Harrap, 1972a). Polyhedral occlusion bodies of nuclear polyhedrosis viruses (NPVs) are known as *polyhedra,* occluded viruses (OVs), or polyhedral inclusion bodies (PIBs). The protein making up the crystalline matrix of an NPV occlusion body is known as polyhedrin. [The original designation was N-polyhedrin to distinguish it from the matrix protein (C-polyhedrin) of cytoplasmic polyhedrosis viruses (CPVs), which are members of the virus family Reoviridae and form morphologically similar occlusion bodies in the cytoplasm of infected insect cells. The shortened version *polyhedrin* is usually employed when a distinction is unnecessary. The word *nuclear* in the NPV designation is valuable for distinguishing NPVs from CPVs.]

Figure 1-2 is an electron micrograph of a single occlusion body (also known as a polyhedron) of AcMNPV. AcMNPV is known as an "M" NPV since the nucleocapsids are enveloped in groups and embedded as such in the polyhedrin matrix. (The letter "M" refers to *multiply embedded* nucleocapsids in occlusion bodies; the budded forms of MNPVs, however, are usually enveloped individually.) BmNPV appears to be an SNPV (singly embedded). However, there has been some controversy concerning this feature—perhaps because different laboratories have worked with different BmNPV isolates in the past. The molecular basis for the difference between an MNPV and SNPV is not known nor is it known whether this distinction is taxonomically relevant.

Polyhedra also have an external surface coat which has been called the polyhedron outer envelope (a term that is confusing because the coat is not a membrane envelope). This outer coat is also referred to as a calyx—a term that we use. The calyx is rich in carbohydrates and is thiol-linked to the polyhedrin matrix by proteins (Whitt and Manning, 1988; Gombart et al., 1989; Zuidema et al., 1989). The calyx is thought to increase the stability of occlusion bodies.

Thus, two biochemically and morphologically distinct virus forms are produced during infection: BV and OV. Although it is thought that the nucleocapsids are identical in these two virus forms, the membranes clearly differ biochemically. The most distinctive difference observed to date is the presence of the viral-encoded glycoprotein, gp64, which is found in BV but not OV. During secondary infection gp64 is intimately involved in virus entry into cells via the process of adsorptive endocytosis (Volkman and Goldsmith, 1985). [Adsorptive endocytosis appears to be the major (but not exclusive) route of entry of BV into cultured cells. Enveloped virions liberated from occlusion bodies enter cells by a different route (Keddie and Volkman, 1985; Volkman, 1986). (See discussion of infection process that follows.)] Other differences include the presence of an O-glycosylated protein, gp41, and the protein p74 in OV, but not BV (J. Kuzio, G. Williams, and P. Faulkner, personal communication).

GVs and Nonoccluded Baculoviruses

In addition to nuclear polyhedrosis viruses (NPVs), there is a second type of occluded baculovirus, the granulosis viruses (GVs). In contrast to NPVs, which

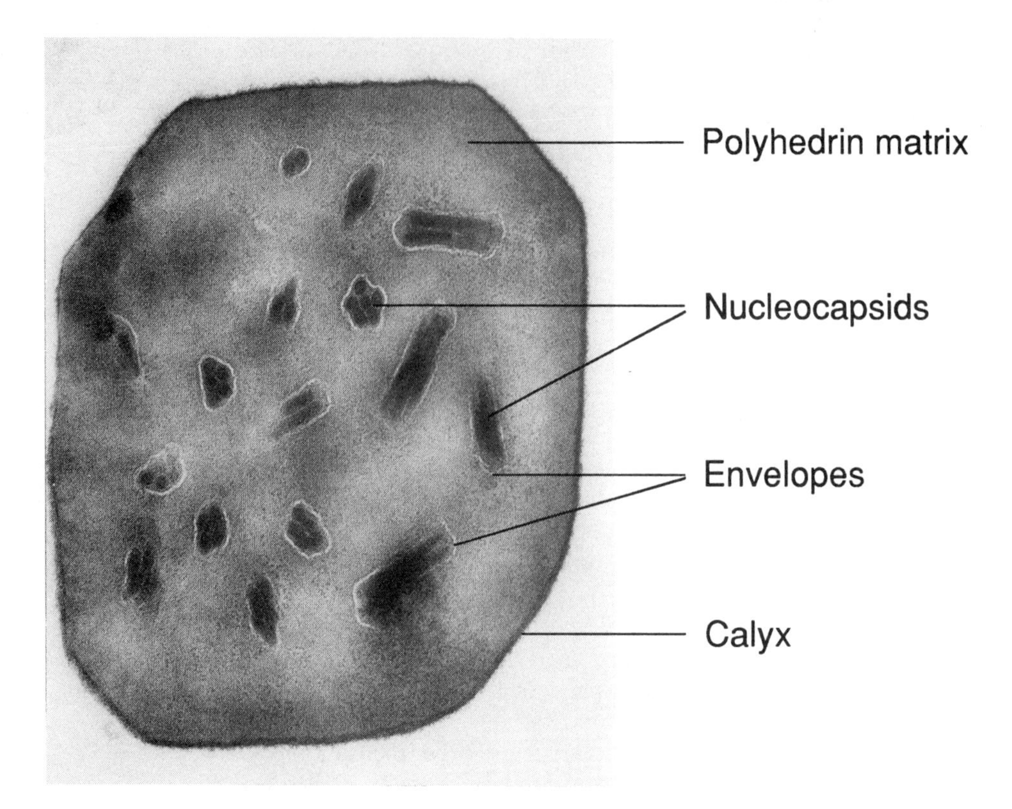

FIGURE 1-2 *The occluded form of AcMNPV.* A single occlusion body or polyhedron of AcMNPV contains numerous virions embedded in a crystalline protein matrix composed primarily of polyhedrin. Virions embedded within the matrix usually comprise more than one nucleocapsid within a common lipid bilayer membrane envelope, which is derived from intranuclear membranes. The paracrystalline structure is surrounded by a carbohydrate-rich layer known as the calyx or outer polyhedron membrane. [Electron micrograph (magnification 56,250X) courtesy of Dr. Jean R. Adams, USDA, Beltsville, MD. Reprinted with permission from *Atlas of Invertebrate Viruses.* © CRC Press, Inc., Boca Raton, FL.]

form large polyhedral occlusion bodies that may be up to 5 μm in diameter and comprise many virions, GVs have only a single virion embedded in a very small occlusion body. The presence of numerous GV occlusion bodies in the nucleus of infected cells results in a granular appearance in the light microscope. The matrix protein of a GV is known as granulin. It should be noted that some baculoviruses do not synthesize an occluded form; these viruses constitute the third basic type of baculovirus and are referred to as *nonoccluded* baculoviruses. There is some taxonomic controversy concerning whether the nonoccluded viruses should be classified as a subgroup of the genus baculovirus, although they clearly belong to the virus family Baculoviridae.

PROGRESS OF INFECTION IN THE INSECT

Distinct Primary and Secondary Phases

The natural cycle of infection by AcMNPV or BmNPV in insect larvae is summarized in Figure 1-3. Insect larvae ingest PIBs as contaminants of their food. The crystalline polyhedrin matrix is solubilized in the alkaline midgut of the insects, releasing embedded virions (Harrap and Longworth, 1974). The virions enter midgut cells by

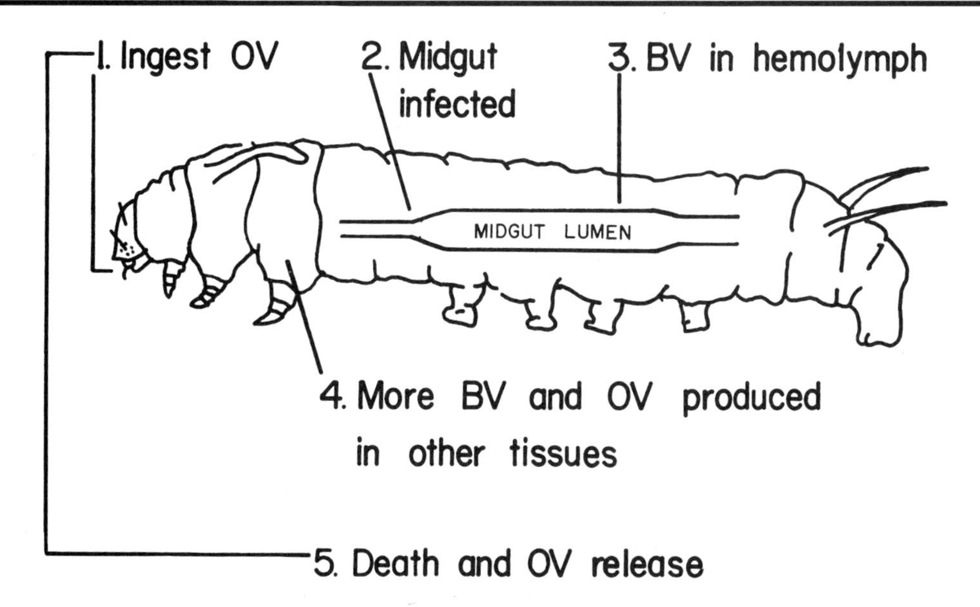

FIGURE 1-3 *Baculovirus infection of insect larvae.* Reprinted with permission from Miller (1989), © ICSU Press.

fusion with the membrane of the microvilli (Granados and Williams, 1986). Infection in the polarized midgut cells results in BV release from the basement membrane side of the cell (Keddie et al., 1989). This BV can gain access to the hemocoel and is transported via the hemolymph to other tissues in the insect (insects have an open circulatory system). BV released from the midgut also infects the epithelial cells of tracheoles, which provide oxygen to the midgut, spreading the infection along the tracheal network (Keddie et al., 1989).

Different NPVs exhibit different tissue tropisms during the secondary phase of infection (Tanada and Hess, 1984). Within larvae, AcMNPV infects a relatively broad spectrum of tissues including hemocytes, hypodermis, and fat body cells.

During a typical AcMNPV or BmNPV infection, the insect continues to feed during most of the infection process, which takes approximately five to seven days. The integument becomes swollen and changes in luster. Both BV and PIBs are produced in most of the tissues infected during the secondary phase. Infection of a late instar larva probably involves approximately 10 generations of virus. Eventually, the insect becomes lethargic and stops feeding. The cuticle melanizes, a polyphenol oxidase-mediated process that results in a discoloration (browning) of the cuticle. The musculature disintegrates, and the larva becomes a cuticular sac of milky fluid containing PIBs. Larval disintegration is sometimes described as *melting* or *wilting*. A substantial portion (e.g., ca. 25%) of the dry weight of the liquified carcass is polyhedra. The cuticle eventually ruptures releasing the PIBs into the environment. Polyhedra are relatively stable in the environment, although they exhibit significant sensitivity to UV light. They are naturally dispersed by a variety of routes (Evans, 1986) and may eventually be consumed by another permissive insect host, thereby reinitiating the infection cycle.

Infection Process and Development of Expression Vectors

The biology of the infection process in insect larvae underlies the utility of baculoviruses as expression vectors (Miller, 1981). The *polyhedrin* gene (*polh*) is nonessential for replication in cell culture (Smith et al., 1983a) yet is expressed at very high levels during a separate and final phase of infection (see the following section, Progress of Infection in Cell Culture). Thus, the basic design of the first baculovirus gene expression vectors was replacement of *polh* with the heterologous gene under *polh* promoter control (Miller et al., 1983b; Smith et al., 1983b; Pennock et al., 1984; Maeda et al., 1985). One advantage of substituting the heterologous gene in place of *polh* is that such recombinant viruses replicate normally as BV in cell culture and can be distinguished visually from wild-type (wt) virus by their occlusion-negative (occ^-) phenotype.

Such occ^- recombinant viruses are not efficient at infecting larvae by the natural oral route of infection and do not persist in the environment; these features have some benefits from environmental and recombinant DNA safety perspectives (Miller, 1981). However, the options for using occ^- viruses for heterologous protein production in larvae are more limited. Transfer plasmids for the construction of occlusion-positive (occ^+) vectors have been developed recently (see Chapters 5, 6, and 7).

PROGRESS OF INFECTION IN CELL CULTURE

For replication in cell culture, the infection cycle can be considered to occur in three basic phases: early, late, and very late. Briefly, these three phases correspond biologically to: (1) reprogramming the cell for virus replication, (2) producing BV, and (3) producing OV.

The kinetics of the infection process and the molecular events occurring during infection have been characterized using AcMNPV and the *Spodoptera frugiperda* (fall armyworm) cell line, IPLB-SF-21, as the model baculovirus-insect cell system. The description of events below will therefore refer to observations in this system. Other baculoviruses, such as BmNPV, may differ, particularly in the kinetics of the infection process.

Early Phase

Infection in cell culture, as in insect hemolymph, is mediated by the budded form of the virus entering by adsorptive endocytosis (Volkman and Goldsmith, 1985; Volkman, 1986). Nucleocapsids migrate through the cytoplasm to the nucleus where they interact end-on with nuclear pores, thereby gaining entry to the nucleoplasm (Granados and Williams, 1986). In the nucleus, the core appears to be released from the "capped" end of the capsid structure. The release of the core may be enabled by a capsid-associated protein kinase, which can phosphorylate the basic p6.9 core protein *in vitro* (Wilson and Consigli, 1985). It is not clear whether p6.9 is released from the DNA following capsid uncoating; some evidence suggests that the viral DNA adopts a nucleosomal structure early in infection (Wilson and Miller, 1986). Viral RNA can be detected 30 minutes after inoculation indicating that arrival of viral DNA in the nucleus and initiation of early transcription are rapid processes (Chisholm and Henner, 1988).

Infected cells undergo significant changes during the first 6 hours of infection, a time period that constitutes the early phase of infection and precedes viral DNA replication. Cytoskeletal rearrangements occur, and the host chromatin disperses within the nucleus, which enlarges during this period. Some of these cytoplasmic and nuclear changes may be brought about by components of the viral inoculum (i.e., proteins in the infecting virions), while others are controlled by newly produced proteins from early viral genes [see, for example, Charlton and Volkman (1991) for cytoskeletal changes]. (The cascade of viral gene expression is described in more detail in Chapter 2.)

Late Phase

The early phase is followed by the late phase, a period of extensive viral DNA replication, late gene expression, and BV production. The late phase extends from

6 hours postinfection (h pi) to approximately 20 to 24 h pi. Logarithmic production of BV occurs from approximately 12 to 20 h pi, decreasing in rate thereafter (Knudson and Harrap, 1976; Lee and Miller, 1979). During this phase, a distinct electron-dense structure known as the virogenic stroma forms in the nucleus (Harrap, 1972c; Kelly, 1981) (see Fig. 1-4A). Capsid sheaths appear to assemble in pockets at the edge of the virogenic stroma and are filled, while associated end-on with the stroma, with a nucleoprotein core to form nucleocapsids (Bassemir et al., 1983; Fraser, 1986). During the late phase, progeny nucleocapsids leave the nucleus, possibly by a variety of routes, and travel through the cytoplasm (Raghow and Grace, 1974; Bassemir et al., 1983) to the cytoplasmic membrane where they interact end-on with regions of this membrane having peplomerlike structures. Nucleocapsids usually bud individually from the cytoplasmic membrane although occasionally more than one nucleocapsid is observed in a BV.

Very Late Phase

The very late phase or occlusion-specific phase begins around 20 h pi. Electron microscopy reveals the elaboration of membrane envelope segments within the nucleus during this occlusion process (Harrap, 1972c; Chung et al., 1980) (see Fig. 1-4A). The origin or mechanism of synthesis of these *de novo* membranes remains unclear. Nucleocapsids interact end-on with these membranes and eventually become enveloped, either individually (SNPVs) or in groups (MNPVs) (Bassemir et al., 1983; Fraser, 1986). Envelopment of the nucleocapsids appears to be an essential prelude to embodiment within the polyhedrin matrix of an occlusion body (see Fig. 1-4B). During the occlusion phase, the production of infectious BV is greatly reduced, if not terminated (Lee and Miller, 1979).

The first light microscopic indication of this occlusion phase is the formation of several PIBs in the nucleus. As they accumulate, the PIBs often form a ring around the inside of the nuclear membrane. (The region between the virogenic stroma and the inner nuclear membrane is known as the *ring zone*.) Eventually, the nucleus becomes virtually filled with occlusion bodies. In a typical wt AcMNPV infection of healthy and fully permissive cells, an average of approximately 70 PIBs accumulate per nucleus. The crystalline nature of the polyhedra gives a highly refractive appearance to groups of infected cells or plaques.

Accumulation of p10 in Fibrous Material

As occlusion proceeds in the nucleus, large arrays of fibrous material (fibrillar structures) begin to accumulate, primarily in the nucleus but sometimes in the cytoplasm as well (see Fig. 1-4B). The bulk of this fibrous material is a single ca. 10 kDa polypeptide known as p10 (Van Der Wilk et al., 1987). Disruption of the p10 gene results in the disappearance of the fibrous material, but the formation of polyhedra containing normal enveloped virions is not disrupted (Williams et al., 1989). The function of fibrous material remains unclear, but it has been suggested to play a role in the lysis of the host cells (Williams et al., 1989) and may be important in controlled cellular disintegration in larvae. Infected cells in culture will eventually lyse. The lysis process does not appear to be an active one, and the time of lysis appears to depend on the health of the cells and the freshness of the media and/or oxygen levels. Without changing media or oxygenating cells, lysis usually begins about 60 h pi, and by 72 h pi, most cells are no longer synthesizing proteins and are in the process of dying and/or lysing. The occlusion period thus extends for approximately two days, during which time the major occlusion phase proteins, primarily polyhedrin and p10, are synthesized.

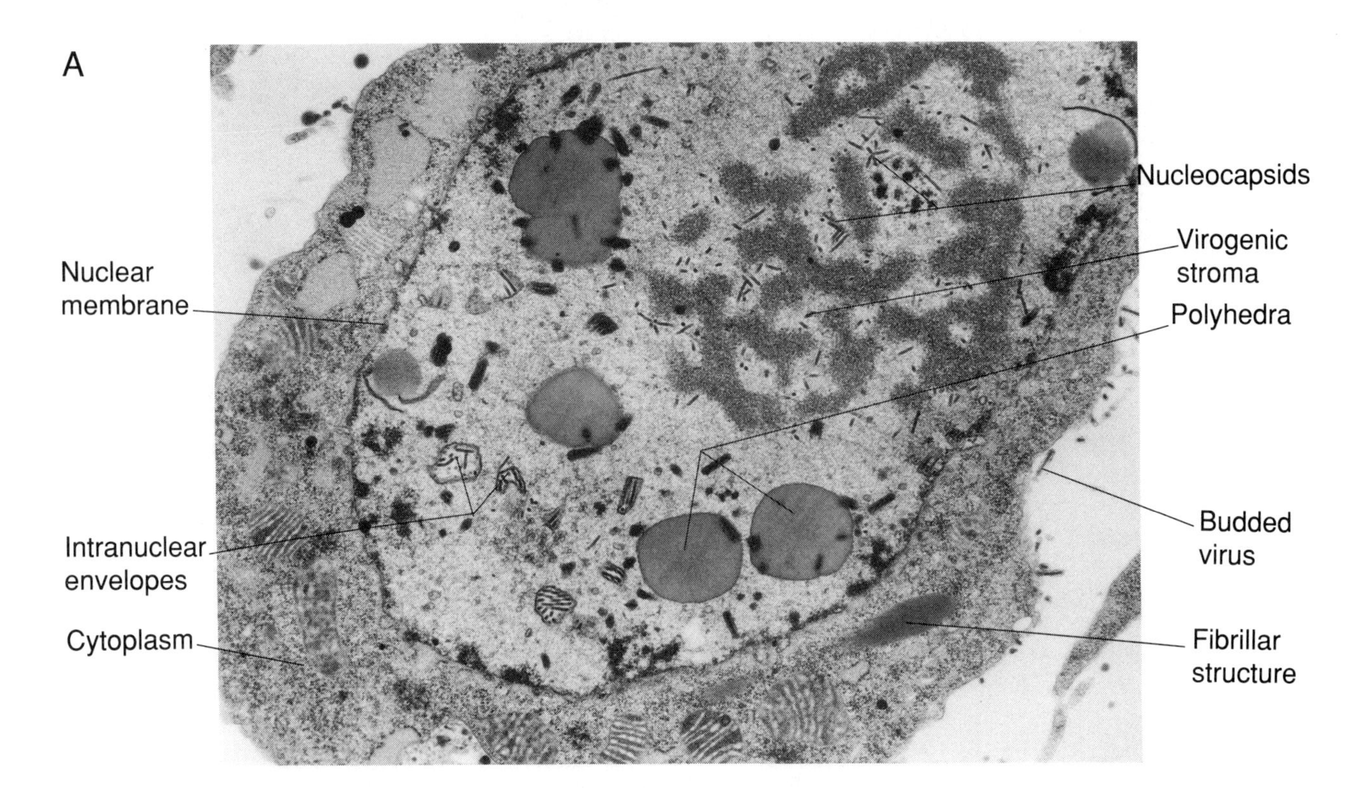

FIGURE 1-4 *Very late stages in AcMNPV infection of S. frugiperda cells.* **A**. Nucleocapsid assembly occurs in association with the virogenic stroma. During the very late phase, the nucleocapsids are enveloped within an intranuclear envelope, and the resulting virions eventually become embedded within polyhedra. A small amount of fibrillar structures may be seen in the cytoplasm. This micrograph appears to capture events occurring in the transition between late and very late phases, as the polyhedra appear to be immature. Budded virus can be seen in the extracellular space.

Effect of Serial Passage of Virus

One important ramification of the lack of infectivity of OV in cell culture is that, during growth in cell culture, there is no selective pressure to maintain viruses that produce OV, whereas there is selective pressure to maintain viruses that produce BV. Upon prolonged serial passage of BV in cell culture, including passage at low multiplicities of infection (MOIs), the baculovirus stock will eventually degenerate into viruses that are defective in PIB formation. The most frequently observed viruses are those that form only a few (less than ten, usually an average of only two or three) PIBs per cell. These viruses are therefore known as *FP* (few polyhedra) mutants. FP mutant plaques may be mistaken as occ$^-$ plaques during screens for recombinant viruses.

Studies of FP mutants at the electron microscopic level indicate that at least some of these mutants are defective in the envelopment of the nucleocapsids in the nucleus during the occlusion phase [for a review, see Miller (1986)]. Also, some FPs are known to produce a higher titer of BV (Potter et al., 1976; 1978). Higher BV titers may or may not be related to the block in nuclear envelopment, but elevated BV levels do explain the selective advantage of FPs for growth in cell culture. The nature of FP mutations are discussed further in Chapter 2.

It may be noted that serial passage at low dilutions may have other effects on the virus. Mutants with deletions in the region of the viral *egt* gene, (O'Reilly and Miller, 1989) arise spontaneously and predominate in serially passaged virus stocks

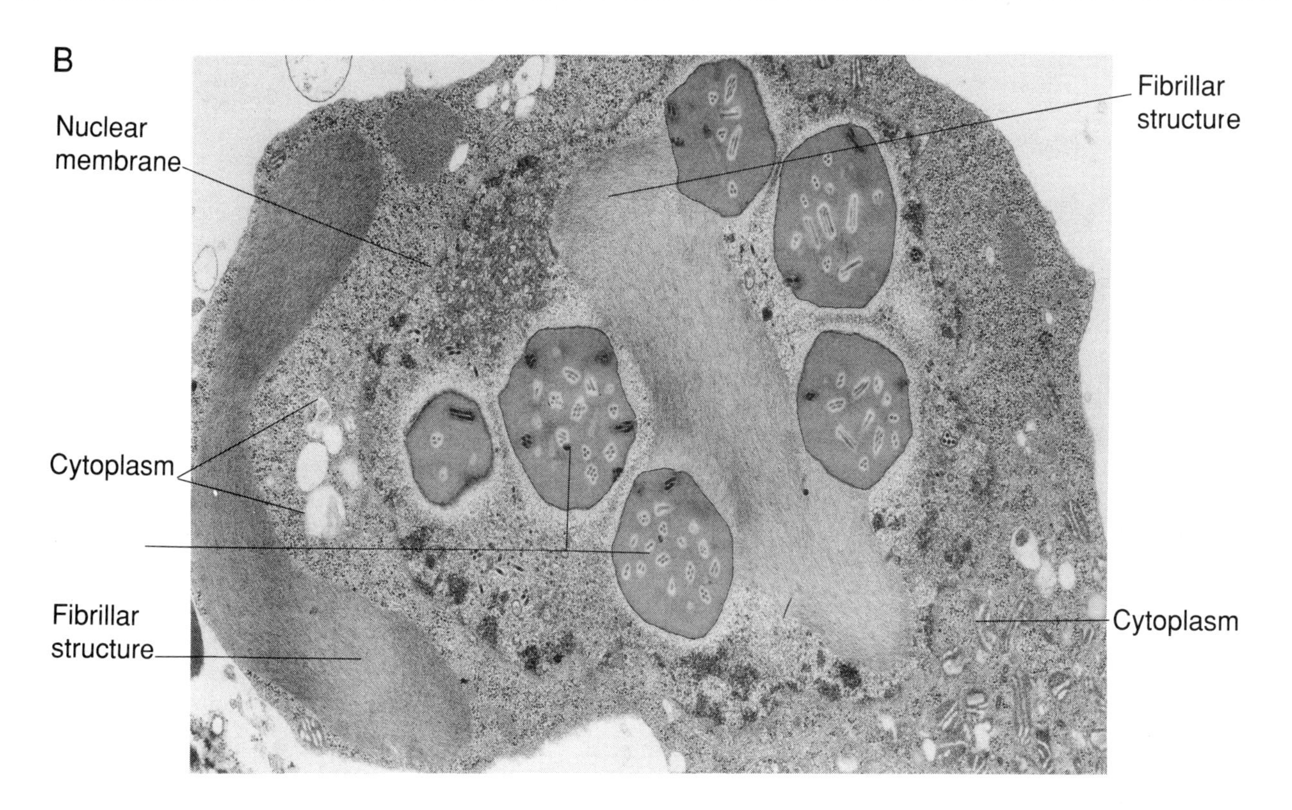

B. Mature polyhedra of AcMNPV contain numerous enveloped virions and most virions contain multiple nucleocapsids. Also visible are fibrillar structures in both the nucleus and cytoplasm. The nuclear membrane remains intact during the infection. [Electron micrographs (magnification: **A.** 15,800X; **B.** 17,556X) kindly provided by Greg Williams, Queens University, Kingston, Ont.]

(Kumar and Miller, 1987). The *egt* gene encodes a UDP-glucosyltransferase, which is specific for ecdysteroids (insect-molting hormones) and allows the virus to block the molting of its insect host. It is not clear why such mutants predominate in serially passaged stocks.

Serial passaging of undiluted virus stocks (i.e., high MOIs) results in the accumulation of defective interfering particles (Kool et al., 1991). The presence of significant proportions of these particles in a virus stock is detected in restriction fragment profiles of DNA prepared from virus particles. The defective interfering particles lack approximately 43% of the genome from 1.7 to 45 map units (mu) on the AcMNPV genome (see Fig. 2-1) so that restriction fragments from this region are present in submolar quantities. These particles are not infectious but interfere with wt virus replication (Kool et al., 1991). Because the region deleted from the genomes of these particles includes *polh,* significant declines in the level of heterologous gene expression occur if care is not taken in the passaging of virus stocks. Low MOI (dilute) passage of viruses can minimize the appearance of defective interfering particles in virus stocks. However, extended passage of viruses in cell culture, even with low dilution passage, results in FP and *egt* mutations (Kumar and Miller, 1987).

2

GENE ORGANIZATION, REGULATION, AND FUNCTION

THE PHYSICAL MAP AND hr REGIONS

The physical map of the 128 kbp genome of *Autographa californica* nuclear polyhedrosis virus (AcMNPV) is presented in Figure 2-1. Since the genome has not been totally sequenced, the map units provided for restriction sites should be considered approximate. The origin, orientation, and fragment designations of this map were derived by consensus and are accepted by most baculovirus laboratories [see Vlak and Smith (1982)]. Numerous genotypic variants of AcMNPV are used, and most differ only slightly from each other (see Chapter 5).

hr *Regions*

One notable feature of the physical map is the presence of numerous EcoRI sites at five locations in the genome. These regions contain two to eight reiterations of an imperfect palindromic sequence containing an EcoRI site at the center of each palindrome. The regions are homologous to each other and are therefore known as homologous regions or *hr* regions. When originally identified (Cochran and Faulkner, 1983), only five *hrs* were distinguished and were numbered *hr*1 to *hr*5. Subsequent analysis and sequencing of each region (Guarino et al., 1986) revealed that EcoRI-L was flanked by two and four reiterations of the homologous palindrome; these two regions are now referred to as *hr*4L and *hr*4R (left and right respectively). The *hr* regions act as enhancers for some early promoters in transient expression assays (Guarino and Summers, 1986b; Guarino and Summers, 1987; Nissen and Friesen, 1989). It is possible that *hr* sequences also play a role in viral DNA replication although no direct evidence has established such a function. At the current time, the origin(s) of DNA replication remain unknown.

THE DISTRIBUTION OF GENES

Figure 2-1 also shows the arrangement of functionally characterized genes; the direction of each open-reading frame (ORF) is shown by arrows. Available information concerning the biological function of characterized genes is summarized in Table 2-1. Additional ORFs have been sequenced but further information concerning their function is not currently available. Regions of the AcMNPV genome that have been sequenced to date are denoted by the bars at the bottom of Figure 2-1.

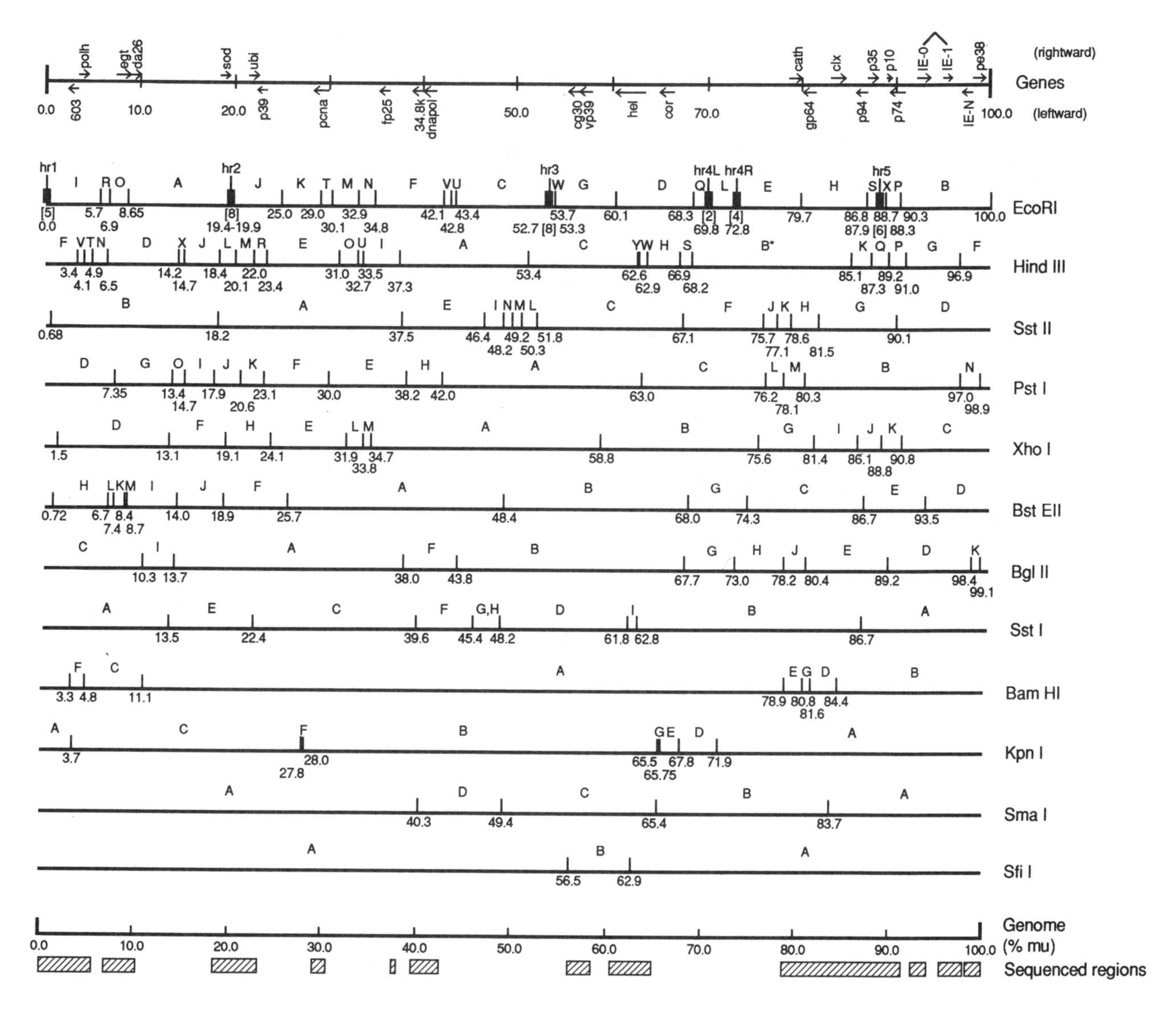

FIGURE 2-1 *Restriction map of AcMNPV genome.* This map shows the positions of the cleavage sites of a variety of different restriction endonucleases on a linear map of the AcMNPV (E2) genome. The positions of the cleavage sites are given in % map units (mu). A scale in mu is presented at the bottom of the figure. The entire genome is 128 kbp so that 1 mu = 1.28 kbp. The approximate locations of regions of the genome that have been sequenced are indicated at the bottom, while the locations and directions of transcription of characterized genes are presented at the top. The HindIII fragments are labeled as they were in Vlak and Smith (1982). However, it seems likely that HindIII-B (marked with an asterisk in the figure) is in fact larger than HindIII-A. In AcMNPV L-1 and C-6, an extra HindIII site occurs within HindIII-B, dividing it into fragments B1 and B2. Cochran et al. (1982) report the existence of a small XhoI fragment of ≈350 bp (XhoI-N). However, its location has not been clearly established, and it is not marked on this map.

Spacing of ORFs

ORFs are usually very closely spaced along the genome; regulatory regions seem to be kept to a minimal length. Of the genes characterized with respect to sequence and protein size thus far, the shortest spacing is 2 bp (between the *vp39* and *cg30* ORFs), and the largest spacing is approximately 350 bp (between *IE-N* and *pe38*),

TABLE 2-1 Characterized Genes of AcMNPV: Location, Transcriptional Regulation, and Function

Location	*Class*	*Designation*	*Function and Key References*
2.6–3.1	L	*603*	Nonessential (Gearing and Possee, 1990)
3.2–3.8	VL	*polh*	Polyhedrin, OV matrix protein, (Smith et al., 1983b; Matsuura et al., 1987)
8.4–9.7	E	*egt*	Ecdysteroid UDP-glucosyltransferase, blocks molting of larval host (O'Reilly and Miller, 1989)
9.7–10.2	E	*da26*	Nonessential (O'Reilly et al., 1990)
19.0–19.4	L or VL	*sod*	Cu/Zn superoxide dismutase homolog, nonessential (Tomalski et al., 1991)
22.0–22.2	L	*ubi*	Ubiquitin homolog, essential gene (Guarino, 1990)
22.2–22.8	E & L	*p39*	Cofractionates with nuclear matrix (Guarino and Smith, 1990)
29.7–30.4	E	*pcna*	Proliferating-cell nuclear-antigen homolog (*etl*), stimulates DNA replication and late gene expression (Crawford and Miller, 1988; O'Reilly et al., 1989)
36.0–36.8	L or VL	*fp25*	Hypervariable region, mutations result in FP phenotype and are selected by serial passage (Fraser et al., 1983; Beames and Summers, 1989)
39.0–39.9	L or VL	*p34.8*	Related to an entomopoxvirus gene (Wu and Miller, 1989; Yuen et al., 1990)
40.0–42.3	E	*dnapol*	DNA polymerase homolog (Tomalski et al., 1988)
56.1–56.9	E	*cg30*	Contains zinc fingerlike motif and leucine zipper (Thiem and Miller, 1989b)
56.9–58.0	L	*p39*	Major capsid protein (Thiem and Miller, 1989a)
60.3–63.4	E	*hel*	Helicase homolog, required for DNA replication (ts8) (Lu and Carstens, 1991)
65.2–65.5	L	*cor*	Protaminelike basic 6.9K protein (p6.9) associated with DNA in core (Wilson et al., 1987)
79.7–80.4	?	*v-cath*	Homolog of cysteine proteases, unknown function (Kuzio and Faulkner, personal communication)
80.4–81.8	E & L	*gp64*	Major envelope glycoprotein of BV involved in virus entry (Whitford et al., 1989)
82.6–83.4	L or VL	*clx*	Associated with the calyx of OV (Oellig et al., 1987; Gombart et al., 1989; Zuidema et al., 1989)
85.1–87.1	E	*p94*	Nonessential (Miller and Miller, 1982; Friesen and Miller, 1987)
87.2–87.9	E	*p35*	Blocks apoptosis (Friesen and Miller, 1987; Clem et al., 1991)
88.9–89.4	VL	*p10*	Nonessential, may affect lysis (Kuzio et al., 1984; Williams et al., 1989)
89.4–90.0	L	*p74*	Essential for OV infectivity (Kuzio et al., 1989)
ca. 92.5	E & L	*IE-0*	First exon of *IE-1* (Chisholm and Henner, 1988)
95.6–96.9	E & L	*IE-1*	*IE-1* main exon, transregulator of early promoters in transient expression assays (Guarino and Summers, 1986a; Guarino and Summers, 1987)
97.7–98.7	E	*IE-N*	Zinc fingerlike and leucine zipperlike motifs, modulates self and *IE-1* expression (Carson et al., 1991a)
99.0–99.8	E	*pe38*	Zinc fingerlike motif, leucine zipper, highly homologous to CG-30 (Krappa and Knebel-Mörsdorf, 1991)

E, L, and VL indicate that a gene is transcribed in the early, late, or very late phases of infection respectively.

with the exception of *hr* regions, the largest of which is approximately 800 bp. Some less well-characterized but sequenced ORFs are known to overlap [e.g., *da18* and *da26* (O'Reilly et al., 1990)]. It would not be surprising if some baculovirus DNA sequences encode more than one protein.

The compact nature of baculovirus genomes is exemplified by the observation that the UAA translational termination signal of an ORF often overlaps the AA<u>UAA</u>A polyadenylation signal of the transcript. Similarly, promoter elements of

ORFs may be found in the coding sequence of an upstream ORF (e.g., *vp39* and *cg30* or *p26* and *p10*). The sequences between many ORFs are extremely A+T rich, which may be related to promoter and transcriptional termination functions.

Functional Organization of ORFs

Thus far, there seems to be little or no clustering of functionally or temporally related genes. Some clustering is observed, but it is not yet clear if it is functionally significant. For example, three genes (*IE-1, IE-N,* and *pe38*), which have been implicated as having regulatory roles in gene expression, are located in the 95–100 mu region. However, another gene (*cg30*) with homology to *IE-N* and *pe38* is located immediately downstream of a virus structural protein at 72 mu. Virus structural genes appear to be dispersed on the genome (e.g., *gp64, cor, polh,* and *vp39*). Three genes (*clx, p10,* and *polh*) have been functionally correlated with the occlusion phase and are located in a quadrant of the genome encompassing the 80–100 and 1–5 mu region. But early genes encoding regulatory proteins as well as other genes of diverse function are interspersed among these occlusion phase genes. Furthermore, a fourth gene, *fp25,* also appears to be involved only in occlusion and is located around 36 mu.

ORFs are found along both strands of the DNA, although usually only one of the two strands in a specific location encodes a gene product. There is no obvious rationale for gene direction although the direction of each one may be important relative to flanking genes (see discussion on mechanisms of gene regulation later in this chapter).

There also appears to be no clustering of genes belonging to the same transcriptional class. Indeed, genes transcribed at late times are often interspersed with genes transcribed at early or very late times. It has been proposed that this organization may have a regulatory role in gene expression (Friesen and Miller, 1986). There is now evidence that the interspersion of different temporal transcriptional units plays a gene regulatory function, probably via an antisense RNA mechanism (Ooi and Miller, 1990) (see discussion on mechanisms of gene regulation later in this chapter). Transcriptional mapping of different regions of the AcMNPV genome has revealed a complex pattern of overlapping transcripts in many of the regions characterized, which suggests that baculovirus gene organization may be intimately associated with subtle gene regulatory mechanisms. [For an early review, see Friesen and Miller (1986)].

NONESSENTIAL GENES

Some genes of AcMNPV are essential for the replication of the virus, but many other genes appear to be *nonessential* as judged by the ability of viruses with null mutations in the given gene to replicate in both cell culture and in insects. It is likely that the nonessential genes provide some advantage for growth or survival of the virus under specific conditions such as an alternate host species, a particular tissue and/or cell type, or survival under harsh field conditions.

Genes or characterized ORFs that are known by genetic disruption studies to be nonessential for AcMNPV growth in both cell culture and in insects include: the 603 ORF (Gearing and Possee, 1990), *egt* (O'Reilly and Miller, 1989), the *da26* ORF (O'Reilly et al., 1990), *sod* [encoding a superoxide dismutase homolog (Tomalski et al., 1991)], *pcna,* formerly known as *etl* [encoding a proliferating-cell nuclear antigen homolog (Crawford and Miller, 1988; O'Reilly et al., 1989)], *clx* [encoding a protein associated with the polyhedron outer envelope or calyx (Zuidema et al., 1989)], and *p10* [encoding the p10 protein that constitutes the fibrillar structures (Kuzio et al., 1984; Williams et al., 1989)]. In addition, a study of the 94K ORF (Friesen and Miller,

1987) was originally pursued due to a retrotransposon insertion in this ORF (Miller and Miller, 1982); the retroelement has little or no effect on BV replication or OV production (Miller and Miller, unpublished observations), but bioassays of PIBs in larvae need to be performed to assess any possible *in vivo* role. Thus seven or eight of the 25 AcMNPV genes that have been functionally characterized are nonessential according to existing assays for gene function.

Functions of Some Nonessential Genes

At least four of these eight nonessential genes (*egt, pcna, clx,* and *p10*), however, exhibit an observable phenotype. The product of *egt* is secreted from the cell and transfers the sugar moiety from a UDP-sugar to ecdysone, the hormone governing insect molting (i.e., 20-hydroxyecdysone). This allows the virus to block the molting of its insect host during infection (O'Reilly and Miller, 1989).

The *pcna* gene is expressed early in infection and stimulates DNA replication as well as late gene expression in cell culture (Crawford and Miller, 1988; O'Reilly et al., 1989). Although *pcna* is not essential for virus growth in the cell lines or larvae tested, it may be essential for replication in some alternate hosts or in some specific tissue types.

The *clx* gene product, also known as the polyhedron envelope-associated protein, is involved in securing an outer carbohydrate layer onto the surface of polyhedra. It is thought that the calyx provides additional stability to PIBs outside the insect (Whitt and Manning, 1988; Gombart et al., 1989; Russell and Rohrmann, 1990).

The *p10* protein is produced very abundantly at very late times pi and forms vast arrays of fibrouslike material in the nucleus of infected cells. Deletion of *p10* is reported to prevent insect cell lysis at very late times pi (e.g., 72 h pi) (Williams et al., 1989). A role for *p10* in the melting or disintegration of the larval host has also been proposed, but the function(s) of this abundant protein remain obscure.

Based on transient expression assays, the *da26* ORF was proposed to play a role in late gene expression (Guarino and Summers, 1988), but genetic disruption of the ORF results in no observed phenotypic difference in cell culture or in insects (O'Reilly et al., 1990). Thus the function of this ORF remains unclear.

OCCLUSION-RELATED GENES

Genes that are known to be involved specifically in OV formation are *polh,* the *fp25* gene, and *clx.* Without *polh,* there is no OV formation [the occ$^-$ phenotype (Smith et al., 1983a)]. The *clx* gene is nonessential (see preceding discussion) for the formation and infectivity of PIBs but plays a subtle role in PIB stability. Disruption of *fp25* gives rise to the FP phenotype (Fraser et al., 1983; Beames and Summers, 1989). FP mutants arise spontaneously in cell culture or in insects, usually by the insertion of an insect transposable element into the *fp25* locus (Fraser et al., 1983; Kumar and Miller, 1987). In cell culture, FP mutants are selected by extended serial passage of the virus at low MOIs because they produce more progeny BV per cycle than wt virus [reviewed by Miller (1986)]. FP mutants also arise in stocks of virus that are serially passaged in insect larvae by injection of BV into the hemocoel (Potter et al., 1978). If virus is passaged through larvae by oral feeding of PIBs, FP mutants arise but do not predominate in the stock.

The FP defect may be due to a dysfunction of the nuclear envelopment process, which prevents the nucleation of the polyhedrin matrix [reviewed by Miller (1986)]. Levels of polyhedrin produced in FP mutant-infected cells appear to be normal or only slightly reduced (Miller, unpublished observations). Conditions favoring the formation of FP mutant build-up (i.e., extended serial passage of BV) should be avoided in expression vector work so that FP mutants do not obscure recombinant virus selection. (See Chapter 12 for guidelines on the propagation and maintenance of virus stocks.)

Another OV-related gene, *p74*, appears to be associated with the envelope of occluded virions (J. Kuzio, G. Williams, and P. Faulkner, personal communication) and plays an essential role in the infectivity of OV in the midgut, the primary site of virus infection in larvae (Kuzio et al., 1989). Mutants disrupted in *p74* replicate in cell culture, but PIBs of *p74* mutants fail to infect larvae when administered by feeding.

GENES ENCODING VIRION STRUCTURAL PROTEINS

Genes known to encode virion structural proteins include *vp39*, *cor*, and *gp64*. The major capsid protein, which forms the basic shell of the rod-shaped nucleocapsid (Fig. 1-1), is encoded by *vp39* (Thiem and Miller, 1989a). *cor* encodes a 6.9 kDa protein, which contains primarily arginine (40%), serine (18%), threonine (13%), and tyrosine (13%) (Kelly et al., 1983; Wilson et al., 1987). This highly basic, protamine-like protein is associated with the DNA within the capsid structure (Tweeten et al., 1980). The major glycoprotein of BV is encoded by *gp64* (Whitford et al., 1989). A monoclonal antibody, AcV1 (Hohmann and Faulkner, 1983), specific for gp64, is able to neutralize the bulk of BV infectivity, but does not neutralize the infectivity of the virions released from OV by alkaline dissolution (Volkman et al., 1984; Keddie and Volkman, 1985; Volkman and Goldsmith, 1985). BV appears to enter cells primarily by receptor-mediated endocytosis although a low level of infectivity may be achieved though membrane fusion. Membrane fusion is the primary method of entry by virions released from OVs in the midgut. Thus, it is likely that gp64 is specifically involved in BV interaction with receptors or the cell surface.

The 34.8 kDa protein, p34.8, may be a virus structural protein but its role has not been established. Initial genetic disruption studies suggest that the gene encoding this protein may be essential (Wu and Miller, 1989). This protein shares sequence homology with an entomopoxvirus protein originally identified as spheroidin, the protein of the occluded form of these viruses (Yuen et al., 1990). However, there is currently some question whether this entomopoxvirus gene indeed encodes spheroidin (Hall and Moyer, 1991).

DNA REPLICATION FUNCTIONS

Genes that appear to be involved in viral DNA replication include *dnapol*, *hel (ts8)*, and *pcna*. The *dnapol* gene (Tomalski et al., 1988) encodes a 114 kDa protein with homology to other DNA polymerases (e.g., DNA polymerases of herpesviruses, poxviruses, and adenoviruses). It is likely that this gene encodes the AcMNPV-induced, aphidicolin-sensitive DNA polymerase described previously (Miller et al., 1981). Enzymatic studies suggest that not only is a new DNA polymerase activity induced during infection but a host DNA polymerase activity is also stimulated or induced (Miller et al., 1981; Mikhailov et al., 1986). By analogy to mammalian proliferating-cell nuclear antigens, viral PCNA (Crawford and Miller, 1988; O'Reilly et al., 1989) may serve as a processivity factor for one of the DNA polymerases. However, it is not essential for replication in at least two cell lines. Therefore, host cell PCNA may duplicate this function, or the viral-induced DNA polymerase(s) can function in the absence of a *pcna* product. The *hel* gene encodes a 143 kDa protein that contains a region-sharing sequence homology with other proteins having ATP-dependent helicase activity (Lu and Carstens, 1991). The phenotype of the *ts8* mutant virus used to localize and define this gene is a block in DNA replication at the nonpermissive temperature (Gordon and Carstens, 1984), consistent with a role of the *hel* product in DNA replication. The origin(s) of baculovirus DNA replication have not yet been identified, but some speculation suggests a role for the *hr* regions (see preceding physical map discussion).

GENES RELATED TO CELL FUNCTION

The gene encoding p35 (Friesen and Miller, 1987) is required to block apoptosis, a specific type of programmed cell death that is induced during baculovirus infection in some cell lines (Clem et al., 1991). Mutant viruses lacking functional *p35* grow poorly in the IPLB-SF-21 cell line because apoptosis is not blocked and cells die before efficient BV or OV production can occur. Apoptotic cell death is an active process in which the cell surface blebs off apoptotic bodies, membrane-bound vesicles containing cytoplasm and nuclear fragments. In wt AcMNPV-infected SF-21 cells, a transient blebbing of the cell membrane, visible by light microscopy, occurs during the late phase of infection. This feature is very prominent during infection by *p35* mutants.

The *ubi* gene encodes a ubiquitin homolog (Guarino, 1990). It is an essential gene but its mode of action remains to be determined. Since ubiquitins are known to be involved in the turnover of proteins, it will be important to determine the role of this ubiquitin in virus replication since the stability of heterologous gene products might be affected.

The gene downstream of *ubi* is *p39*, which encodes a 39K protein (p39) (Guarino and Smith, 1990) that is expressed both early and late in infection. Although *p39* has not been studied functionally at a genetic level, p39 has been reported to cofractionate with the nuclear matrix [results of Wilson reported in Guarino and Smith (1990)]. However, the procedures used to isolate the nuclear matrix may also incorporate other components of the infected baculovirus cell, including virogenic stroma, so that the actual location of p39 within the nucleus remains to be established.

A gene (*v-cath*) adjacent to *gp64* encodes a protein with sequence homology to cysteine proteases (J. Kuzio and P. Faulkner, personal communication). Initial studies indicate that the *v-cath* gene is nonessential. Protease activity of the gene product has not yet been detected. Cysteine protease activity has been detected in the extracellular medium of infected cells very late in infection (Vernet et al., 1990; Yamada et al., 1990). However, cysteine proteases are also known to be associated with uninfected cells and may simply be released into the extracellular fluid of infected cells upon lysis.

REGULATORY GENES

Several genes that appear to be involved in gene regulation have been identified. These genes include those encoding IE-0, IE-1, IE-N, PE-38, and CG30. *IE-0* is actually an exon of one of the two forms of IE-1 (Chisholm and Henner, 1988). *IE-0/IE-1* is the only known spliced gene in the AcMNPV genome to date. IE-1 is able to transactivate some early promoters in transient expression assays (see discussion on mechanisms of gene regulation later in this chapter) (Guarino and Summers, 1986a; 1987). The *IE-1* exon alone is sufficient for transactivation activity. The activity of IE-1 and IE-0/IE1 differ subtly in transient expression assays (Kovacs et al., 1991). The other three proteins, IE-N, PE38, and CG30, (Thiem and Miller, 1989b; Carson et al., 1991a; Krappa and Knebel-Mörsdorf, 1991) all share common unique structural motifs: an unusual, double zinc-fingerlike motif and a C-terminal leucine zipper. The zinc-fingerlike motif ($CX_2CX_{19\text{-}27}CXHX_2CX_2CX_{11\text{-}17}CX_2C$) is present in all three genes except that the last CX_2C motif of IE-N is CXC. The IE-N leucine zipper is also less clearly structured than those found in PE38 and CG30. The unusual zinc-fingerlike motif is found in other proteins, which are thought to have DNA-related functions (Freemont et al., 1991). Thus, baculoviruses appear to have a family of related genes that are probably involved in gene regulation. Whether members of this family function in early viral, late viral, or host gene regulation remains to be established.

MECHANISMS OF GENE REGULATION

Baculovirus genes are transcribed in a regulated cascade. Three basic phases of transcription can be distinguished: early, late, and very late [for an earlier review see Friesen and Miller (1986)]. Most viral genes are transcribed primarily during one phase although some genes are transcribed in two or possibly three phases. Early genes may be defined as those genes that are transcribed in the absence of any viral gene expression (Friesen and Miller, 1986; Crawford and Miller, 1988; Tomalski et al., 1988; Nissen and Friesen, 1989). Late and very late genes are dependent on early viral gene expression and on DNA replication (Rice and Miller, 1986; Wilson et al., 1987; Thiem and Miller, 1989a). Transcription of very late genes, those expressed primarily during the occlusion phase, is activated approximately 10 to 12 hours after the initiation of late gene transcription (Thiem and Miller, 1990).

Early Gene Transcription

The transcription of many early genes is strongly dependent on the product of *IE-1* when introduced into cells in the form of naked DNA (i.e., in transient expression assays) (Guarino and Summers, 1986a; 1987). However, there is no requirement for viral protein synthesis for early viral gene transcription in a normal virus infection (Crawford and Miller, 1988; Tomalski et al., 1988; Nissen and Friesen, 1989). One hypothesis that might explain this discrepancy is that IE-1 enters the cell as a component of the virions (Miller, 1988b). Although *IE-1* is transcribed throughout infection, steady-state levels of *IE-1* mRNA are highest late in infection (Chisholm and Henner, 1988; Huh and Weaver, 1990a; Kovacs et al., 1991).

Transactivation of some early promoters by IE-1 in transient expression assays can be stimulated by *hr* sequences (see physical map section for discussion on *hr* sequences), particularly when the *hr* sequences are placed in *cis* upstream or downstream of the early promoter/reporter gene fusion (Guarino and Summers, 1987; Nissen and Friesen, 1989). Extracts of cells transfected with *IE-1* cause a shift in the mobility of *hr*5-containing DNA fragments (Guarino and Dong, 1991), consistent with an IE-1-mediated enhancer role for *hr*5 and a transactivator function for IE-1. IE-1 appears to have a helix-loop-helix motif and possibly a leucine zipperlike sequence (Guarino and Summers, 1987). The function of the *IE-0* exon, encoding an additional 54 amino-acid N-terminus, is not yet clear, but the *IE-0*/*IE-1* and *IE-1* products have different transacting properties in transient expression assays (Kovacs et al., 1991). IE-1 functions in transient expression assays in the absence of this exon, and the *IE-1* exon has its own promoter. IE-1 also has a negative regulatory effect on *IE-N* and *IE-0* expression in transient expression assays (Carson et al., 1991b; Kovacs et al., 1991).

The product of *IE-N* appears to augment subtly (2.5-fold) the activity of IE-1 and has an autoregulatory effect on its own expression in transient expression assays (Carson et al., 1988; 1991b). *IE-N* is located in a back-to-back orientation to *pe38*, and the transcripts of these two genes comprise the major mRNAs found very early in the infection (e.g., 1 h pi) (Krappa and Knebel-Mörsdorf, 1991). Genetic analysis of the function of *IE-1, IE-N,* and *pe38* has not been reported.

Early gene transcription is thought to be mediated by host cell RNA polymerase II because no viral gene expression is required and early gene transcription is sensitive to α-amanitin (Grula et al., 1981; Huh and Weaver, 1990b), a potent inhibitor of eukaryotic RNA polymerase II. Correct transcriptional initiation from early viral promoters can be observed in uninfected cell extracts that support *in vitro* transcription (Hoopes and Rohrmann, 1991). These data are consistent with the view that early transcription is mediated by host RNA polymerase II. If virion-associated proteins do play a role in the process, their activity is probably limited to increasing or decreasing the transcriptional initiation frequency of host RNA polymerase II on viral and/or host promoters. Thus, the response of individual early viral gene transcription to cycloheximide treatment will probably show considerable variation, as has been reported (Huh and Weaver, 1990a).

Late and Very Late Gene Transcription

Late and very late viral gene transcription, however, is insensitive to α-amanitin (Grula et al., 1981; Huh and Weaver, 1990b). Some biochemical evidence suggests that a new RNA polymerase activity is present in virus-infected cells (Fuchs et al., 1983; Yang et al., 1991). Thus, even though baculovirus mRNAs are reported to be capped (Jun-Chuan and Weaver, 1982) and most appear to be polyadenylated (Lübbert and Doerfler, 1984; Rohel and Faulkner, 1984; Friesen and Miller, 1986), the enzyme(s) responsible for transcription is likely to be a virus-modified host RNA polymerase (e.g., a modified RNA polymerase III), a new RNA polymerase with virus-encoded subunits, or some combination of these two possibilities. The genes encoding other factors and/or RNA polymerase subunits involved in late and very late gene expression have not yet been identified, although temperature-sensitive mutants with phenotypes consistent with defects in such functions have been described (Partington et al., 1990).

Late and very late transcription is dependent on early viral gene expression and on DNA replication (Rice and Miller, 1986; Wilson et al., 1987; Thiem and Miller, 1989a; Huh and Weaver, 1990a); both cycloheximide and aphidicolin block the transcription of late and very late genes. The nature of the coupling between DNA replication and late gene transcription remains to be defined. Disruption of the viral-encoded PCNA delays DNA replication as well as late gene expression (Crawford and Miller, 1988; O'Reilly et al., 1989). The *ts8* mutant, having a mutation in a gene (*hel*) with homology to helicases, is defective in both DNA replication and late gene expression (Gordon and Carstens, 1984).

Late and Very Late Promoters—The Importance of TAAG

The late and very late promoters are distinctive and unusual with regard to their location. The primary determinant of both late and very late promoter activity is the tetranucleotide TAAG, which is located at the transcriptional start point of all known late and very late transcripts (Rohrmann, 1986; Possee and Howard, 1987; Wilson et al., 1987; Thiem and Miller, 1989a). TAAG sequences are relatively rare in the AcMNPV genome and are found primarily at late or very late transcriptional start points. Late and very late promoters are distinguished from each other by their relative activity during the late and very late phases. The *polh* promoter is an example of a very late promoter: it exhibits low activity during the late phase (i.e., 6 h pi–18 h pi) but becomes highly active beginning about 18 h pi. By 27 h pi to 48 h pi, approximately 20% of the total polyadenylated RNA in the cell is *polh* mRNA (Adang and Miller, 1982; Rohel et al., 1983). In contrast, the late *vp39* promoter is most active between 12 h pi and 24 h pi (Thiem and Miller, 1990).

The *polh* promoter has been characterized most extensively using both deletion (Possee and Howard, 1987) and linker-scan mutational analyses (Rankin et al., 1988; Ooi et al., 1989). The site of transcriptional initiation is within the sequence TAAATAAGTATT, which is highly conserved among different baculovirus *polh* genes and is very similar to the sequence found at the transcriptional start site of *p10* RNA (TAATAAGAATT). As discussed above, the TAAG portion of this sequence is absolutely essential for transcriptional initiation to occur; linker scan substitutions affecting the TAAG site decrease reporter gene expression approximately 2000-fold and lower steady-state levels of RNA to undetectable levels (Ooi et al., 1989). Additional promoter determinants are found in the untranslated leader region [i.e., between the TAAG sequence and the translational initiation codon (ATG) of the polyhedrin ORF]. The low steady-state RNA levels in such linker-scan mutants are due to less efficient transcriptional initiation, not an alteration in RNA turnover rates (Ooi et al., 1989). Thus, the major determinant of *polh* transcription is located at the transcriptional start point, and additional determinants are found between the RNA start point and the translational initiation codon (i.e., in the region specifying the untranslated leader-RNA region).

Influence of Sequences Upstream of TAAG

Nucleotides immediately upstream of the TAAG site may play a subtle role in regulation. The A of the ATAAG sequence at the *polh* RNA initiation point may influence levels and possibly timing of this promoter (Ooi et al., 1989). All strong late and very late RNAs initiate from an ATAAG sequence. Weaker late RNAs have been observed to initiate from TTAAG and GTAAG sites (Nissen and Friesen, 1989; Thiem and Miller, 1989a; Tomalski et al., 1991). However, the influence of this 5′ nucleotide cannot alone specify strength because linker insertion mutations of the nucleotides to the 5′ side of the *polh* TAAG site (i.e., converting, in the process, the ATAAG sequence to a GTAAG sequence) only decrease transcription approximately 3-fold (Ooi et al., 1989). Insertion of linkers further upstream of the TAAG site (e.g., 10–20 nucleotides upstream) *increases* (1.5-fold) the level of transcription. The "LSXIV" promoter (P_{XIV}) is the strongest of this series of linker-modified promoters (Rankin et al., 1988; Ooi et al., 1989) and is used in some vectors for increased levels of gene expression (Wang et al., 1991). The temporal regulation of P_{XIV} is similar to the *polh* promoter.

Polh *Promoter Regulation*

The phenotypes of the characterized *polh* promoter linker-scan mutants are consistent with the view that *polh* transcription is transactivated primarily through sequences corresponding to the untranslated leader region. These sequences are responsible for the very high levels of expression at very late times in infection (Ooi et al., 1989). There is no substantive evidence of a repressor binding site controlling temporal regulation of *polh* transcription. Thus, the lack of significant *polh* transcription during the late phase is unlikely to be due to repressor control. The direct correlation of steady-state levels of RNA with reporter protein levels suggests that *polh* expression is regulated primarily at the transcriptional rather than the translational level. However, translational regulation cannot be ruled out, particularly because all late and very late RNAs have the same 5′ end (i.e., AAG), which might influence their translation.

The *p10* promoter has been less extensively characterized, but it appears to be regulated in a similar fashion to the *polh* promoter with regard to the location and nature of the promoter determinants (Qin et al., 1989; Weyer and Possee, 1989). The nucleotides flanking the AcMNPV *polh* and *p10* TAAG sites are remarkably similar

(Rohrmann, 1986) and may reflect a role for these sequences in very late versus late temporal regulation. However, clear evidence for such a role did not emerge from the linker-scan analyses of the *polh* promoter (Ooi et al., 1989).

TAAG sequences also appear to be the primary determinant of transcriptional initiation from late promoters (Thiem and Miller, 1990). The *vp39* promoter actually has three transcriptional initiation sites, each starting within a TAAG site. When each TAAG site is deleted, transcriptional initiation from that region of the "promoter" ceases.

A hybrid late/very late promoter was constructed by fusing the two distal TAAG sequences of the *vp39* promoter to the *polh* promoter (i.e., the TAAG and 50 bp specifying the untranslated leader region of *polh*) (Thiem and Miller, 1990). All three TAAG sites largely retained their original temporal identity in this fusion promoter. The two distal *vp39*-derived TAAG sites were clearly regulated as late start sites, exhibiting strong activity during the late phase. During the very late phase, the *polh*-derived TAAG site was preferentially activated. The presence of *vp39* late promoters upstream of the *polh* TAAG site may have provided some "late" character to the *polh* promoter, but it nevertheless behaved primarily as a very late promoter. These results argue strongly for unusual and highly compact promoters with temporal regulation largely dictated by sequences immediately flanking the TAAG sites. Hybrid late/very late promoters are likely to find application in expression vector work.

Moving and Flipping Promoters

The late and very late promoters are compact and remain functional even when moved to different locations in the genome. The *polh* promoter, for example, retains full temporal and quantitative function after being "flipped" in the opposite orientation in the genome (Ooi et al., 1989). The *p10* and *cor* promoters, when moved to the *polh* region, appear to be fully active (Hill-Perkins and Possee, 1990; Weyer et al., 1990). Two *polh* promoters can be placed back-to-back and retain almost full function (Emery and Bishop, 1987; Wang et al., 1991). Promoters placed in tandem can boost transcriptional initiation and/or alter temporal regulation (Thiem and Miller, 1990; Wang et al., 1991).

Temporally Complex Promoters

Some genes are expressed in more than one phase of the infection process. The *p39*, *p35*, and *gp64* genes, for example, have both early and late transcriptional start sites; the late start sites comprise TAAG sequences (Guarino and Summers, 1986a; Nissen and Friesen, 1989; Whitford et al., 1989). Similarly, steady-state levels of transcripts of *IE-1* and the *p26* ORF continue to increase throughout infection suggesting that these genes are transcribed at least through the early and late phases of infection (Chisholm and Henner, 1988; Huh and Weaver, 1990a; Kovacs et al., 1991).

Influence of Gene Arrangement on Transcriptional Regulation

Many regions of the baculovirus genome are transcribed into multiple, overlapping RNAs. These overlapping RNAs can be in the antisense as well as sense directions relative to the ORF. Late transcripts are often found as a series of 5′ coterminal RNAs with differing 3′ termini [reviewed by Friesen and Miller (1986)]. One explanation of such overlapping late RNAs is the failure of the transcriptional machinery to efficiently recognize polyadenylation signal and/or to efficiently terminate transcription at late times pi. The failure to terminate transcription or to efficiently polyadenylate, however, may be related to gene regulation mechanisms.

There is now evidence to support the view that the arrangement of different temporal classes of genes on the baculovirus genome can play a role in gene regulation. This evidence comes from analysis of transcription of the two ORFs flanking the *polh* region; both of these ORFs are transcribed late in infection from the strand opposite the one used for *polh* transcription (see *polh* region map, Fig. 7-1). The steady-state level of the common 3.2kb mRNA, which is sense for flanking ORFS but includes antisense *polh* sequences, declines rapidly during the late phase. The rapid decline is directly tied to the onset of *polh* transcription (Ooi and Miller, 1990; 1991). The decline may be due to *polh* RNA hybridizing to the 3.2 kb antisense RNA and thereby destabilizing it, but other explanations such as transcriptional interference are also possible. What is clear is that transcription of neighboring genes can influence the regulation of upstream and/or downstream genes.

Effects of TAAG Elements on Heterologous Gene Expression

The mechanisms of late and very late gene expression in baculovirus-infected cells may have several ramifications for baculovirus gene expression vectors. If TAAG sequences in an appropriate context are present within the heterologous gene, transcription may initiate within the gene. If present in the antisense strand, the antisense RNAs may interfere with expression of the heterologous gene. Although such internal transcriptional initiation has not been reported, few researchers have analyzed both the sense and antisense transcription of their heterologous gene. Since we do not yet know what contexts of the TAAG sites are most active, it is difficult to predict the effects of such TAAG sequences on heterologous gene transcription. See Chapter 9 for further discussion of these issues.

3

VIRUS-HOST INTERACTIONS

VIRUS EFFECTS ON HOST GENE EXPRESSION

AcMNPV infection results in a shut-off of host gene expression. A decline in the steady-state levels of host mRNAs begins approximately 12 h pi. By 24 h pi, steady-state levels of cellular mRNAs such as actin and histone are quite low (Ooi and Miller, 1988). A decline in host protein synthesis is clearly observed at 18 h pi, and shut-off is virtually complete by 24 h pi (Carstens et al., 1979; Dobos and Cochran, 1980; Kelly and Lescott, 1981; Miller et al., 1983c). By 24 h pi to 36 h pi, gene expression is primarily, if not exclusively, viral-specific.

The mechanism(s) by which host RNA and protein levels are down-regulated are not known. The decline in host protein synthesis probably requires late gene expression (Ooi and Miller, 1988), and genes that influence the time of viral DNA replication and late gene expression (e.g., *pcna*) also influence the decline in host protein synthesis (Crawford and Miller, 1988).

Host chromatin structure appears to remain largely intact (Clem et al., 1991) and retains a nucleosomal structure (Wilson and Miller, 1986) throughout a wt virus infection. Viral DNA, however, adopts a very unique nucleoprotein structure late in infection (Wilson and Miller, 1986). It is possible that the unique nucleoprotein structure of viral DNA is due to the association of the protaminelike p6.9 protein, which must occur prior to virion assembly. However, this unique nucleoprotein structure and its compartmentalization in the virogenic stroma could also be important with regard to viral versus host gene regulation.

The decline in host protein synthesis might explain the apparent decline in the function of the endoplasmic reticulum (ER) during the very late phase of infection (Jarvis and Summers, 1989). The rate of movement and the amount of glycosylation of some heterologous gene products, particularly those such as tissue plasminogen activator (tPA), which require extensive N-glycosylation for secretion, appear to decline between 24 h pi and 48 h pi.

Several other host cell functions decline during the very late phase or are unable to keep pace with the high rate of heterologous protein synthesis at that time. A decline in the amount of phosphorylation of SV40 T-antigen and p53 was reported during the period between 24 h pi and 48 h pi, even though steady-state levels of the proteins were increasing rapidly during this time (O'Reilly and Miller, 1988). Also, intron splicing, a process that does not appear to be particularly efficient in baculovirus-infected cells, is reported to decline in efficiency at very late times pi in infected *B. mori* cells (Iatrou et al., 1989).

For proteins requiring extensive post-translational modification, expression during the late phase using late or late/very late hybrid promoters may increase yields and the efficiency of post-translational modification.

BACULOVIRUS HOST RANGE

Block to Infection of Nonpermissive Insect Cells

Baculoviruses are generally considered to have a relatively narrow host range, each baculovirus being able to infect efficiently only a few taxonomically related insect species (Gröner, 1986). In the most comprehensive and controlled survey of AcMNPV insect host range (Bishop et al., 1988), second or third instar larvae of a wide variety of insect species were challenged with up to 10^6 PIBs per insect; the LD_{50} (amount of virus causing 50% mortality in treated population) of the most susceptible species tested (*Trichoplusia ni*) was 10^2 PIBs per larva. Nonpermissive lepidoptera included all 17 butterfly species tested (belonging to five different families) and 58 moth species (belonging to 11 different families). Only six species were found to be permissive, and these species belonged to two different lepidopteran families, Noctuidae and Sphingidae. Many other species within these two families, including some members of the *Autographa* and *Spodoptera* genera, were not affected, even at 10^6 PIBs per larva. Thus, AcMNPV, which is considered to have a relatively broad host range for a baculovirus, nevertheless has a highly restricted host range.

AcMNPV enters cells of taxonomically diverse insects and can express genes under "early" promoter control. This was demonstrated by using a recombinant AcMNPV (vLCL1galcat) carrying *E. coli lacZ* under the control of the polyhedrin promoter P_{polh} and the CAT gene (*cat*) under the control of the Rous sarcoma-virus long-terminal repeat (RSV-LTR). The RSV-LTR is active in taxonomically diverse organisms. The *cat* was expressed relatively early in permissive cells and was expressed to approximately the same level in permissive and nonpermissive cells from two dipteran species, *Drosophila melanogaster* and *Aedes aegypti* (Carbonell et al., 1985). In contrast, expression of *lacZ*, under *polh* promoter control, was efficient in permissive cells but not observable in nonpermissive cells.

The restriction to AcMNPV infection in many, if not all, nonpermissive insect cells is, therefore, not its ability to enter those host cells but rather its inability to replicate its DNA and express late functional gene products. This has been confirmed, in part, by other work showing restrictions to late gene transcription (Rice and Miller, 1986), but a great deal of additional work is still needed to delimit the factors controlling the ability of different insect cells to support AcMNPV DNA replication and late gene expression.

Block to Infection of Mammalian Cells

The nature of the block to baculovirus infection appears to be different in insects and mammals (Carbonell et al., 1985; Carbonell and Miller, 1987). Although electron microscopy indicates that AcMNPV nucleocapsids enter mammalian cells (Volkman and Goldsmith, 1983), probably through a phagocytic process, the viral DNA rarely, if ever, reaches the nucleus in an expressible form (Carbonell and Miller, 1987). Thus, AcMNPV has some block to entering and/or expressing genes in the nucleus of mammalian cells. AcMNPV is also unable to transform BALB/c 3T3 cells or primary cultures of Syrian hamster embryos (Hartig et al., 1989) and does not persist in human cells (Hartig et al., 1991). Additional earlier studies concerning the safety of using baculoviruses as pesticides have been considered previously (Summers et al., 1975; Summers and Kawanishi, 1978; Gröner, 1986).

Recombinant baculovirus vectors are therefore expected to have no adverse effects on mammals. As with any recombinant organism, however, precautions must be taken concerning their distribution, and recombinant viruses must be treated as if they were potential biohazards. Those working with vectors carrying potent toxin genes, or vectors carrying all or major portions of other viral genomes, particularly genes encoding membrane-bound fusogenic proteins, should exercise caution in handling these vectors. Consideration should also be given to the possibility of baculoviruses being carried from the laboratory by invertebrates. An effective program of insect pest control must be implemented. Occ^- viruses are less able to survive outside insect hemolymph or culture media than occlusion-positive viruses. The survival capacity of occluded viruses, however, is quite remarkable. They are known to survive for 20 or more years in soil. Containment requirements for baculovirus expression vector work are considered in Chapter 10.

4

SUMMARY OF BACULOVIRUS FEATURES RELEVANT TO EXPRESSION VECTORS

Several features of the baculovirus system that are particularly advantageous are presented next. For each advantage, balancing exceptions or pitfalls are noted.

EUKARYOTIC ENVIRONMENT FOR PROTEIN PRODUCTION

Using a eukaryotic system for expressing a eukaryotic gene can be particularly important in obtaining biologically active proteins. The insect baculovirus expression system provides a eukaryotic environment that is generally conducive to the proper folding, disulfide bond formation, oligomerization, and/or other post-translational modifications required for the biological activity of some eukaryotic proteins. Post-translational modifications that have been reported to occur in baculovirus-infected insect cells include signal cleavage, proteolytic cleavage, N-glycosylation, O-glycosylation, acylation, amidation, phosphorylation, prenylation, and carboxymethylation. The sites of such modifications are usually at identical positions on the proteins produced in insect and mammalian cells.

The precise nature of some of the post-translational modifications observed in insect cells, however, differs from those in mammalian cells, usually in subtle ways. Similarities and differences of insect and mammalian post-translational modification systems are discussed in Chapter 15. Owing to the high levels of expression that can be achieved in this system coupled with the decline in host cell functions at very late times in infection, the efficiency of post-translational modification may decline at very late times in infection. This decline in efficiency can generate heterogeneous protein products that differ in the extent of post-translational modification. Such differences can be minimized, as suggested in Chapter 6, if uniform products are required. In general, however, the insect baculovirus expression system mimics a vertebrate cell system quite remarkably with regard to protein post-translational modification. [See Luckow (1991) for a recent review of the range of genes that have been expressed and the spectrum of post-translational modifications that have been observed with the baculovirus expression vector system. Appendix 3 includes an updated list of all genes expressed as of this manual's press date. However, it does not include information on the post-translational modification of the gene products.] For genes encoding glycosylated and secreted proteins, there is a greater likelihood of obtaining biologically active products in the baculovirus expression vector system than in prokaryotic expression systems.

EXCEPTIONALLY HIGH EXPRESSION

The most distinguishing feature of baculovirus expression vectors is the potential for very strong expression of a heterologous gene. The highest expression levels reported using baculovirus expression vectors is 25%–50% of the total cellular protein. Such levels are equivalent to *polh* expression and correspond to approximately 1 gram of protein product per 10^9 cells (e.g., a 1 liter culture).

All heterologous proteins, however, are not produced at the same level as polyhedrin, and levels approaching 25% of the total cellular protein have been achieved in few cases. Most of these cases involved expression of structural genes of other virus families, the products of which are quite stable. Most heterologous proteins are produced at levels ranging from 10 mg to 100 mg per 10^9 cells. Nevertheless, in the cases where different eukaryotic expression systems have been compared, the baculovirus system has usually outperformed the other expression systems in overall protein production. Yields of a heterologous gene product must be determined empirically; it is difficult to provide any guidelines for how a gene will behave in the expression system unless similar genes have been expressed previously.

EXPRESSION DURING THE OCCLUSION PHASE

The use of very late promoters (e.g., the *polh* and *p10* promoters), which are activated and strongly transcribed during the unique occlusion phase of virus replication, provides a clear advantage for baculovirus-based expression systems. Because the occlusion phase is distinct from the late phase encompassing BV formation, expression of the heterologous gene interferes minimally with BV production, and there is very little selective pressure on the virus to mutate toward heterologous gene deletion or inactivation.

Very late expression may be particularly advantageous for the expression of heterologous genes with a deleterious effect on essential cell functions. However, no specific examples of the production of cytotoxic gene products have been described to date. (There are occasional verbal reports that vectors carrying some genes have been difficult or impossible to obtain, presumably due to a negative effect of the cell or surrounding cells that would compromise plaque formation.) One reported application of the feature of very late baculovirus expression negating cell toxicity effects, however, is substituting selenomethionine into heterologous proteins (Chen and Bahl, 1991c). This normally irreversibly cytotoxic amino acid analog can be provided in the media during the very late phase of infection and extensively incorporated into heterologous proteins for the purpose of isomorphous replacement for X-ray diffraction studies.

The negative aspect of expressing genes during the very late phase of virus infection is that the post-translational modification capacity of cells appears to decline during the very late phase. In addition, expression during the late or very late phase of a lytic virus infection means that the cell will die, and large-scale protein production must be run in batches rather than in continuous culture.

EXPRESSION AT 27° C

Baculoviruses are normally propagated at 27° C. This temperature is likely to be permissive for most temperature-sensitive mutants derived from organisms that are normally cultured at 37° C. Thus, the baculovirus expression vector system can be used to produce large quantities of active protein encoded by a temperature-sensitive allele of a gene (Reynisdottir et al., 1990). This is not possible in expression vector systems where the vector itself is propagated at 37° C.

The obvious disadvantage of gene expression at 27° C is that the baculovirus system may not be appropriate for the expression of some genes encoding cold-sensitive proteins.

CAPACITY FOR LARGE INSERTIONS

The rod-shaped nucleocapsids of baculoviruses can extend to accommodate larger viral DNA genomes, and it is likely that a baculovirus vector can accommodate an additional 100 kbp of DNA or more. The largest reported insertion thus far is approximately 15 kbp, but the limit has not been reached or, to our knowledge, been challenged. The number of genes that might be expressed simultaneously using baculovirus vector systems is also likely to be high. Using currently available vector systems, one can develop a vector that expresses three heterologous genes.

To make very large insertions, however, it may be necessary to construct a series of transfer plasmids that allow the building of the insert in successive increments. This limitation has more to do with the fragility of large DNAs *in vitro* rather than the vector system *per se*. Also, if more than three genes are to be expressed, the user would need to construct additional transfer plasmids.

UNSPLICED GENES (e.g., cDNAs) ARE EFFICIENTLY EXPRESSED

All characterized very-late genes of baculoviruses are unspliced but efficiently expressed. Since many researchers isolate cDNAs, the baculovirus system is particularly useful for expressing such genes. For those researchers who wish to express genes containing introns, we recommend isolating the cDNA version for high-level expression purposes. However, the baculovirus expression system does carry out at least some splicing, and its usefulness in identifying and obtaining specific gene products from multigene families has been noted (Iatrou et al., 1989).

SIMPLICITY OF TECHNOLOGY

Baculovirus vectors are helper-virus independent and therefore relatively simple to use. Constructing a recombinant baculovirus is considerably faster and simpler than constructing a cloned, high-expressing, recombinant eukaryotic cell line (e.g., cloning a methotrexate-amplified recombinant CHO cell line).

Nevertheless, there are significant time requirements: the baculovirus expression system clearly takes a longer time to develop than a typical *E. coli* expression system. The primary time requirement of the baculovirus expression system, after construction of the transfer plasmid, lies in successive plaque purifications; generally, each plaque purification takes five days to a week, and three plaque purifications are recommended to ensure a pure recombinant plaque. Additional time is required for cotransfections and virus stock production. Thus, the development of a baculovirus vector will take at least six weeks from the time the heterologous gene has been cloned in the transfer plasmid.

II

CHOOSING A TRANSFER PLASMID AND PARENT VIRUS

Part II of the manual is intended to help you make an informed decision concerning the design and construction of your baculovirus vector. You will need to decide which virus, host, transfer plasmid, and methods to use in constructing and identifying your recombinant vector.

The first decision you will need to make, discussed in Chapter 5, is which host and virus to use. The two basic virus options are *Autographa californica* nuclear polyhedrosis virus (AcMNPV) or *Bombyx mori* nuclear polyhedrosis virus (BmNPV). Your choice will be influenced by whether you anticipate producing your protein in insect larvae. For those expecting to use only cell cultures, AcMNPV is currently the preferred virus because of the diversity of available transfer plasmids, protein yield, and growth properties of cells supporting AcMNPV replication. For those choosing to use insect larvae in their work, the relative advantages of working with larvae supporting AcMNPV or BmNPV are discussed.

Chapter 5 also includes information concerning the differences among available viruses and information concerning insect cell lines that have been used extensively for baculovirus expression vector work. We think that an AcMNPV-based vector and a *S. frugiperda* cell line (SF-21 or SF-9) are the best options for the majority of basic gene expression work. This host/cell system is particularly useful for scale-up of protein production in cell culture.

Once you have chosen a virus and a host cell line, you will need to decide what type of transfer plasmid you will use to construct your virus vector. All baculovirus vectors are constructed using allelic replacement to insert heterologous genes into the virus genome. Chapter 6 therefore starts with a discussion of the basic features of allelic replacement such as factors affecting efficiency. The chapter continues with a discussion of the basic differences among the many transfer plasmids available and how to choose a plasmid that will suit your needs. Important differences among the available transfer plasmids include: (1) the type of promoter used; (2) the number of different genes that can be inserted for simultaneous transfer and expression; (3) provision of a signal sequence and/or another N-terminus for secretion or expression as a fusion protein; (4) the site in the virus genome to which the heterologous gene is directed; and (5) the presence or absence of a marker gene for recombinant identification purposes.

The choice of transfer plasmid will significantly influence the method used to identify the recombinant and may also dictate the nature of the parent virus used in the allelic replacement event. For some transfer plasmids, for example, it may be helpful to use a parent virus carrying the *E. coli lacZ* gene to more easily identify recombinants following allelic replacement. The factors pertaining to the choice of transfer plasmid and parent virus for construction of the recombinant virus will need to be considered carefully.

Chapter 7 contains specific information concerning a number of different transfer plasmids that were selected for their utility as well as their diversity with respect to expression vector preferences and/or needs. The information includes restriction maps, polylinker sequences, sources from which the plasmids can be obtained, and so on. Chapter 8 provides information concerning the different recombinant viruses that are available to assist in the identification of appropriate recombinant vectors.

Finally, Chapter 9 discusses factors you may want to consider in tailoring your gene to suit the transfer plasmid and obtain optimum gene expression in the baculovirus expression system. Factors discussed include translational initiation context, 5′ and 3′ untranslated sequences, introns, and signal sequences.

5

CHOICE OF VIRUS AND HOST SPECIES

Although there are over 500 baculoviruses reported in the literature, the focus of this book is limited to AcMNPV and BmNPV because these viruses are currently the most developed and have the most useful properties as expression vectors. Initial sequence comparisons suggest that BmNPV may share as much as 90% sequence identity with AcMNPV (but neither genome has been sequenced entirely at this time). However, the two viruses differ in host range, and the different hosts or host cell lines offer unique advantages for particular applications of the expression vector system. This chapter first discusses the relative merits of each system and then presents information concerning the origins of and differences among available virus variants. Finally, this chapter provides information concerning the different insect cell lines available for vector propagation and protein production.

CHOOSING A PARENT VIRUS: AcMNPV VERSUS BmNPV

Although the basic technology for using AcMNPV- or BmNPV-based vectors is very similar, AcMNPV-based systems have a number of advantages for most common applications. The cell lines supporting AcMNPV replication are superior in growth characteristics and expression levels than cell lines supporting BmNPV replication. BmN4 cells, which are normally used for BmNPV cell expression work, are usually passed every 4 to 5 days at a dilution of 1:2 (Maeda, 1989a). In contrast, SF cells used for the propagation of AcMNPV typically double every 18 to 24 hours. Thus, vector construction, propagation of virus stocks, and overproduction of recombinant proteins are all likely to be more time consuming in the BmNPV system.

In addition, the range of transfer plasmids and parent viruses available for the AcMNPV-based system is much greater than that for the BmNPV-based system. Thus, with AcMNPV vectors, researchers have the options of expressing genes in occlusion-minus (occ$^-$) or occlusion-positive (occ$^+$) viruses, expressing genes using a variety of promoters active at different stages of infection, using vector systems designed for rapid identification of recombinants, and so on. In contrast, the currently available BmNPV vectors are restricted to the expression of a single gene, alone or as a polyhedrin fusion, in an occ$^-$ virus.

However, a BmNPV-based system can offer certain advantages if the overproduction of recombinant proteins in insect larvae is desired. The principle advantage to using BmNPV is that larvae of the insect host *Bombyx mori* (silkworm) are approximately 10 times larger than the preferred larval hosts for AcMNPV and can attain weights of approximately 5 g. Because of their large size, they can be injected with the budded form of the virus, and 0.5 ml of hemolymph can be easily collected. In contrast, common hosts for AcMNPV-based vectors, such as *Trichoplusia ni* (cabbage looper), *Heliothis virescens* (tobacco budworm), and *Spodoptera frugiperda* (fall armyworm), achieve average maximum weights of approximately 300 mg to 500 mg. Injection and hemolymph collection from these smaller larvae is possible, but collection yields only approximately 50 µl per larvae and can be rather tedious if large numbers of insects need to be processed. Thus, use of a BmNPV-based vector in *B. mori* represents an attractive protein production system, particularly for proteins that are secreted into the hemolymph. There are other attractive features of the silkworm system. The larvae exhibit no cannibalistic behavior and can therefore be reared in groups. They do not migrate far, the adults cannot fly, and they are unable to survive in the field (Maeda, 1989a). A popular organism for research and teaching, *B. mori* eggs are available from a variety of suppliers and researchers. In Asia, silkworms have been mass-cultured for silk production for centuries; automated rearing procedures are used currently in Japan. One potential problem to consider is that many research colonies of *B. mori* are reared on fresh mulberry leaves, which are available during only certain seasons of the year.

However, insect size is not the only parameter to consider when evaluating larval production options, and there are circumstances when an AcMNPV-based system may be preferable. For example, researchers who need very large quantities of protein, particularly protein that is not secreted, may wish to process thousands of insects for protein production. Clearly, it is not practical to infect this number of insects by injection. AcMNPV-based systems have a distinct advantage in these circumstances: it is feasible to generate occ^+ recombinant viruses to allow infection by contamination of the larval diet with occluded virus.

Note also that recent evidence suggests that it may be possible to use the tobacco hornworm, *Manduca sexta,* as a host for the AcMNPV-mediated production of recombinant proteins (Gretch et al., 1991). *M. sexta* is a considerably larger insect, achieving weights of up to 10 g. However, it only supports limited AcMNPV replication, and it is not clear yet whether *M. sexta* is an optimal larval system.

In summary, whereas AcMNPV-based vectors are currently preferable for cell-culture-based expression work, *B. mori* may have advantages for certain larval expression applications. Researchers trying to decide which virus system to employ should consider whether they anticipate producing their protein in insect larvae. The relative merits of protein production in cell culture versus insect larvae are addressed in the introduction to Part IV. However, the salient points are summarized here.

The potential advantages of larval production include: (1) higher efficiency of certain types of post-translational modifications; and (2) less costly production. Post-translational modifications occurring intracellularly are likely to be more heterogeneous in infected larvae due to expression in a variety of cell types. This may or may not be desirable, depending on the application. Potential disadvantages include the need to tend living insects, and the greater difficulty of recombinant protein purification. The magnitude of the latter problem will depend on whether the protein is secreted.

Large-scale production in cell culture has the advantages of being: (1) the same system used in the initial characterization of the kinetics and levels of synthesis of the protein; and (2) "cleaner" than insect larvae with regard to recombinant protein purification, especially if serum-free medium is used. There may also be more uniformity in post-translational modification because only a single cell type is supporting synthesis. A significant disadvantage to this approach is the potentially high cost. Commercially prepared media may cost $10 to $25 per liter. In addition, if large-scale production is required, specialized equipment (e.g., fermenters) will be needed.

Thus, infection of insect larvae can provide a cost-effective alternative to the scale-up of recombinant protein production in cell culture. However, it is difficult to provide definitive general recommendations on whether a protein should be scaled-up in insect larvae or in cell culture. It is difficult also to predict *a priori* whether large-scale production will be necessary at all because some proteins are expressed at levels up to 500 mg/l of cells (ca. 10^9 cells per liter) whereas others are expressed at levels of only 5 mg/l of cells. Obviously, once a recombinant virus has been generated, one can easily address these questions empirically by carrying out small-scale larval infections, and so on. It is important to bear in mind from the outset that your initial choice of expression system will dictate the options for scale-up of protein production later. Researchers who do not anticipate a need to produce very large quantities of recombinant protein are advised to use an AcMNPV-based vector and focus on expression in cell culture.

AcMNPV AND BmNPV VARIANTS

Different AcMNPV genotypic variants are used by different baculovirology laboratories. All of these clonal variants are derived from one original isolate (Vail et al., 1971) although their interim histories differ. The original isolate contained multiple variants that were distinguishable by restriction endonuclease digestion (Lee and Miller, 1978; Vail et al., 1982). The commonly used variants (e.g., L-1, E2, C-6, and HR3) differ by one or a few minor differences in the restriction map. L-1 is used by Miller's group, E2 by Summers's group, HR3 by Faulkner's lab, and C-6 by Possee and colleagues. L-1 and C-6 differ from E2 by the presence of an additional HindIII site in the HindIII-B fragment, while both C-6 and HR3 differ from L-1 and E2 by a small insertion in HindIII-L. By consensus, a common physical map and fragment nomenclature for the AcMNPV genome was adopted by most AcMNPV laboratories (Vlak and Smith, 1982). (See Chapter 2.) Doerfler's group, however, adopted a different fragment nomenclature for the "E" isolate, which differs from other common AcMNPV variants by the presence of a transposable element insertion around 81 map units (mu) (Schetter et al., 1990). The variants may also differ in some of the sites of less commonly used restriction enzymes. The absence of Bsu36I sites, a feature which has some value in expression vector work, has been verified only for the L-1 and C-6 variants. There are no known phenotypic differences among the commonly used AcMNPV variants.

There are several recombinant AcMNPVs that are very useful in some vector design strategies. These viruses are described in detail in Chapter 8.

The two principle laboratories involved in the development of BmNPV vectors are those of Dr. Susumu Maeda (University of California-Davis) and Dr. Kostas Iatrou (University of Calgary). The physical map of the cloned virus used by Maeda and his colleagues has been published recently (Maeda and Majima, 1990). The differences between the BmNPV isolates used by these two laboratories are not completely established.

CHOICE OF CELLS

Properties of cell lines that are important in baculovirus expression vector work include: (1) growth rate (i.e., doubling time); (2) ability to support virus replication and plaque formation; (3) ability to grow in suspension for ease of cell production; and (4) ability to support recombinant protein synthesis, post-translational modification, and secretion (if required). The cells that are most commonly used with AcMNPV-based vectors are SF-9 and SF-21AE cell lines. Both cell lines grow well (doubling times of approximately 18 to 22 hours) in both monolayer and suspension culture. They both originated from IPLB-SF-21 cells, which were derived from *S. frugiperda* pupal ovarian tissue at the USDA Insect Pathology Laboratory (IPLB) at Beltsville, Maryland (Vaughn et al., 1977). Unless there is a specific need to indicate which of these cell lines is used, they will be called SF cells. SF-21AE cells were adapted in England (AE means adapted in England) to TC-100 media by H. Stockdale in the early 1970s and were distributed widely by D. Knudsen, now at Colorado State University (Fort Collins, Colorado). The SF-9 cell line is a clonal cell line isolated from SF-21AE in 1983 by G. Smith and C. Cherry. This line has been entered in the ATCC cell culture collection (ATCC CRL 1711).

Detailed studies of the differences between SF-9 and SF-21AE cells have not been reported, and it is not clear if there is a particular advantage to using one cell line versus another. Some researchers prefer SF-21AE cells, claiming higher levels (ca. twofold) of expression than observed in SF-9 cells and vice versa, depending on the laboratory. It is not clear whether such differences are due to the nature of the protein produced, the health of the cells in that laboratory, or the cell line *per se*. Researchers should be cautious of the source of the cell line they choose because mishandling of the cells may change the growth and health characteristics of even a clonal cell isolate.

Other cell lines that support the replication of AcMNPV are available. One cell line used in many AcMNPV laboratories for certain types of research is TN-368, a cell line established originally from minced adult ovaries of *T. ni* (Hink, 1970). This cell line grows well as a monolayer and supports virus replication and protein synthesis to similar levels and similar kinetics as observed in SF-21 cells. However, TN-368 cells tend to clump in suspension culture and, in our hands, we find that this cell line is not as versatile as SF-21-based cells. Another *T. ni* cell line, BTI-TN-5B1-4 (commercially known as "High 5"), was established at Boyce Thompson Institute (Ithaca, New York) and is reported to exhibit "up to one log higher levels of protein expression compared to SF-9" (Invitrogen Corporation advertisements). However, this cell line does not grow in suspension, making it inconvenient for large-scale cell culture. In addition, there is currently little information available concerning the kinetics of viral gene expression in these cells, the efficiency of different types of post-translational modification, and so on. Thus, at the present time, we hesitate to recommend this cell line for general use.

A cell line that supports AcMNPV replication and protein synthesis with similar kinetics as SF cells has been established from *Mammestra brassicae* (cabbage moth) (King et al., 1991). Higher levels (three to four times) of expression were observed in *M. brassicae* than in SF cells on a per-cell basis. *M. brassicae* cells are substantially larger, which may account in part for this difference. Again, more information on the growth properties and post-translational modification capacity of this cell line is necessary before we could recommend its use as a general cell system. Those interested in optimizing gene expression, especially at the industrial level, may be interested in investigating expression properties of a variety of different cell lines. Some cell lines may express a particular gene better than others. Information concerning established insect cell lines is available elsewhere (Hink, 1976; 1980; 1989).

Although there are differences in cell lines, we recommend that most users of the AcMNPV-based vector system use SF cells because these are well-tested and have excellent growth and handling properties. These cells have been reported to produce some proteins at levels approaching 20% or more of the total cell protein, and it is unlikely that any other generally available cell line will provide more than threefold higher yields of protein than these. Extensive cell-line comparison studies are probably not worthwhile for most research laboratories.

Two *B. mori* cell lines are commonly used for BmNPV vector work. Dr. K. Iatrou uses "Bm5" cells originally isolated by Dr. T.D. Grace, whereas Dr. S. Maeda uses a clonal cell line, BmN4, which he cloned from "BmN" cells. BmN cells also had their origin in Grace's laboratory. Grace supplied Dr. D. Knudsen with two cell lines, BmN(1) and BmN(5), in 1976. Knudsen sent these cells to Dr. L. Volkman (University of California at Berkeley), who provided Maeda with a BmN culture. Thus, there is some ambiguity as to whether the Bm5 and BmN4 cell lines were derived from the same original source, but it is likely that they were. It is also likely, but not certain, that these lines were derived from pupal ovarian cells. Dr. Maeda indicates that other laboratories are now establishing new cell lines of *B. mori* [mentioned in Maeda (1989a)] so that better cell lines may be available in the future.

6

CHOICE OF TRANSFER PLASMID

Baculovirus genomes are large (e.g., 100 kbp–200 kbp) and usually contain one or more recognition sites for known restriction endonucleases. When AcMNPV was being developed as an expression vector, there were no known restriction endonucleases that lacked recognition sites in this genome, so allelic replacement was adopted to insert foreign genes into the genome (Smith et al., 1983b; Pennock et al., 1984; Maeda et al., 1985). Allelic replacement remains the preferred method for heterologous gene insertion. Thus, this chapter begins with a discussion of several important aspects of the allelic replacement reaction, the general properties of transfer plasmids that facilitate foreign gene placement and expression, and the factors that may influence the frequency of recombinant virus formation.

ALLELIC REPLACEMENT

In allelic replacement strategies, one starts by inserting the foreign gene into a transfer plasmid so that it is downstream of the required viral promoter and flanked on both sides by viral sequences that will target the gene and promoter to a particular region in the viral genome (see Fig. 6-1). The plasmid and parental viral DNA are cotransfected into insect cells, and enzymes in the cells recombine the DNAs. This primarily involves homologous recombination. However, occasionally nonhomologous recombination occurs; but it is comparatively rare as long as the homologous recombination event does not affect the viability of the virus. Thus, the primary recombination reactions occur between the regions of the plasmid DNA flanking the foreign gene and their homologous counterparts in the viral genome.

An allelic replacement reaction actually involves two distinct homologous recombination events: one in each of the two regions of viral DNA flanking the foreign gene in the transfer plasmid (Fig. 6-1). These two reactions occur independently. The most common type of recombinant virus found in virus stocks obtained following cotransfections is the *single-crossover* recombinant, in which the entire plasmid DNA has integrated into the viral genome. These recombinants are inherently unstable because they contain duplications of portions of the viral genome (i.e., the flanking regions present in the transfer plasmid DNA duplicate existing regions in the virus genome). Single-crossover recombination does not disrupt the function of the viral gene at the site of the insertion if the plasmid contains sufficient flanking sequence to provide an intact copy of the viral gene at one of the ends of the insertion. Depending on where a second homologous recombination occurs, the

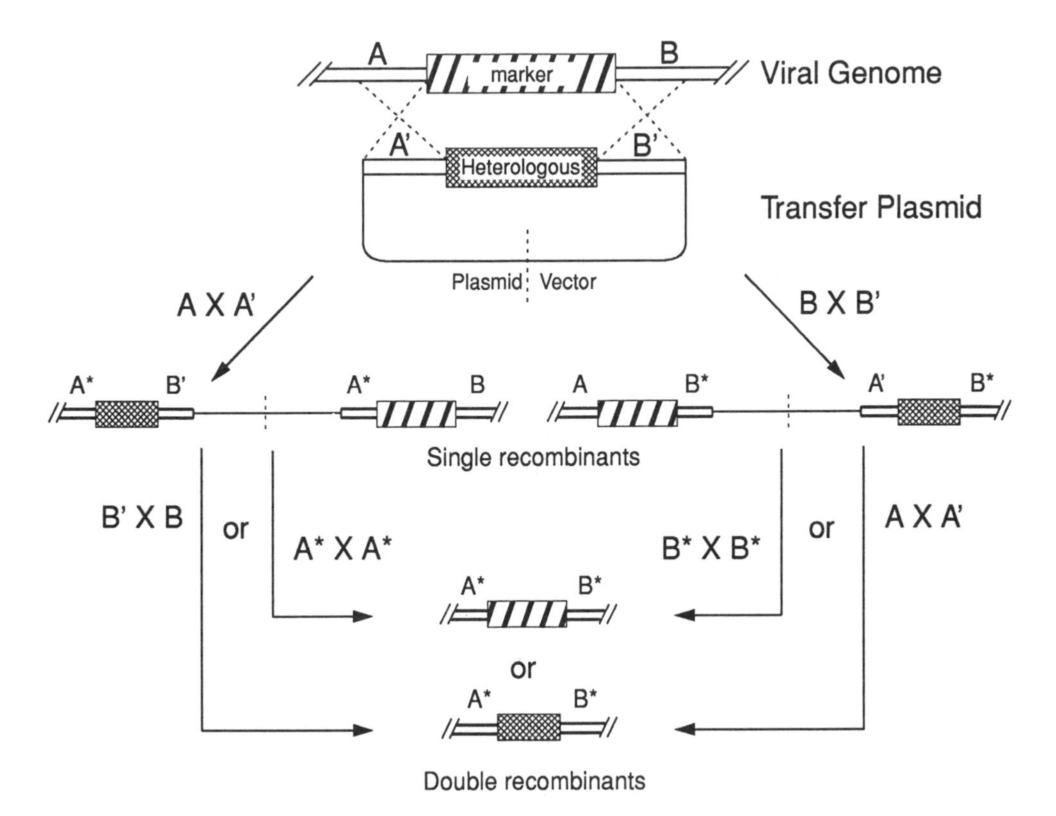

FIGURE 6-1 *Illustration of allelic replacement reaction.* The structures of the parent virus and the transfer plasmid are depicted at the top of the figure. The marker gene could be *polh*, *lacZ*, or some other gene. A and B represent the flanking sequences present in both viral and plasmids DNAs in which homologous recombination takes place; A and B are the sequences in the viral genome, whereas A′ and B′ are the same sequences in the transfer plasmid. The center of the figure shows the two possible outcomes of single crossover recombination events; A* and B* are sequences derived from recombination between A and A′ or B and B′, respectively. The bottom of the figure shows the structures of the recombinants obtained when the single recombinants illustrated in the center of the figure recombine the second time.

single-crossover recombinant may give rise to either a true allelic replacement (i.e., a *double-crossover* recombinant) or the starting (parent) virus (Fig. 6-1).

Because of the prevalence and instability of single-crossover viruses in stocks derived from allelic replacement, it is best to choose a vector construction design that will allow you to distinguish double recombinants from single recombinants and parent virus. The most direct way to do this is to start with a parent virus with a nonessential "marker" gene conferring a visible plaque characteristic. For example, one might use a parent virus that makes occ$^+$ plaques, or a virus carrying the *E. coli lacZ* gene that forms blue plaques in the presence of an appropriate chromogenic indicator of β-galactosidase. The transfer plasmid would be designed so that the marker gene is replaced with the heterologous gene upon allelic replacement. Double crossovers are thus distinguished from single crossovers and parent virus by the absence of the marker gene (occ$^-$ or white plaques, in the examples cited). In this regard, it is therefore better to have a marker gene in the parent virus than to have a marker gene in the transfer plasmid.

The nature of the allelic replacement reaction thus shapes the strategies that can be employed in virus vector construction and recombinant virus identification. These strategies, in turn, influence the design of the plasmids that are used to transfer the foreign gene into the viral genome. Plasmids used in the allelic replacement reaction are referred to either as transfer plasmids or transplacement plasmids. Some allelic replacement strategies rely heavily on the nature of the parent virus used for recombinant virus construction. Other strategies use both plasmids carrying an identifiable marker gene (e.g., a plasmid-carried *lacZ* or *polh* gene) combined with a screen for the loss of a virus-borne gene function (e.g., occ$^+$ or *lacZ*).

Construction strategies that rely solely on the presence of a plasmid-borne marker must be used with caution because recombinant viruses expressing this marker will consist mainly of single-crossover recombinants. For example, carrying *lacZ* on the plasmid and screening only for a blue plaque phenotype will yield primarily single-crossover, blue plaque recombinants. These virus stocks must therefore be rescreened for appropriate double-crossover recombinants using some other technique such as restriction profiles or DNA hybridization characteristics. If the effort is not made to obtain a double-crossover recombinant, the stock of virus generated from a plaque-purified single recombinant virus will probably eventually be taken over by double-crossover recombinants that have lost the foreign gene of interest. In general, we recommend using a screening strategy that will allow direct screening for double-crossover recombinants.

The relative proportion of double recombinants in the mixture of viruses resulting from cotransfection of transfer plasmid and viral DNAs influences how much effort it will take to screen for the recombinant virus of interest. In most allelic replacement reactions involving 2 kb to 4 kb of viral DNA flanking each side of the heterologous DNA in the transfer plasmid, double-recombinant viruses will probably represent 0.2% to 1% of the total virus stock. Thus, methods for rapidly and easily detecting a rare virus in a mixture of viruses are important.

A technique that substantially improves the proportion of recombinants in the virus stock is linearization of the viral DNA at the target site (Kitts et al., 1990). The DNA genomes of the AcMNPV variants C-6 and L-1 (and perhaps others) lack recognition sites for the restriction endonuclease Bsu36I (CC'TNAGG). Recombinant AcMNPV DNAs constructed with a Bsu36I site at the target site and linearized at this site were 15- to 150-fold less infectious than uncut circular viral DNA. However, when cotransfected with transfer plasmids, the proportion of recombinant viruses was increased 10- to 30-fold compared to circular parental viral DNA. (To some extent, there is a trade-off in the yield of viruses versus the proportion of recombinants.) The experiments demonstrating this effect used a plasmid-carried *lacZ* gene and scored percent recombinants as blue plaques; thus, the percent recombinants reported included both single- and double-crossover recombinants. However, linearization of the parental DNA should decrease the proportion of single-crossover recombinants obtained. Because the *E. coli lacZ* gene contains a Bsu36I site, recombinant AcMNPVs (at least C-6- and L-1-based viruses) carrying the *lacZ* gene at the target site can be linearized with Bsu36I to increase the proportion of recombinants. One of the disadvantages of this approach is that recombinant viruses are not visibly distinguishable from viruses in which the Bsu36I ends were rejoined to form a nonfunctional marker gene (e.g., a nonfunctional *lacZ* gene).

Parent viruses with unique restriction sites at positions outside of the marker gene but within the region to be allelically replaced should improve this linearization technique. For example, Invitrogen is currently marketing a linearized AcMNPV DNA in which a unique Sse8387I site (CCTGCA'GG) is placed upstream of *polh*. Linearization should not disrupt *polh*, and thus one should be able to distinguish double crossovers from rejoined breaks as long as portions of *polh* have not been removed during the rejoining event. PharMingen has recently begun to market linearized DNA that contains a deletion extending into the 1629 ORF downstream from *polh*. This deletion is lethal and must be complemented by the transfer plasmid to allow the recovery of viable virus. Thus, the majority of virus plaques obtained should represent recombinant viruses.

Some investigators have inquired about the prospects of having a baculovirus with a unique restriction site at the proper location downstream of a strong promoter to avoid the allelic replacement step. Even if a unique restriction site were placed in the baculovirus genome at a preferred site, the efficiency of inserting a single heterologous fragment and virus recircularization might not be as high as desired, unless a forced cloning approach (e.g., two different cohesive ends on the viral DNA and matching ones on the fragment to be inserted) were possible. This lower efficiency would be due to the large size of the viral DNA and, consequently, the comparatively low local concentration of the two ends of the viral DNA. The other problem is that the unique restriction site would probably not be a commonly used one since AcMNPV has multiple cleavage sites for most common enzymes. The transfer plasmid pEV/cc (Miller et al., 1986) was designed to allow the construction of a virus that can be linearized downstream of the *polh* promoter to allow direct cloning of a fragment. The approach makes use of the ClaI methylating enzyme and DpnI, which cuts at symmetrically methylated GATC sites. However, it is necessary to insert the heterologous gene fragment by blunt-end ligation, and this plasmid has not found wide use.

In summary, allelic replacement is generally used to introduce heterologous genes into baculovirus genomes. In designing a vector construction strategy, it is important to choose markers on the virus and plasmid that allow the distinction of double-crossover recombinants from the more frequent single-crossover recombinants. Using a loss-of-function marker facilitates this distinction. The overall proportion of virus recombinants may be increased by using a viral DNA that is linearized at the target site.

CHOOSING A PROMOTER

The "workhorse" promoter of the baculovirus expression system has been the *polh* promoter—a very strong promoter that is expressed very late in infection and is normally responsible for the synthesis of polyhedrin, the nonessential occlusion-body matrix protein (see Part I for the biology of this system). The *polh* promoter has many excellent features, and a *polh*-based promoter is recommended for standard applications of the vector system. However, the *polh* promoter is not the only very strong, very late promoter in baculoviruses. The *p10* promoter is another very strong, very late promoter that is increasingly used with the development of *p10* promoter-based transfer plasmids and parent viruses, which are convenient for vector construction and identification. Also, some researchers are exploring the possibility of expressing genes earlier in the infection process, the shut-off of host protein synthesis, and the decline of host functions involved in post-translational modifications. Thus, late promoters and even early promoters are being used in some baculovirus expression work.

This section begins with a discussion of the nature of the *polh* promoter, noting key features to consider when choosing transfer plasmids employing this promoter. There are many available transfer plasmids that use this promoter. Some have unique properties, such as being able to express more than one heterologous gene (e.g., dual-gene transfer plasmids) or express genes as fusions to other N-terminal sequences. Thus, we also describe within the *polh* promoter topic area (1) the use of plasmids carrying the *polh* gene for constructing occ^{+} vectors; and (2) the use of plasmids supplying an ATG for expression of genes as fusions.

A discussion of the comparative advantages of using the *polh* or *p10* promoter follows. The last topics of this section consider the possible disadvantages of strong, very late promoters and the pros and cons of alternative promoters or stable insect cell-based expression systems.

The polh Promoter

Although polyhedrin is nonessential for virus replication in cell culture, it is responsible for a visible plaque characteristic—the presence of polyhedra in the nucleus of infected cells (occ$^+$). The presence or absence of polyhedra remains one of the most frequently employed and useful markers in vector construction strategies. Although the *polh* promoter is quite small (50 bp) and can be moved to other locations in the genome, most available transfer plasmids use the *polh* promoter in its natural locus, either in its original orientation or in reverse orientation, to take advantage of this marker system.

An important aspect of the *polh* promoter relating to the choice and design of transfer plasmids is the unusual position of the promoter relative to the transcriptional start site (see Chapter 2 for a more complete discussion and documentation of this subject). Briefly, sequences specifying *polh* transcription (i.e., promoter sequences) are found within sequences encoding the untranslated leader region of *polh* mRNA.

In making the first series of nonfusion transfer plasmids for AcMNPV *polh* promoter-based expression, Smith et al. (1983b) constructed pAc373, which had an oligonucleotide with a BamHI site inserted at –7 with respect to the *polh* translational initiation codon (ATG, corresponding to +1, +2, +3). This vector became widely used as an expression vector before it was realized that the *polh* leader region comprised the promoter and that the nucleotides between –7 and +1 contributed to optimum expression. (Matsuura et al., 1987; Ooi et al., 1989). There is a twofold increase in expression if the five to seven nucleotides upstream of the ATG are retained in the promoter (Matsuura et al., 1987; Ooi et al., 1989).

Plasmids such as pAcYM1 (Matsuura et al., 1987), pEVmXIV (Wang et al., 1991), and pBM030 contain the entire *polh* promoter, including the +1 A of the original *polh* ATG. The levels of protein production using such a +1 vector can reach the same levels as those achieved for *polh* expression (Matsuura et al., 1987); levels of some stable recombinant proteins are reported to reach 20% to 50% of the total cellular protein. However, very few genes are expressed at this high level.

Luckow and Summers (1989) used a different transfer plasmid design. In pVL941, the *polh* translational initiation codon was changed from ATG to ATT by site-directed mutagenesis. Foreign genes could be inserted at a restriction site placed within the former N-terminal polyhedrin coding sequences. Plasmids such as pVL1393 and pBlueBac2 (see Chapter 7) are derivatives of pVL941 with a multicloning site to facilitate foreign gene insertion. Like the +1 vectors above, the levels of transcripts in these vectors are equivalent to those observed for *polh* RNA. The translational start codon of the foreign gene in these vectors is displaced further downstream from the transcription control regions. Recent studies have found, however, that translation can initiate at the ATT of this type of vector (Beames et al., 1991). Such initiation results in the fusion of polyhedrin N-terminal sequences to some of the foreign gene products produced by these vectors. The placement of genes out-of-frame to the ATT should eliminate contaminating gene fusion products (Beames et al., 1991); however, the influence of upstream translational initiation on downstream initiation at the foreign ATG has not yet been fully investigated. This type of problem may not be limited to pVL941 derivatives. It is possible that undesired fusion proteins will be generated with any vector in which a codon that can function as a translational start site (e.g., ATT, ACG) occurs upstream of and in-frame with the foreign ATG.

The 50 bp *polh* promoter can work with equal efficiency in either orientation with respect to the AcMNPV genome (Ooi et al., 1989). In addition, two *polh* promoters or two very late promoters (e.g., *polh* and *p10* promoters) can be placed in back-to-back orientation and both operate at nearly optimum efficiency to drive very high levels of expression of two different genes (Emery and Bishop, 1987; Weyer et al., 1990; Wang et al., 1991). The spacing and sequence of these promoter arrangements can influence the levels of expression (Wang et al., 1991). Examples of transfer plasmids with back-to-back promoters are p2XIV VI$^-$ and pAcUW3 (Chapter 7); each have downstream multicloning sites that allow the simultaneous insertion and expression of two different heterologous genes.

Similar transfer plasmids that allow the coinsertion of both a heterologous gene and *polh* to yield an occ$^+$ viral phenotype (Emery and Bishop, 1987; Wang et al., 1991) have been constructed. Occ$^+$ viruses are not only easily identifiable in a background of occ$^-$ viruses but they are also useful in large-scale protein production in insect larvae because the virus can be administered orally by contamination of the insect diet (see pSynXIV VI$^+$ series plasmids and pAcUW2B in Chapter 7). It is recommended that such occ$^+$ plasmids be used with an occ$^-$ parent virus carrying a *lacZ* gene in place of *polh* (e.g., vSynVI$^-$gal [Wang et al., 1991]) to allow selection of the double recombinant as a white, occ$^+$ virus. If such a *lacZ*-containing virus is used as the parent virus, double-crossover recombinant screening techniques can be employed and/or the Bsu36I linearization method can be used if desired.

The baculovirus expression system has an excellent track record for expressing genes as nonfusion proteins (i.e., for expressing genes using their natural, translational initiation codon and N-terminal sequence), but some proteins are expressed to higher levels as fusions. Several sets of transfer plasmids are available to supply a translational initiation codon (ATG) and allow the downstream insertion of heterologous genes in different reading frames. All such currently available transfer plasmids use the *polh* promoter to drive heterologous gene transcription, but most differ in the nature of the N-terminal sequence and multicloning sites provided. Some plasmids, such as pAc360 (Chapter 7), are designed to fuse the heterologous gene to the polyhedrin N-terminus. The presence of the polyhedrin N-terminal sequences has been shown to improve the level of expression of several proteins (Luckow and Summers, 1988a). The reasons for higher levels of expression for these fusions have not been firmly established; it may be that the heterologous protein is stabilized by the N-terminal sequences. Other transfer plasmids are designed to fuse the heterologous gene to a short peptide and provide an ATG in a good context. Examples include pAcC4 and the pSynXIV VI$^+$X3 series of transfer plasmids (Chapter 7). The pSynXIV VI$^+$X3 series plasmids are also designed to be compatible with signal peptides in targeting to the ER. The efficiency of these plasmids in gene expression and protein secretion, however, has not been fully tested and may differ from protein to protein, depending on flanking amino acid contexts.

Although N-terminal fusions may increase the stability of some proteins, the biological activity of many proteins is adversely affected by additional sequences at their N-terminus. Most researchers prefer to have lower yields of heterologous protein than risk having biologically inactive or inappropriate proteins. Those experiencing poor expression or wishing to have proteins for other purposes such as antibody production may, however, choose to try fusion vectors in the hope of improving yields.

Some users may wish to express their genes as N-terminal fusions to have a translational initiation codon in a good context for initiation. The importance of the AUG context is discussed in Chapter 9. Briefly, the context of the AUG does have an impact on the levels of gene expression (Luckow and Summers, 1988a). Comparisons of the contexts of initiation AUGs of several highly expressed late baculovirus genes reveal a strong preference for A at –3 and a bias against G at –1 and –2. Rather than selecting an N-terminal fusion vector, you may wish to consider site-specific mutagenesis of the context of the natural AUG of the heterologous gene if problems are encountered with expression levels. Alternatively, you may wish to use one of the fusion vectors just to see whether changing the context of the upstream untranslated leader sequences and the AUG context might significantly increase expression prior to committing the time to site-directed mutagenesis. See Chapter 9 for a further discussion of the possible effects of flanking sequences on gene expression.

Invitrogen is currently developing a transfer plasmid, pBac*HIS*, in which the heterologous gene will be fused to a short leader peptide from the bacteriophage T7 gene 10 protein. The leader peptide is designed to facilitate purification of the fusion protein. After purification, it may be removed by enterokinase cleavage.

The p10 *Promoter*

The *p10* protein is produced abundantly very late in infection and forms large fibrillar structures in infected cells (Van Der Wilk et al., 1987; Williams et al., 1989) (see Part I). The function of this protein is not known although it is reported that lysis of cells infected with *p10* mutant viruses is delayed (Williams et al., 1989). If this is true, there may be some advantage to using a virus lacking the *p10* gene as a base for vector construction. Mutations in the *p10* gene do not affect budded virus formation nor do they have a visible effect on occluded virus formation (see Part I). Thus, vector designs in which the *p10* gene is replaced by the heterologous gene do not confer an identifiable phenotype on the double recombinants as do plasmids designed for *polh* replacement.

The promoter of the *p10* gene is regulated in a fashion similar to the *polh* promoter (Weyer and Possee, 1988; Qin et al., 1989; Weyer and Possee, 1989; Weyer et al., 1990). The relative strengths of the two promoters also appear to be quite similar. Like the *polh* promoter, the *p10* promoter has an essential TAAG sequence at its transcriptional start point, and the region from around the initiation point to the ATG is sufficient to drive high-level transcription (Weyer et al., 1990).

Recently, *p10*-based plasmids and viruses that allow easier screening for recombinant viruses have been developed (Vlak et al., 1988; Williams et al., 1989; Vlak et al., 1990). All the *p10*-based plasmids described in Chapter 7 (pAcAS2, pAcUW1, pEP252) are employed with a parent virus carrying *lacZ* at the *p10* locus. Replacement of *lacZ* with the heterologous gene allows the selection of double recombinants as white plaques in a background of blue plaques.

The *p10* promoter has also been substituted for the *polh* promoter in the *polh* locus and found to drive normal *polh* expression (Weyer et al., 1990). It has also been placed back-to-back with the *polh* promoter. One such vector is pAcUW2B (see Chapter 7), which allows coexpression of *polh* and a heterologous gene.

Pros and Cons of Using Other Promoters

Although the very late *polh* and *p10* promoters are very effective promoters for high-level expression of heterologous genes, some investigators have noted a decline in the level of post-translational modifications such as phosphorylation (O'Reilly and Miller, 1988) and N-glycosylation (Jarvis et al., 1990b) at very late times postinfection. Note, however, that most users have been quite satisfied with the levels of post-translational modification obtained using the very late promoters. We generally recommend the use of a *polh*- or *p10*-based plasmid in initial work to establish the expression capability of the system. In most cases, the recombinant virus generated will prove perfectly adequate. However, if you perceive a need for earlier expression to allow additional time for post-translational modification and secretion, a few appropriate transfer plasmids are available.

Two AcMNPV late promoters, the promoter of the major capsid protein vp39 and the promoter of the cor gene encoding the p6.9 core-associated protein, are used in some transfer plasmids (e.g., plasmids pc/pS1 and pAcMP1). (See Chapter 7.) These promoters drive the strong expression of genes encoding major-virus structural proteins (Tweeten et al., 1980; Wilson et al., 1987; Thiem and Miller, 1989a; Hill- Perkins and Possee, 1990) and express reporter genes maximally between 12 h pi to 24 h pi (Hill-Perkins and Possee, 1990; Thiem and Miller, 1990). In contrast, AcMNPV *polh* promoter-driven expression is maximal between 24 h pi and 60 h pi. From 12 h pi to 24 h pi, expression levels from the late *vp39* promoter exceed those from the *polh* promoter, but by 48 h pi, expression levels from the *polh* promoter exceed those from the *vp39* promoter (Thiem and Miller, 1990). A hybrid *vp39/polh* fusion promoter, referred to as $P_{cap/polh}$, is active at both late and very late times (Thiem and Miller, 1990). (See pc/pS1, Chapter 7.)

Although you should expect a lower yield (e.g., approximately fivefold lower) of total protein using a vector expressing only during the late phase, the advantage in such a promoter is the possibility of improving the uniformity of post-translational processing, particularly if the proteins are harvested by 36 h pi. These features, however, have not been rigorously tested with a variety of different foreign gene products and/or post-translational modifications. It is possible that the total amount of product that is post-translationally modified may actually increase using such vectors. The rate of movement of proteins through the secretory pathway, for example, appears to decline at very late times (Jarvis et al., 1990b), and some proteins move slowly through the ER even under normal conditions. The products of genes expressed from a late promoter will have an additional 12 hours, approximately, to move through the ER and Golgi.

A possible drawback in using vectors with late promoters is that expression of the heterologous gene will occur during the period of budded virus synthesis. If the heterologous gene product interferes with budded virus formation or is cytotoxic, it may be difficult to isolate a recombinant virus expressing the gene, and/or it may be difficult to maintain a stable stock of the recombinant virus. This is the main reason why we recommend that new users first try vectors with very late promoters. If the gene product does not appear to be toxic, then exploring the features of late promoters is recommended if post-translational modification uniformity is important.

The use of promoters that are expressed during the early phase (pre-DNA replication stage) of infection might provide an additional few (up to 6) hours of foreign gene expression. However, the potential problem of interference with virus replication would be expected to be even more severe. Candidate early promoters also are not particularly strong, and the study of the potential of these promoters to drive foreign gene expression is only at an early stage. The early promoters that have been used to drive marker gene (*cat* and/or *lacZ*) expression in recombinant viruses are the *pcna* (also known as *etl*) (Crawford and Miller, 1988), *p35* (Nissen and Friesen, 1989), *egt* (O'Reilly and Miller, 1989), and *da26* (O'Reilly et al., 1990) promoters. Other early promoters have been described but their ability to drive foreign gene expression within the viral genome has not yet been described. Continuing work may attempt to use combination early and late and/or very late promoters. Because heterologous gene expression at earlier times in the infection may result in cytotoxicity or poor virus yields, we do not recommend this as the initial approach in baculovirus expression work.

Gene Expression in Stable Insect Cell Cultures

Interest in obtaining stable insect cell lines that express foreign genes without killing the cell stems from (1) the possibility of using continuous rather than batch-culture protein production methods; and (2) the decline in post-translational modification potential in baculovirus-infected cells at very late times pi (see Chapter 15). The overall levels of expression from stable cell expression systems are unlikely to approach the levels of expression from a regular baculovirus expression system, except for those cases in which the gene is poorly expressed using the baculovirus system and post-translational modification of the protein is very slow. Nevertheless, you should be aware that such stable, insect cell-expression systems are being developed. The techniques used in stable cell expression work are not covered here, and the strengths and weaknesses of the available systems are only discussed briefly.

A stable cell-expression system, using *Drosophila melanogaster* cells and an inducible metallothionein promoter to drive heterologous gene expression, has been described (Johansen et al., 1989). One of the most valuable aspects of the *Drosophila* system is that the number of copies of the heterologous gene in the cell can be established almost immediately by manipulating the ratio of the two plasmids. This is a characteristic of *Drosophila* cells (and possibly other insect cells). Thus, unlike mammalian cells, a long period of plasmid amplification is not required to establish a strongly expressing cell line. The metallothionein promoter is tightly regulated in these cells, minimizing cell instability from heterologous gene expression.

A stable *S. frugiperda* cell expression system using the early *IE-1* promoter of AcMNPV has also been described (Jarvis et al., 1990a). Stably transfected cells expressed the *lacZ* at approximately one-thousandth the level observed in a regular baculovirus expression system. In the case of tissue plasminogen activator (*tPA*) gene expression, however, the *IE-1* promoter/SF-9 cell-based stable expression system had expression levels similar to the regular baculovirus expression system. The *tPA* gene is not expressed particularly well in the regular baculovirus expression system (i.e., 1 mg/l, which actually compares reasonably well to tPA production in other expression systems), and movement of tPA through the ER is slow (Jarvis et al., 1990a). The tPA was processed through the ER more quickly and was more efficiently secreted in the stable cell expression system. With further development, an *S. frugiperda*-based cell expression system will probably compare well to *Drosophila* and mammalian expression systems. Use of a strong and inducible promoter would improve the value of this system.

In summary, the *polh* and *p10* promoters are excellent promoters because of their strength and the very late timing of their expression. The transfer plasmids available for the *polh* promoter allow for convenient screens for double recombinants and are more versatile than currently available *p10* promoter-based plasmids. Options for *polh* promoter-based plasmids include dual-gene expression vectors and gene fusion vectors. Promoters that drive gene expression earlier in the infection process may be considered for nontoxic proteins that are slowly processed in baculovirus-infected cells.

IDENTIFYING BACULOVIRUS RECOMBINANTS

The first part of this chapter described allelic replacement reactions, emphasizing the need to obtain double- rather than single-crossover recombinant viruses and the relative merits of using the presence or absence of gene markers as screening methods. The basic conclusion was that screening for the loss of a gene from the parental viral genome was a more powerful (more direct) method for identifying double recombinants than screening for the gain of a gene via plasmid transfer. The generality of this conclusion depends on the precise construct being used for allelic replacement. In most cases, the generality holds true and should be considered a major determinant in choosing a transfer plasmid and a parent virus to use in recombinant vector construction. This section, therefore, considers some of the different methods available for the identification of recombinant baculoviruses. It discusses some of the pros and cons of the different marker genes that are currently used in baculovirus vector construction and considers some of the subtleties that may affect the usefulness of these genes in designing transfer plasmids and viruses and/or in selecting double-crossover recombinant viruses.

The general types of identification methods are those based on: (1) occlusion phenotype; (2) plaque color (e.g., *lacZ*-based systems); (3) PCR amplification; and (4) heterologous gene-specific traits, including DNA hybridization methods, antibody screening methods, and enzyme-activity detection methods. The latter identification methods will probably not influence significantly the choice of vector design whereas the former two will have significant influences.

The Occlusion Phenotype; Occ$^+$ Versus Occ$^-$ Screens

The presence of the *polh* gene, driven by a very late promoter, results in the occ$^+$ phenotype, which is characterized by the presence of multiple, large (i.e., 2 μm – 5 μm in diameter) polyhedra (i.e., polyhedral occlusion bodies) in infected cell nuclei. Plaques of occ$^+$ virus are more refractive than plaques of occ$^-$ virus. If an occ$^+$ virus is used as the parent virus for a transfer plasmid carrying the heterologous gene within the *polh* locus, virus stocks arising from the cotransfection can be screened visually for double-crossover recombinants using the nonrefractive nature of occ$^-$ plaques. This approach is used with many currently available transfer plasmids, and methods for screening occ$^-$ plaques are presented in Chapter 13. Most researchers, however, find this screen difficult, or at best, tedious. The problem is that occ$^-$ plaques can often be difficult to distinguish from discontinuities in the cell monolayers, particularly in the presence of large numbers of occ$^+$ viruses.

One simple way to improve the occ$^-$ screening method is to use a virus carrying *lacZ* at some other location in the viral genome as the parent virus for vector construction. The *lacZ* gene is therefore retained in the recombinant, and the occ$^-$ plaque is distinguished from monolayer discontinuities by its blue color. An example of this is the use of the virus vDA26Z as the parent virus with occ$^-$ transfer plasmids, as described in the following sections.

The converse screen, observing occ$^+$ plaques in a background of occ$^-$, may also be used for recombinant identification. In this method, an occ$^-$ virus is used as the parent virus, and a transfer plasmid carrying the *polh* gene as well as the foreign gene is used for the allelic replacement reaction. Transfer plasmids pAcUW2B and the pSynXIV VI$^+$ series are examples of this type of plasmid. The resulting occ$^+$ recombinant plaques can be rapidly and easily distinguished visually in the background of occ$^-$ virus. However, these occ$^+$ viruses are likely to be primarily single-crossover recombinants, and a further screen for a double-crossover recombinant is necessary. One way to achieve this is to use a virus, carrying *lacZ* at the target site for heterologous gene insertion, as a base for vector construction. For example, a virus carrying *lacZ* in place of *polh,* such as vSynVI$^-$gal, can be used as the parent virus with a transfer plasmid that carries both *polh* and the heterologous gene. The double-crossover recombinant will be occ$^+$ and white.

The advantages of obtaining occ$^+$ recombinants are that they are easy to identify during vector construction and can then be used to orally infect insects. The possible disadvantages of using occ$^+$ virus recombinants are: (1) some constructs may have slightly lower levels of heterologous gene expression due to coexpression of *polh;* (2) high levels of *polh* (ca. 32 kDA on SDS-PAGE analysis) might obscure heterologous gene expression; and (3) more caution needs to be taken in handling and disposing of occ$^+$ virus recombinants because they are orally transmissible by insects.

The lacZ *Gene and Blue Plaque Phenotype*

The *lacZ* gene is a very useful marker because expression of *lacZ* results in the formation of blue plaques in the presence of an appropriate chromogenic indicator. There are three ways in which *lacZ* can be employed for baculovirus vector construction and identification. First, the loss of *lacZ* carried on the parental viral DNA at the target site can be used as a screen for double recombinants. Examples of this approach include the parent virus vSynVI$^-$gal with *lacZ* at the *polh* locus under the control of a synthetic promoter, and AcAS3, AcUW1-*lacZ*, and Ac228z with *lacZ* inserted at the *p10* locus (see Chapter 8).

Second, a *lacZ* gene at a nontarget site in the viral DNA can be retained to aid an occ$^-$ screen for a double recombinant. An example would be the use of the recombinant virus vDA26Z (Chapter 8), which carries *lacZ* under early promoter control in the *da26* locus. This virus is particularly useful as a parent virus for those screening for occ$^-$ plaques because the occ$^-$ plaques remain dark blue and lack occlusion bodies (as described earlier). The possible additional advantage of this approach is that the recombinants have blue plaques and are easily titered in future work. If a need subsequently arises to remove *lacZ*, it can be removed by cotransfection with a plasmid carrying a wt copy of that region of viral genome.

Third, the transfer of the *lacZ* gene from the transfer plasmid to the recombinant can be used to help identify recombinants. Examples of plasmids that use this approach are pBlueBac2 and pAcDZ1 (Chapter 7). While the introduction of *lacZ* provides a striking marker to identify recombinants, it is also important to use a method to observe the loss of a parental virus phenotype to ensure that all selected plaques are double recombinants (as described later in this section).

The nature of the viral promoter controlling *lacZ* expression influences the blueness of the plaques. Very efficient promoters, such as the *polh* and *p10* promoters, provide more expression than required, resulting in large blue areas around the plaques that can interfere with the identification of neighboring white plaques. Conversely, some viral promoters are too weak to provide sufficiently high levels of expression to easily visualize and distinguish blue plaques from white. Promoters that transcribe efficiently but relatively early in the infection process (e.g., the *etl* promoter or the *da26* promoter) provide an appropriate level of *lacZ* expression to allow easy identification and distinction of blue plaques.

Some designs for vector construction use both the transfer of *lacZ* from the transfer plasmid to the virus, as well as the loss of *polh* for double-recombinant virus identification. This is probably the most visually striking method for identifying plaques because all blue plaques appearing on the plate are either single or double recombinants, and they can be quickly screened under a light microscope for their occlusion phenotype. This approach is used with the transfer plasmids pBlueBac2 and pAcDZ1 (see Chapter 7), which express *lacZ* under *etl* or *Drosophila hsp70* promoter control respectively. In the case of pBlueBac2, it is not known if duplication of the *etl* promoter may lead to genomic instability owing to homologous recombination between promoter regions. The presence of the large *lacZ* gene in both these plasmids limits the availability of suitable cloning sites; the plasmids only contain one or two unique cloning sites.

For those considering insertion of heterologous genes into the *p10* region, vector construction designs using *lacZ* generally involve a parent virus with *lacZ* at the *p10* locus. In these cases, both starting and recombinant viruses will be occ$^+$; recombinants are distinguished as white occ$^+$ plaques. In the parent virus AcAS3 (Chapter 8), *lacZ* is under *hsp70* promoter control. The basal activity of the *Drosophila hsp70* promoter directs adequate, but not excessive, *lacZ* expression without heat shock. In the virus AcUW1-*lacZ* (Chapter 8), *lacZ* expression is under *p10* promoter control. In this case, screening is complicated by the excessive levels of β-galactosidase produced by the parent virus plaques. In Ac228z (Chapter 8), *lacZ* expression is also under *p10* promoter control. However, the *lacZ* gene is out of frame with the *p10* ATG, and the blue color of the plaques is not too diffuse.

PCR Amplification

Polymerase chain reaction (PCR) amplification of specific sections of the parent and/or recombinant virus genome provides a powerful method to screen for recombinant viruses. It is best used in combination with one of the screening approaches outlined earlier. The technique can enable a researcher to rapidly confirm the presence of double-crossover recombinants and differentiate them from contaminating wt and single-crossover recombinant viruses. It can be used to confirm that the appropriate gene has been inserted into the recombinant virus at the correct location and in the correct orientation. As long as the appropriate primers are used, the method may be used with all the transfer plasmids and parent virus combinations, irrespective of the marker gene employed. Thus, PCR amplification is an extremely versatile complement to any recombinant baculovirus screening protocol. Recommendations for the choice of amplification primers and a protocol for the PCR amplification reaction are provided in Chapter 13.

Other Identification Methods: Hybridization, Antibody, or Enzyme Activity Screening

DNA hybridization, antibody, and enzyme activity screening have been used by various workers to identify viruses carrying a heterologous gene of interest. Antibody and enzyme-activity screening protocols are obviously restricted to instances where an appropriate antibody or assay is available. These types of screening processes may be used with any transfer plasmid. However, a positive signal only reflects the presence of the heterologous gene in the virus genome; it does not distinguish single recombinants and double recombinants.

Usually, these types of gene identification schemes are employed in conjunction with the loss of a marker (e.g., the occ^+ to occ^- screen). Enzyme activity screens are likely to be the most useful because a positive reaction ensures that the gene product is active. (There is usually not any problem obtaining recombinants with functional genes, but if a positive enzyme assay is not available, it is always best to obtain and test several double recombinants as insurance against something having gone awry in the allelic replacement step.)

Recombinant Virus "Selection" Methods

The baculovirus construction designs described above employ methods for identifying recombinants but not for "selecting" recombinants. Recently, Patel et al. (1992) described a system that allows the selection of recombinant baculoviruses in the yeast *Saccharomyces cerevisiae*. They have generated an AcMNPV derivative that can replicate in both insect and yeast cells. If this virus is used as a parent virus, it is possible to select for the required recombinant in yeast, and then introduce the recombinant into the insect cells. We do not have any personal experience of this system, but it seems that this, or similar approaches, should facilitate the isolation of recombinant viruses.

COEXPRESSION OF TWO OR MORE GENES

Several approaches have been taken to expressing two or more heterologous genes in a single baculovirus-infected insect cell. The two basic approaches are: (1) insert two different genes into a single baculovirus vector and express both upon infection of cells with this dual-expression vector; or (2) coinfect cells with two different viruses, each expressing one or more heterologous genes. The first approach is probably preferable for expressing multi-subunit proteins and is particularly useful for scaled-up production. The second approach can be a useful research tool and can employ existing single-gene expression vectors. The nature, advantages, and disadvantages of these approaches are now discussed in more detail.

Several transfer plasmids have been designed that allow the simultaneous insertion of two different heterologous genes into the *polh* locus using the occ$^+$ to occ$^-$ screen for recombinants (Emery and Bishop, 1987; Wang et al., 1991). These transfer vectors use back-to-back promoters so that transcription of the two genes is driven in opposite directions. Some plasmids use two similar, very late promoters to allow approximately equivalent levels of transcription from the two genes (see, for example, plasmids p2XIV VI$^-$ and pAcUW3 in Chapter 7). Other plasmids use one strong and one weaker promoter to provide differential levels of transcription from the genes (see Wang et al., 1991). A weaker promoter for one gene would be useful in cases where one gene product is needed in lower amounts (e.g., a gene product involved in post-translational modification) or one of the subunits of a protein is produced in excess of the others.

Expression of two genes using the same virus is more advantageous than coinfection with two different viruses because the coinfection method requires a careful balancing of the multiplicities of infection of each virus. Furthermore, the coinfection process requires infection at high multiplicities (i.e., at least five plaque-forming units of each virus per cell). For large-scale production methods that use lower multiplicities, and for infections of larvae, dual expression vectors are essential for efficient simultaneous expression of two genes.

Coinfection with two or more viruses, however, can be useful in some applications such as determining the effect of one gene product on another. Coinfection with multiple viruses can also lead to the expression of multiple gene products. The most striking example of this approach thus far is the coexpression of five different heterologous genes of blue-tongue virus using three different viruses, two of which were expressing two different genes (Loudon and Roy, 1991). The result was an impressive assembly of double-shelled, viruslike particles. The use of expression systems like this will probably lead to many interesting applications in defining complex pathways. The baculovirus expression system is particularly amenable to these types of studies.

7

AVAILABLE TRANSFER PLASMIDS

PLASMID MAPS

This chapter provides information about and maps of selected transfer plasmids for use in constructing a baculovirus expression vector. Chapter 8 discusses the different parent viruses that may be used with these plasmids. The plasmids were selected as representative of the different types of plasmid currently available. Table 7-1 lists the plasmids described and summarizes their uses.

For each plasmid, we provide a circular map that illustrates the component elements of the plasmid and shows the locations of selected restriction sites. Unique restriction sites are indicated in bold, and the position of the polylinker or cloning site (indicated by an asterisk) is shown. Plasmid vector sequences are represented by the thin line on each map, and other elements of the plasmids are distinguished by differential shading. The *p10* and polyhedrin (*polh*) promoter and coding sequences are shaded in black on all maps, except in plasmids that have back-to-back *p10* and/or *polh*-derived promoters. In these cases, the promoter oriented in the opposite direction to normal *polh* transcription is shaded lighter.

TABLE 7-1 Summary of Transfer Plasmids

Description	*Plasmid*
PLASMIDS FOR EXPRESSION OF A SINGLE HETEROLOGOUS GENE	
AcMNPV polh-*based*	
Plasmids that do not supply an ATG	
Expression of heterologous gene alone	pAcYM1
	pEVmXIV
	pVL1393
Coexpression of heterologous gene and marker gene	pAcDZ1
	pBlueBac2
Coexpression of heterologous gene and *polh*	pAcUW2B
	pSynXIV VI$^+$
Plasmids that supply an ATG	
Expression of heterologous gene alone	pAcC4
Expression of *polh*/heterologous gene fusion	pAc360
Expression of signal peptide/heterologous gene fusion	pVT-Bac
Coexpression of heterologous gene and *polh*	pSynXIV VI$^+$X3/2,3,4

TABLE 7-1 Summary of Transfer Plasmids *(continued)*

Description	*Plasmid*
AcMNPV p10-*based*	
Expression of heterologous gene alone	pAcAS2 pAcUW1 pEP252
AcMNPV late promoter-based	
Expression of heterologous gene alone	
cap/polh fusion promoter	pc/pS1
cor promoter	pAcMP1
BmNPV polh-*based*	
Expression of heterologous gene alone	pBM030
PLASMIDS FOR COEXPRESSION OF TWO HETEROLOGOUS GENES	
AcMNPV *polh* and *p10* promoters	pAcUW3
Two AcMNPV P_{XIV} promoters	p2XIV VI^-

For each plasmid, the sequence of the cloning site(s) and a description highlighting key features of the plasmid is also provided. In most cases, lower-case letters represent synthetic DNA sequences that were introduced during the cloning process. The parent virus used with each plasmid by the laboratory of origin is noted, along with the expected phenotype of the recombinant virus if that parent virus is used. However, in many cases, researchers may wish to consider using a

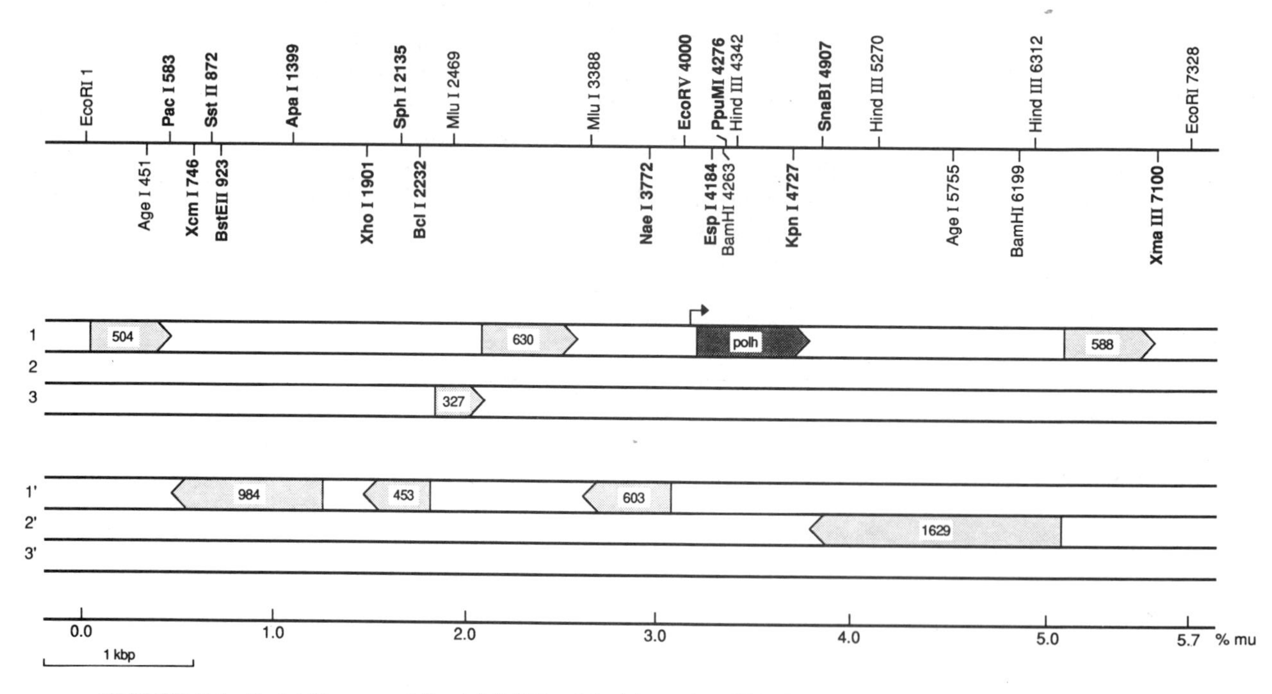

FIGURE 7-1 *Restriction map of the AcMNPV polyhedrin region.* This figure presents a map of the EcoRI-I fragment of AcMNPV. It is based on the sequence data of Possee et al. (1991). Unique restriction sites are in bold; nt 1 is the G of the EcoRI site at 0.0 mu. A scale in mu is presented at the bottom of the figure. Major ORFs in the region are illustrated underneath the restriction map. The arrow indicates the *polh* transcriptional start site; the *polh* coding region is represented by the darkly shaded box. Other ORFs are named according to Possee et al. (1991). The sequence of this region is presented in Appendix 1.

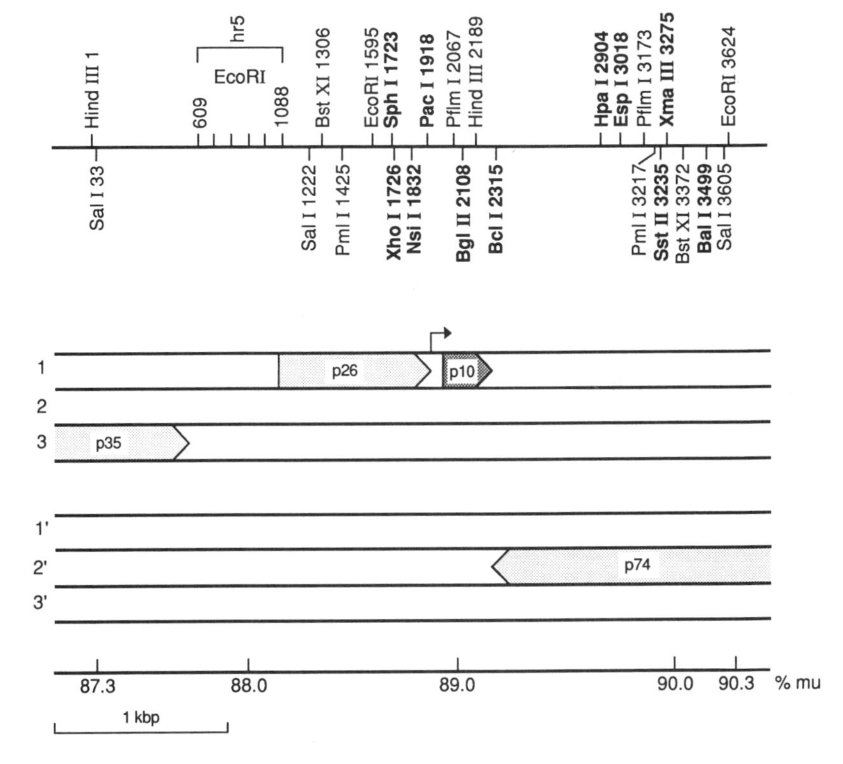

FIGURE 7-2 *Restriction map of the AcMNPV p10 region.* A map of the AcMNPV genome from the HindIII site at 87.3 mu to the EcoRI site at 90.3 mu is presented. A scale in mu is provided at the bottom of the figure. Unique restriction sites are in bold; nt 1 is the first A of the HindIII site at 87.3 mu. The map was prepared from the sequence of this region as compiled from data published by Lübbert and Doerfler (1984), Kuzio et al. (1984), Liu et al. (1986), Friesen and Miller (1987), and Kuzio et al. (1989). Guarino et al. (1986) have also published sequence data for the *hr*5 sequences within this region. Their sequence differs in a number of positions from that of Liu et al. (1986). It is not clear whether this reflects the different AcMNPV strains used. The differences do not alter the restriction map. ORFs in the region are indicated below the restriction map; *p10* coding sequences are represented by the darkly shaded box. The arrow indicates the *p10* transcriptional start site.

different parent virus. In these cases, the phenotype of the recombinant may be changed. Information on available parent viruses and their uses is in Chapter 8. Beneath each plasmid map, the source from which the plasmid may be obtained is indicated. Addresses, telephone numbers, and fax numbers of all sources are provided at the end of this chapter. For some plasmids, it may be necessary to complete and return a biological materials exchange agreement, or similar document, before obtaining the plasmid.

In general, numbering of plasmid maps begins at the plasmid vector/viral DNA junction upstream of the *polh* or *p10* gene. However, numbering on the polylinker sequences is relative to the translational start codon (ATG = +1, +2, +3) of the original *polh* or *p10* open-reading frame. For reference, Figures 7-1 and 7-2 provide restriction maps of the sequences flanking the AcMNPV *polh* and *p10* genes respectively. In addition, the sequences of these regions are provided in Appendices 1 and 2.

Every effort was made to ensure that the nucleotide numbers and restriction sites indicated on the maps are accurate. However, in certain cases, the junctions of the different elements constituting the plasmid have not been sequenced, and the coordinates provided should therefore be viewed as tentative.

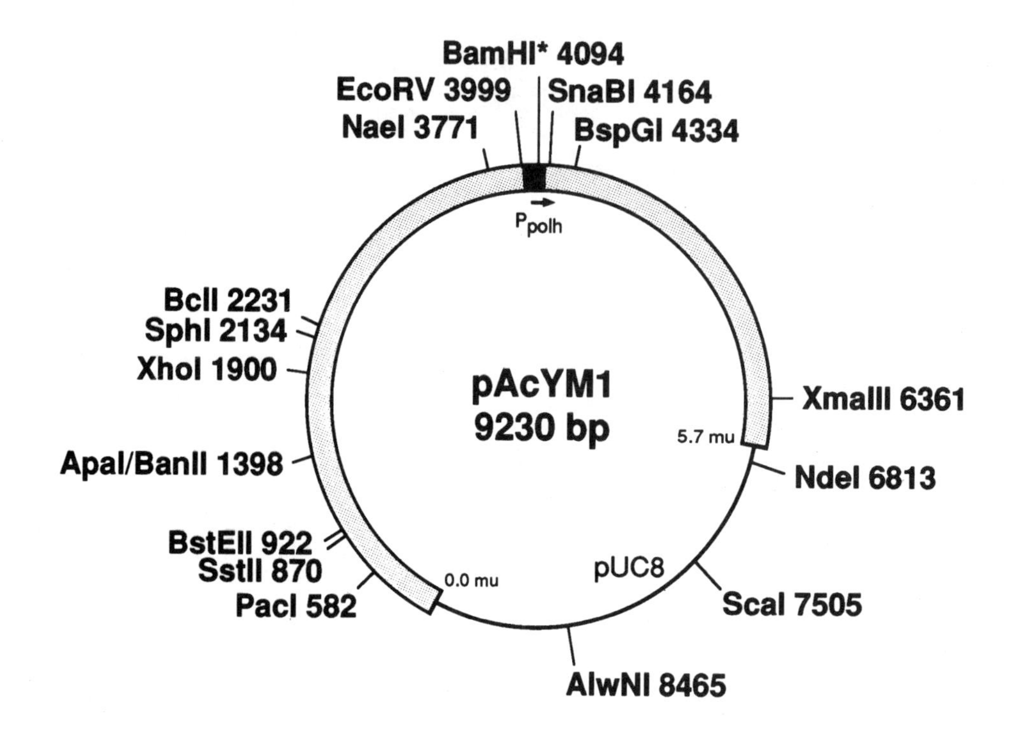

Plasmid	pAcYM1
Promoter	P_{polh}
Parent virus	wt AcMNPV
Plaque phenotype	occ^-
Related plasmids	pAcCL29 series
Selection in *E. coli*	Amp^r
Source	R.D. Possee

pAcYM1

```
      -90        -80        -70        -60       -50                 ,  -30
EcoRV|            |          |          |           * polh mRNA 5 end → |
GATATCATGGAGATAATTAAAATGATAACCATCTCGCAAATAAATAAGTATTTTACTGTTTTCGTA

          -20        -10         +1        +752
           |          |          | BamHI    |
ACAGTTTTGTAATAAAAAAACCTATAAATAcggatccgGTTATT
```

The plasmid pAcYM1 (Matsuura et al., 1987) facilitates the generation of occ⁻ recombinant virus expressing a heterologous gene under the control of the P_{polh} promoter. The entire P_{polh} promoter up to the +1 position is present in this plasmid.

The transfer plasmid pAcCL29 (Livingstone and Jones, 1989) is a derivative of pAcYM1, which includes the intergenic region of phage M13. Infection of transformed *E. coli* cells containing this plasmid with M13 helper phage results in the generation of single-stranded pAcCL29 DNA, facilitating sequence analysis, site-directed mutagenesis, and the generation of strand-specific probes. Further derivatives are pAcCL29-1 and pAcCL29-8 in which the BamHI cloning site has been replaced with a polylinker containing SstI, KpnI, SmaI, BamHI, XbaI, SalI, and PstI cloning sites.

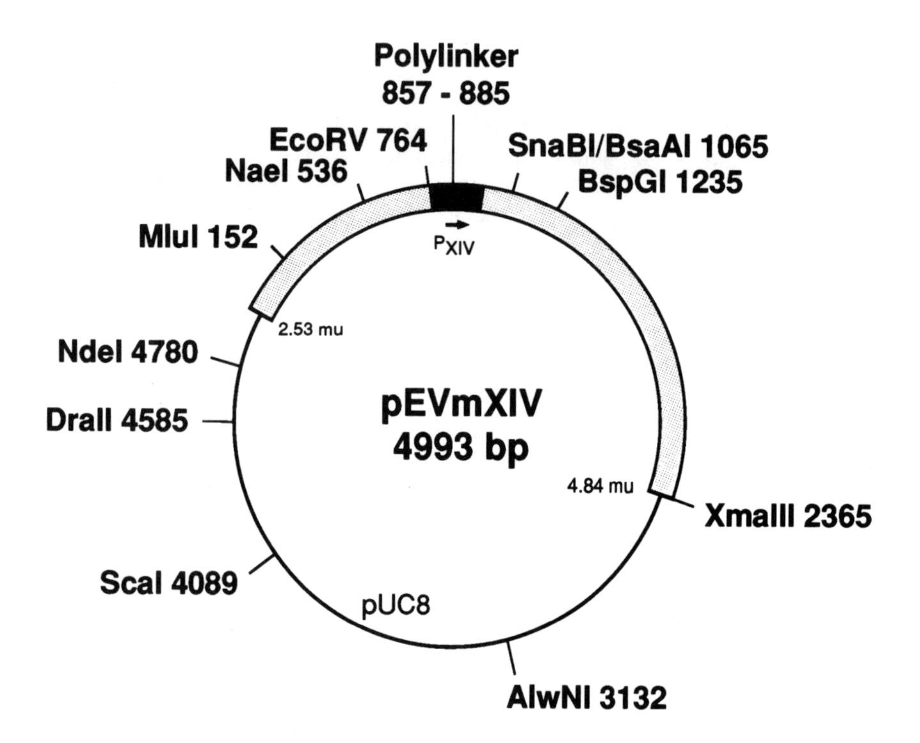

Plasmid	pEVmXIV
Promoter	P_{XIV}
Parent virus	wt AcMNPV
Plaque phenotype	occ^{-}
Related plasmids	pEVmod, pEV55
Selection in *E. coli*	Amp^{r}
Source	L.K. Miller

pEVmXIV

```
    -90         -80         -70          -60          -50               ′    -30
EcoRV|           |           |HindIII    |            * polh mRNA 5′end →     |
GATATCATGGAGATAATTAAAATGccaagcttggCGCAAATAAATAAGTATTTTACTGTTTTCGTA

        -20         -10          +1                         +631       +640
         |           |          BglII     EcoRI      ClaI  |           |
ACAGTTTTGTAATAAAAAAACCTATAAATAgatctcgagaattctagatcgatGGTACCGACT
                                       XhoI       XbaI         KpnI
```

The transfer plasmid pEVmXIV (Wang et al., 1991) allows the generation of an occ⁻ recombinant virus expressing the heterologous gene under the control of P_{XIV}, a modified polyhedrin promoter. P_{XIV} is a linker-scan mutant of P_{polh}, which has been reported to drive transcription ≈50% more efficiently than P_{polh} (Ooi et al., 1989). The HindIII linker is located upstream of the transcriptional start site. The entire P_{XIV} promoter, up to the +1 position, is present in this plasmid.

The plasmid pEVmod is identical to pEVmXIV except that it has a wt P_{polh} (Wang et al., 1991). Another related plasmid is pEV55, which has a wt P_{polh}, but has additional flanking sequences at either end of the cloned polyhedrin gene (Miller et al., 1986). All three plasmids have the same polylinker. Note that the XbaI site overlaps a site for *dam* methylation. This site may not be cleaved efficiently in DNA isolated from *dam*⁺ *E. coli.*

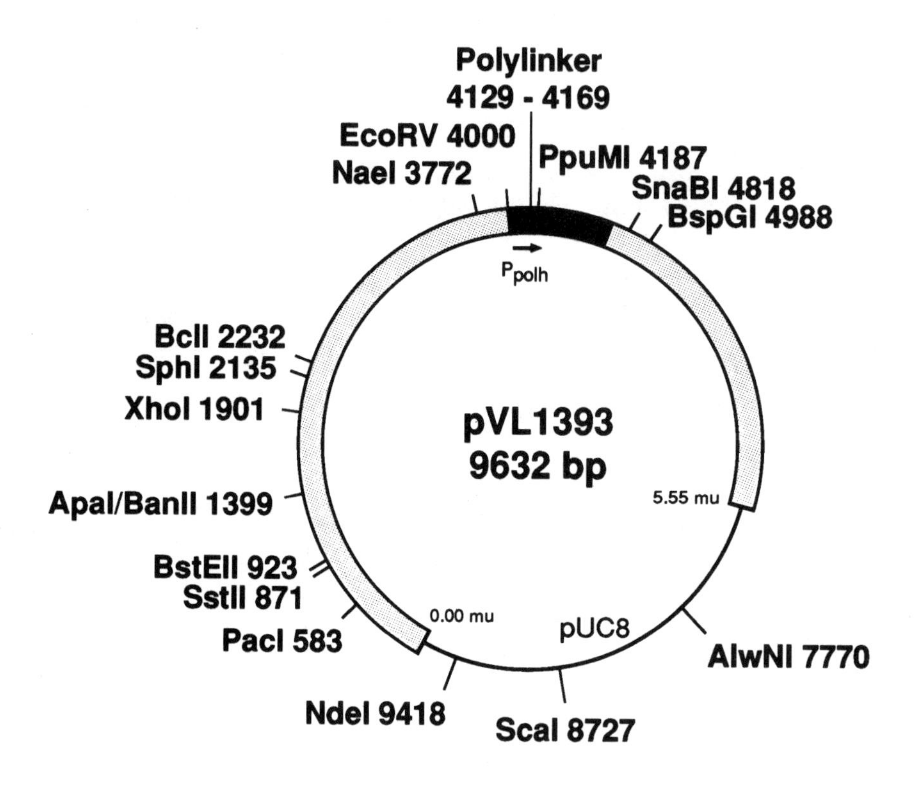

Plasmid	pVL1393
Promoter	P_{polh}
Parent virus	wt AcMNPV
Plaque phenotype	occ^-
Related plasmids	pVL1392, pVL941, p36C, pAcC12, pAcC13
Selection in *E. coli*	Amp^r
Source	Invitrogen

pVL1393

```
    -90         -80         -70         -60         -50
EcoRV |          |           |           |          * polh mRNA 5' end →
GATATCATGGAGATAATTAAAATGATAACCATCTCGCAAATAAATAAGTATTTTACTGTTTTCG

   -30         -20         -10        +1+3       +10         +20         +30
    |           |           |          | |        |           |           |
   TAACAGTTTTGTAATAAAAAAACCTATAAATATtCCGGATTATTCATACCGTCCCACCATC

      +35                                                          +177
       |  BamHI    KpnI        EcoRI        NotI   PstI              |
   GGGCGcggatcccgggtaccttctagaattccggagcggccgctgcagatctgatccTTTC
               SmaI         XbaI            EagI      BglII
```

pVL1392

```
-30         -20         -10        +1+3       +10         +20         +30
 |           |           |          | |        |           |           |
TAACAGTTTTGTAATAAAAAAACCTATAAATATtCCGGATTATTCATACCGTCCCACCATC

   +35                                                          +177
    |       BglII       NotI        EcoRI        KpnI    BamHI   |
GGGCGcggatcagatctgcagcggccgctccagaattctagaaggtacccgGGATCCTTTC
                     PstI   EagI            XbaI         SmaI
```

In pVL1392 and pVL1393 (Luckow and Summers, unpublished data), the *polh* ATG is mutated to ATT, and the polylinker is located at position +35. The plasmids were designed in this way because it was thought that sequences downstream from the *polh* ATG might contribute to optimal promoter activity. Recent evidence obtained with recombinant viruses generated with these plasmids indicates that, despite the mutation at +3, some translation still initiates at the +1 position (Beames et al., 1991). Thus, if you use these plasmids, it is advisable that you ensure that the ATG of the cloned gene is out-of-frame with the ATT at +1. Most translation should initiate at the first ATG of the heterologous gene.

The plasmids pVL1392 and pVL1393 differ from each other only in the orientation of the polylinker. Both are derived from the plasmid pVL941 (Luckow and Summers, 1989) by deletion of a short segment of the AcMNPV/pUC8 junction downstream of *polh*, and insertion of the polylinker in place of the single BamHI cloning site. Other similar plasmids include p36C (Page, 1989), pAcC12, and pAcC13 (Quilliam et al., 1990). The *polh* ATG is altered to ATC and a BamHI site is located at position +33 in p36C. The plasmids pAcC12 and pAcC13 are very similar to pVL1392 and pVL1393, but have different polylinker sequences.

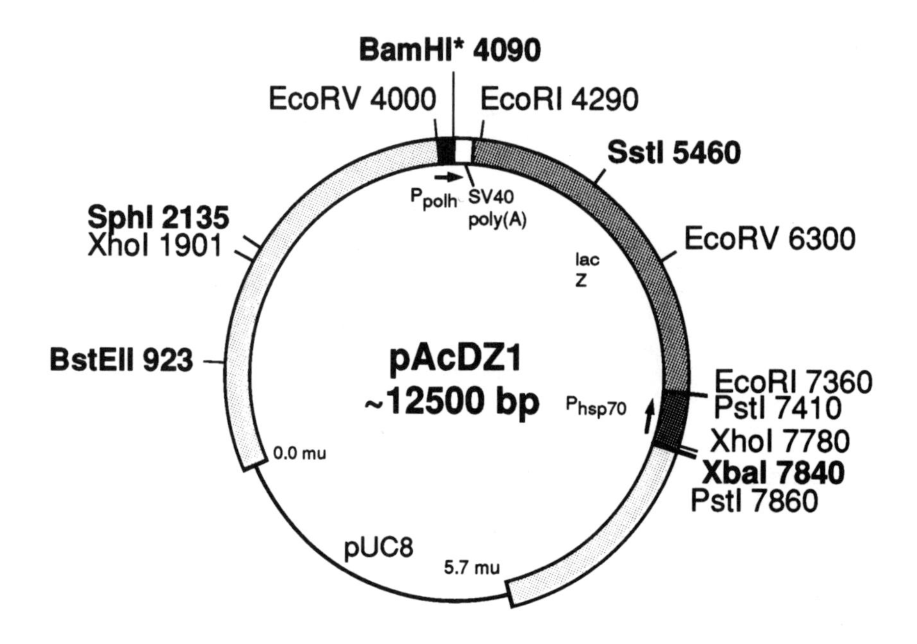

Plasmid	pAcDZ1
Promoter	P_{polh}
Parent virus	wt AcMNPV
Plaque phenotype	occ^-, blue
Related plasmids	pAcDZ6
Selection in *E. coli*	Amp^r
Source	J.M. Vlak

pAcDZ1

```
    -90         -80         -70         -60         -50                    -30
EcoRV|           |           |           |          * polh mRNA 5′end→      |
GATATCATGGAGATAATTAAAATGATAACCATCTCGCAAATAAATAAGTATTTTACTGTTTTCGTAACAGTTT

 -20         -10         -1                                           +670
  |           |           |   BamHI                                     |
TGTAATAAAAAAACCTATAAATccggatcc- ←SV40 poly(A) - lacZ - Phsp70  -GTTTCCC
```

The transfer plasmid pAcDZ1 (Zuidema et al., 1990) is designed to express heterologous genes under the control of P_{polh}. The plasmid also includes *lacZ* under the control of the *Drosophila hsp70* promoter (shaded in the sequence above) as a marker gene to facilitate the identification of recombinants. The SV40 polyadenylation signals after the 3′-end of *lacZ* also include translational stop signals in all three reading frames. Recombinant viruses are occ^- and blue. (The basal activity of the P_{hsp70} is sufficient to produce enough β-galactosidase without heat shock.) It is important to confirm that selected blue plaques are occ^- to avoid the possibility of selecting single-crossover recombinants that would be blue and occ^+.

The plasmid pAcDZ6 is a derivative of pAcDZ1, which includes the same P_{polh} promoter and cloning site, as well as the *lacZ* marker gene. However, pAcDZ6 is ≈2 kbp smaller and is therefore an easier plasmid to use.

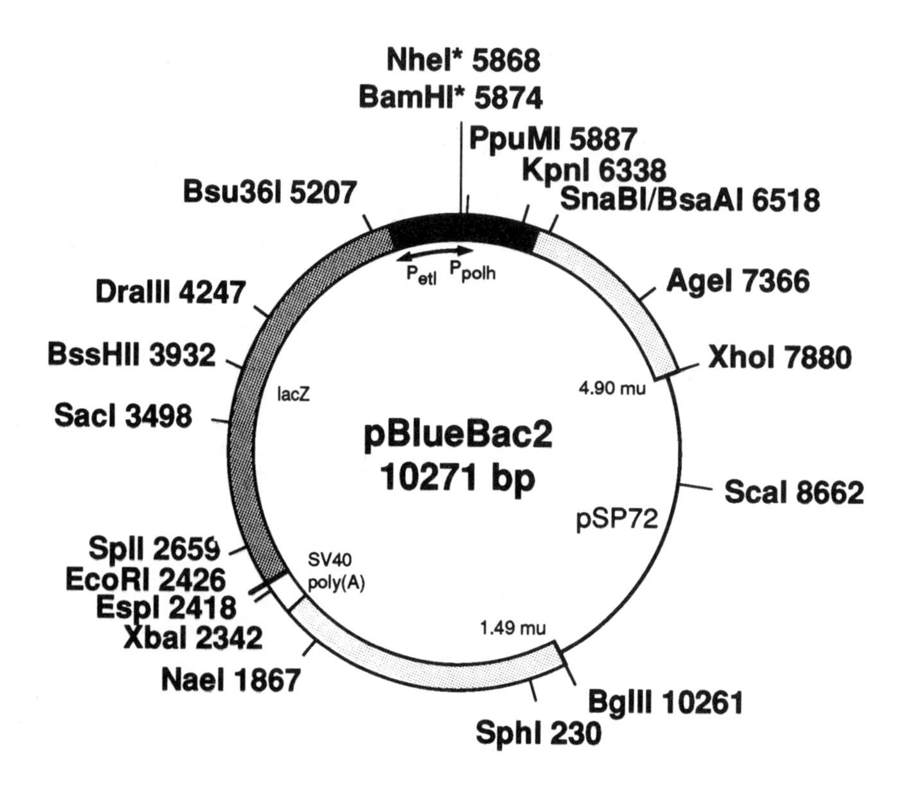

Plasmid	pBlueBac2 (pETL blue)
Promoter	P_{polh}
Parent virus	linearized AcMNPV
Plaque phenotype	occ^{-}, blue
Related plasmids	pBlueBac
Selection in *E. coli*	Amp^{r}
Source	Invitrogen or C. Richardson

pBlueBac2 (pETL blue)

```
    -90         -80         -70         -60         -50                  -30
EcoRV|           |           |           |             * polh mRNA 5'end→ |
GATATCATGGAGATAATTAAAATGATAACCATCTCGCAAATAAATAAGTATTTTACTGTTTTCGTAACAGTTT

  -20         -10         +1+3       +10         +20         +30    +36     +171
   |           |           | |         |           |           |      | NheI |
TGTAATAAAAAAACCTATAAATATtCCGGATTATTCATACCGTCCCACCATCGGGCGTgctagcGGATCCTTT
                                                                      BamHI
```

The *polh* ATG of pBlueBac2 (pETL blue) (Richardson et al., 1992) has been altered to ATT, and heterologous genes are inserted at an NheI or BamHI site located 36 nt downstream. This approach, which was also used with pVL1393 and related plasmids (see pVL1393 map and Chapter 6), was taken because it was thought that sequences downstream from the *polh* ATG might contribute to optimal promoter activity. Recent evidence obtained with recombinant viruses generated with pVL1393 indicates that, despite the mutation at +3, some translation still initiates at the +1 position (Beames et al., 1991). Thus, if using these plasmids, it is advisable to ensure that the ATG of the cloned gene is out-of-frame with the ATT at +1. Most translation should initiate at the first ATG of the heterologous gene.

The plasmid pBlueBac2 also includes the *E. coli lacZ* gene under the control of an early viral promoter, P_{etl} (Crawford and Miller, 1988). Recombinant viruses generated with this plasmid are occ^- and blue. Remember that it is important to verify that selected blue plaques are occ^-, as the majority of blue plaques will represent single-crossover recombinants that still include an intact *polh* gene. Use of parental viral DNA linearized at or near the *polh* locus will minimize the number of single-crossover recombinants obtained.

The plasmid pBlueBac2 was derived from pBlueBac by replacement of the pUC8 vector part of pBlueBac with pSP72 (Promega) sequences. The bacteriophage f1 origin of replication present in pBlueBac was also removed, and the plasmid was altered so that the BamHI site adjacent to the NheI site is unique in pBlueBac2.

In mid-1992, Invitrogen plans to introduce another pBlueBac derivative, pBac*HIS*. Viruses generated with this plasmid should overproduce recombinant proteins fused to a short leader peptide derived from the bacteriophage T7 gene 10 protein. This leader contains a high-affinity, divalent metal-ion binding site and a site for cleavage by enterokinase. The recombinant protein can be purified by chromatography on an Ni^{+2} affinity resin. If desired, the leader can then be removed from the protein by enterokinase cleavage.

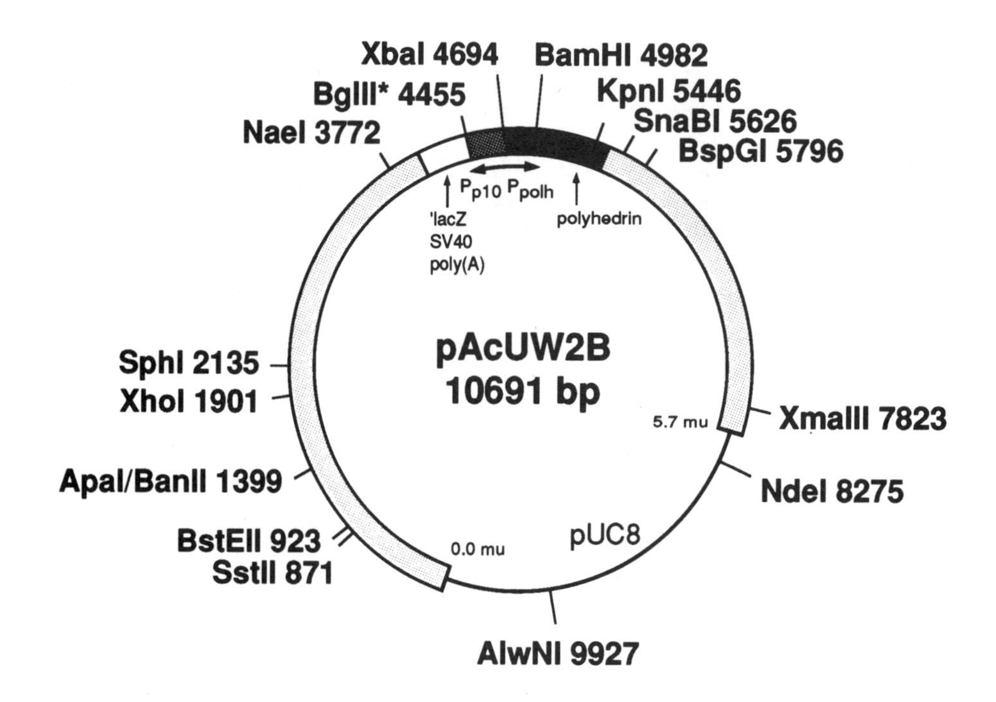

Plasmid	pAcUW2B
Promoter	P_{p10}
Parent virus	AcRP6-SC
Plaque phenotype	occ^{+}
Related plasmids	pAcUW2A
Selection in *E. coli*	Amp^{r}
Source	R.D. Possee

pAcUW2B

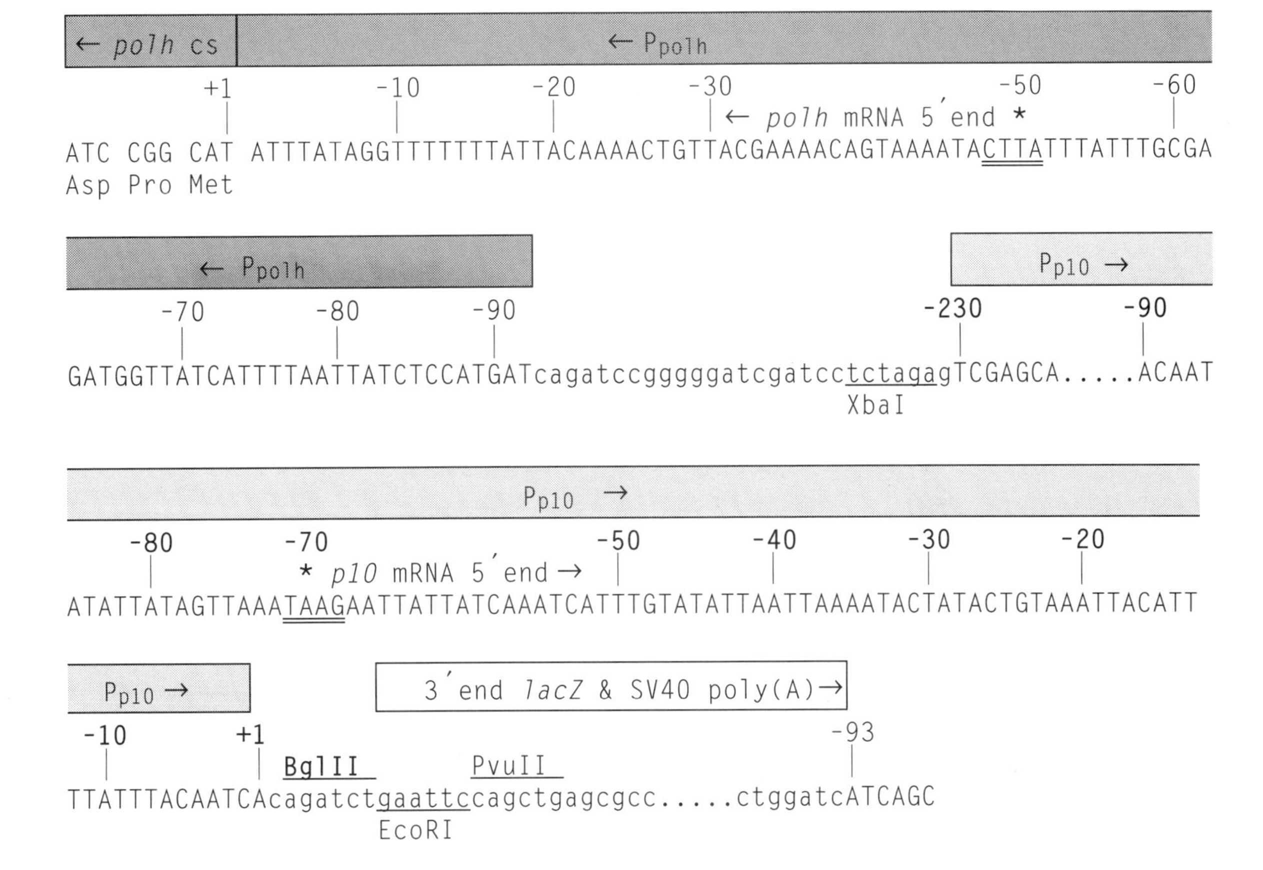

The plasmid pAcUW2B (Weyer et al., 1990) is designed for the expression of a heterologous gene in an occ$^+$ recombinant virus. The heterologous gene is placed under the control of the *p10* promoter, P_{p10}, and the transfer plasmid also includes the wt *polh* promoter and coding sequences (cs). A sequence comprising 322 bp from the 3′-end of the *lacZ* gene, followed by 132 bp of SV40-derived sequences, is located immediately downstream from the BglII cloning site. The SV40 sequence provides polyadenylation signals for the heterologous gene.

The sequence above is presented in the direction of transcription of the heterologous gene, that is, anticlockwise relative to the figure opposite. Numbers in bold refer to P_{p10}, whereas numbers in regular script refer to P_{polh}.

Extra sequences between P_{p10} and P_{polh} derive from the cloning procedure. This plasmid can be used with the parent virus AcRP6-SC, linearized at the introduced Bsu36I site in the *polh* locus (see Chapter 8). Recombinants are identified by screening for occ$^+$ plaques. This screen will not distinguish between single- and double-crossover recombinants. However, the frequency of single-crossover recombinants will be reduced by linearization of the parental viral DNA.

The transfer plasmid pAcUW2A (Weyer et al., 1990) is virtually identical to pAcUW2B. The two plasmids differ only in the relative orientation of the two promoters. In pAcUW2A, P_{p10} and P_{polh} are in tandem, whereas in pAcUW2B they are located back-to-back.

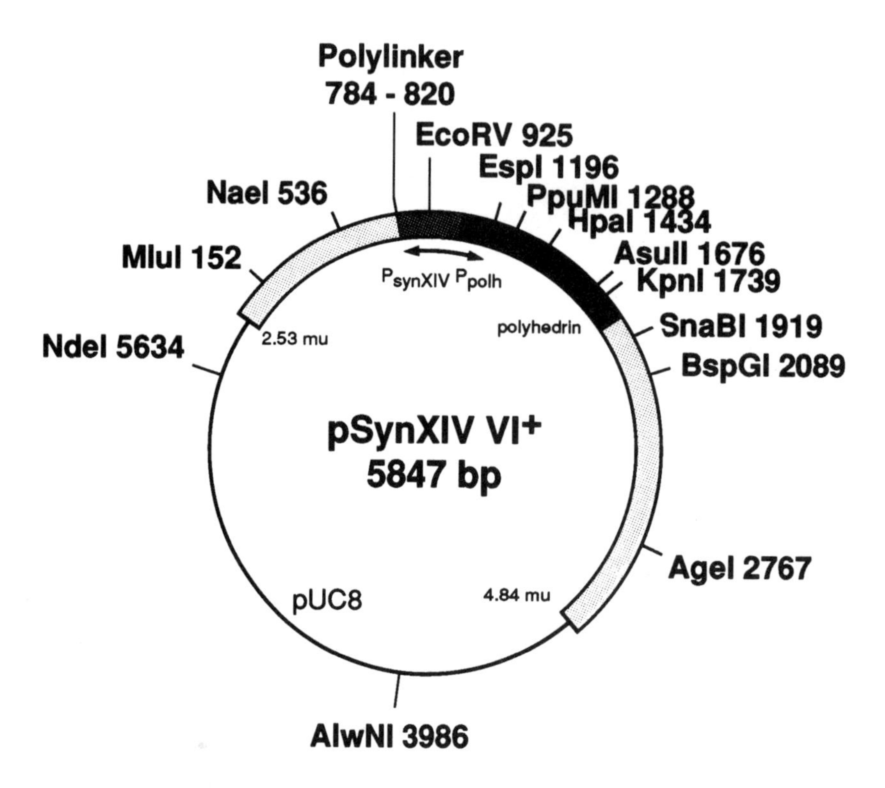

Plasmid	pSynXIV VI$^+$
Promoter	P_{synXIV}
Parent virus	vSynVI$^-$gal
Plaque phenotype	occ$^+$, white
Related plasmids	pSynXIV VI$^+$X3, pSynVI$^+$wtp, pLSXIV VI$^+$X3, pLSXIV VI$^+$
Selection in *E. coli*	Amp^r
Source	L.K. Miller

pSynXIV VI+

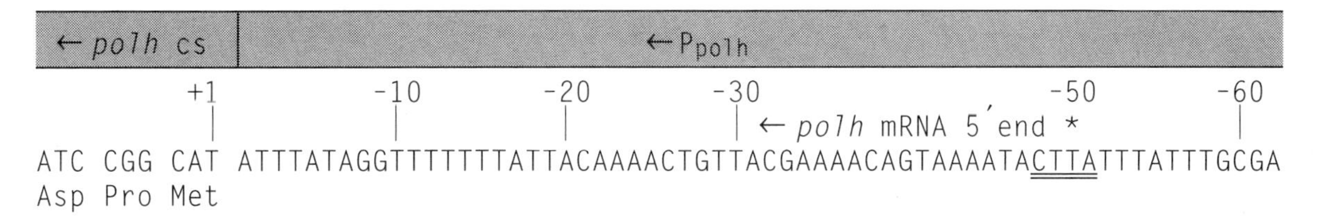

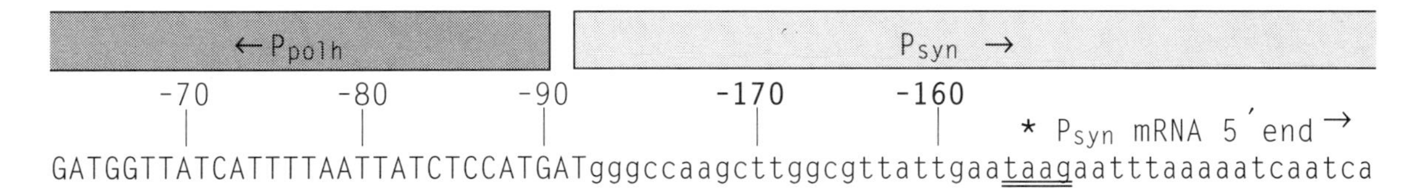

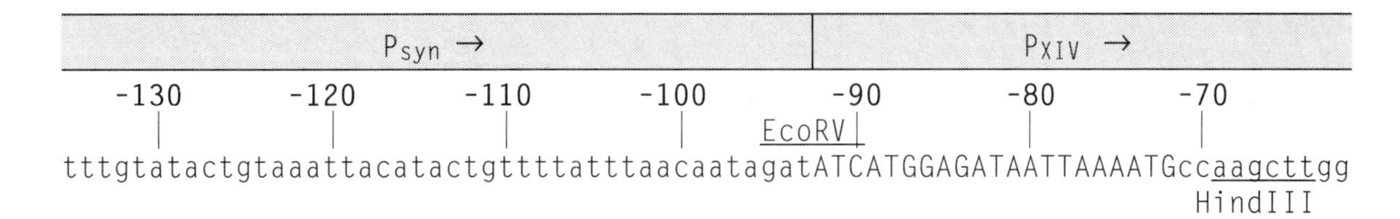

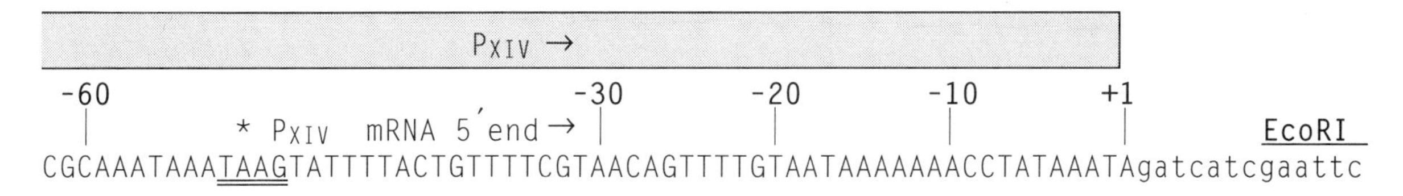


```
                                                        -93
          BamHI       XbaI        BstXI                  |
ctgcagccgggggatccactagttctagagcggccgccaccgcggtggagctccaATCAGC
 PstI             SpeI         NotI     SstII    SstI
```

pSynXIV VI+X3

```
    -10       +1                                                     -93
     |         |      EcoRI      PstI      SalI     polyA             |
ATAAAAAAACCTATAAATAgatcatcgaattctcgagctgcagatctgtcgacccgggaataaagagctccaATC
                                  XhoI      BglII       SmaI        SstI
```

The transfer plasmid pSynXIV VI$^+$ (Wang et al., 1991) allows the expression of a heterologous gene in an occ$^+$ recombinant virus. The heterologous gene is under the control of a hybrid promoter P_{synXIV}, comprised of the artificial promoter P_{syn} and the linker-modified polyhedrin promoter P_{XIV} arranged in tandem. The plasmid also includes the wt *polh* promoter and coding sequences (cs). The sequence is presented in the direction of transcription of the heterologous gene, that is, anticlockwise relative to the figure opposite. Numbers in bold refer to P_{synXIV}, whereas numbers in regular script refer to P_{polh}. This plasmid is best used with the parent virus vSynVI$^-$gal, an occ$^-$ virus with *lacZ* in place of *polh*. Recombinant viruses are white and occ$^+$, in a blue, occ$^-$ background. pSynXIV VI$^+$X3 differs from pSynXIV VI$^+$ only in the sequence of the polylinker (Wang et al., 1991). Other related plasmids are pSynVI$^+$wtp, pLSXIV VI$^+$, and pLSXIV VI$^+$X3 (Wang et al., 1991). These have different promoters directing the expression of the heterologous gene and either one of the polylinkers shown in the sequence. In the nomenclature of these plasmids, XIV refers to the P_{XIV} promoter; VI$^+$ or $^-$ refers to the presence or absence of *polh* sequences in the HindIII-V/EcoRI-I region; and X3 refers to the polylinker used.

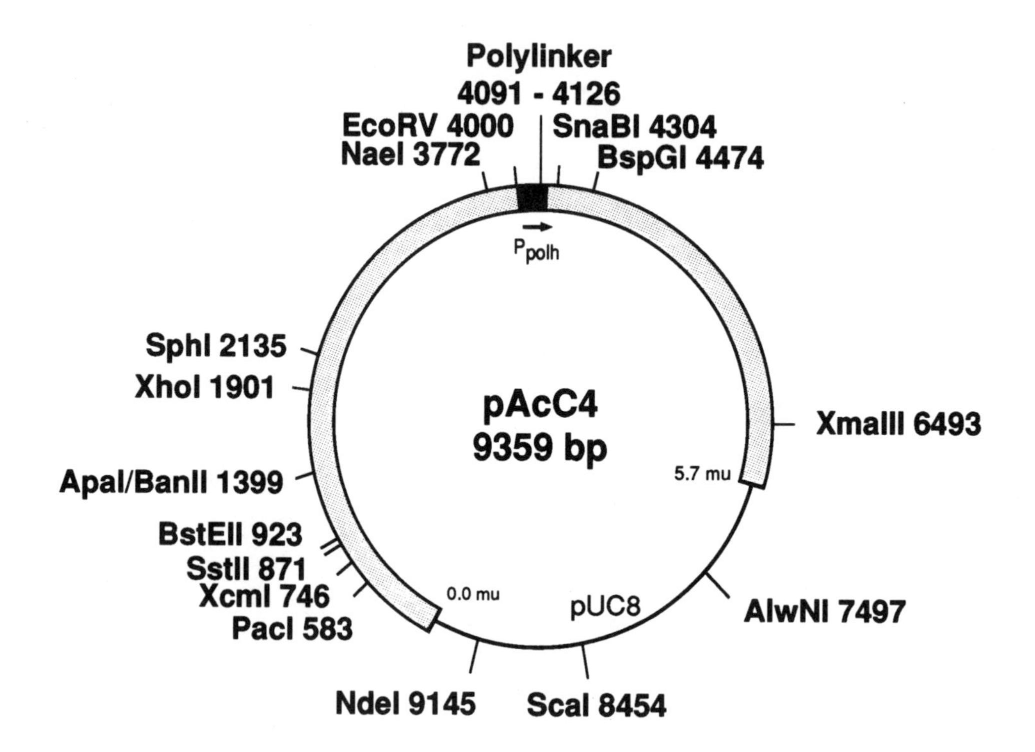

Plasmid	pAcC4
Promoter	P_{polh}
Parent virus	wt AcMNPV
Plaque phenotype	occ^{-}
Related plasmids	pAcC5
Selection in *E. coli*	Amp^{r}
Source	R. Clark

pAcC4

```
       -90            -80            -70            -60            -50                     -30
EcoRV|                 |              |              |             * polh mRNA 5′end→       |
GATATCATGGAGATAATTAAAATGATAACCATCTCGCAAATAAATAAGTATTTTACTGTTTTCGTAACAGTTTT

-20            -10        -3                                                           +654
 |              |          | NcoI           Kpn I       BglII    EcoRI                   |
GTAATAAAAAAACCTATAAccatggcggcccgggtacctgcagatctagaattcggatcctgatcaccgggGGA
                                     SmaI         PstI      XbaI        BamHI
```

pAcC5

```
-20            -10        -3                                                           +654
 |              |          | NcoI                       BamHI        XbaI    PstI      SmaI |
GTAATAAAAAAACCTATAAccatggcggcccggtgatcaggatccgaattctagatctgcaggtacccgggGGA
                                                         EcoRI   BglII      KpnI
```

Plasmids pAcC4 and pAcC5 [O'Rourke and Kawasaki, described in Luckow and Summers (1988b)] allow the expression of heterologous genes that lack an ATG under the control of the *polh* promoter, P_{polh}. These plasmids differ only in the orientation of their polylinkers. In both cases, the polylinker includes an ATG (within the NcoI site) at the normal position of the *polh* ATG. Heterologous genes lacking an ATG may be cloned at the NcoI site or inserted in-frame at one of the cloning sites downstream. Recombinant viruses generated with these plasmids are occ⁻. Note that the XbaI site overlaps a site for *dam* methylation. This site might not be cleaved efficiently in DNA isolated from *dam*⁺ *E. coli*.

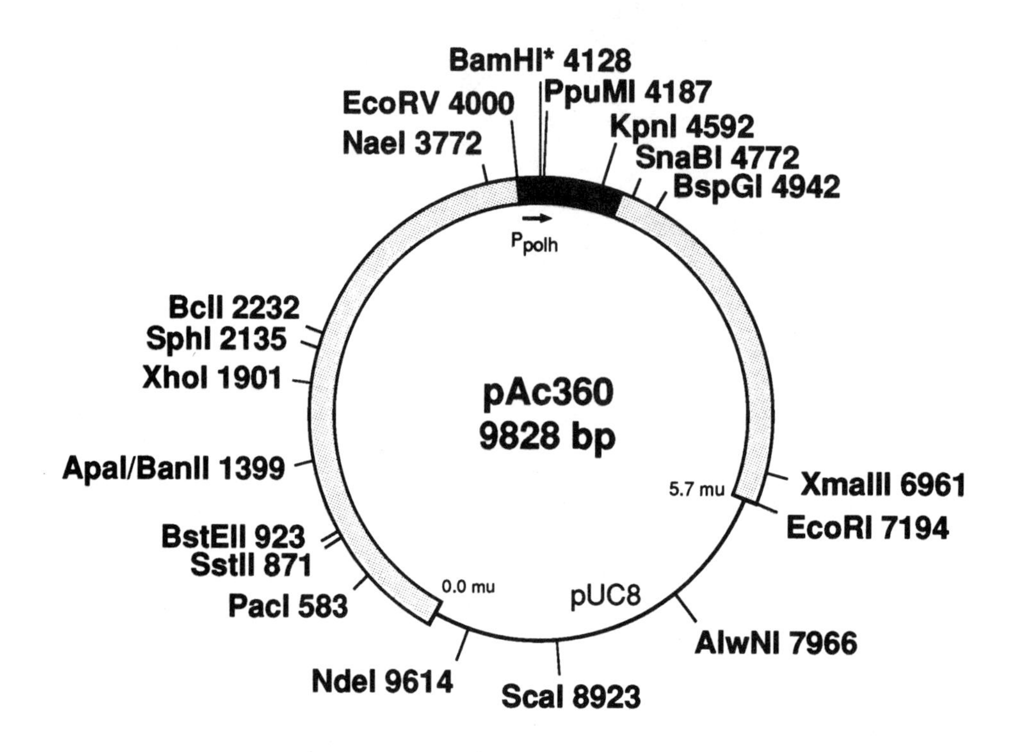

Plasmid	pAc360
Promoter	P_{polh}
Parent virus	wt AcMNPV
Plaque phenotype	occ^-
Related plasmids	pAc311, pAc101, pAc401, pAc436
Selection in *E. coli*	Amp^r
Source	Invitrogen

pAc360

```
     -90       -80       -70       -60       -50       ,     -30
EcoRV|          |         |         |        * polh mRNA 5'end→ |
GATATCATGGAGATAATTAAAATGATAACCATCTCGCAAATAAATAAGTATTTTACTGTTTTCGTAA

        -20       -10        +1          +10         +20         +30
         |         |          |           |           |           |
CAGTTTTGTAATAAAAAAACCTATAAAT ATG CCG GAT TAT TCA TAC CGT CCC ACC ATC
                             Met Pro Asp Tyr Ser Tyr Arg Pro Thr Ile

   +34      +177
    |  BamHI   |
GGG Ccg gat ccT TTC
Gly Pro Asp Pro Phe
```

pAc311

```
-10        +1          +10         +20         +30    +35        +177
 |          |           |           |           |      |   BamHI   |
 ACCTATAAAT ATG CCG GAT TAT TCA TAC CGT CCC ACC ATC GGG GGc gga tcc TTT
            Met Pro Asp Tyr Ser Tyr Arg Pro Thr Ile Gly Arg Gly Ser Phe
```

The plasmid pAc360 (Smith et al., 1983b) is representative of several plasmids designed to facilitate the expression of heterologous genes as fusions with the N-terminal part of *polh*. Other similar plasmids include pAc311 (shown here), pAc101, pAc401, and pAc436 (Luckow and Summers, 1988a; Smith et al., 1983b). These differ in the extent of *polh* sequences added to the heterologous gene and the restriction site used for cloning. In pAc401 and 436, a SmaI cloning site is located at position +4 or +6 respectively, whereas pAc101 uses the natural BamHI site at nt 171. All of these plasmids give rise to occ⁻ recombinant viruses.

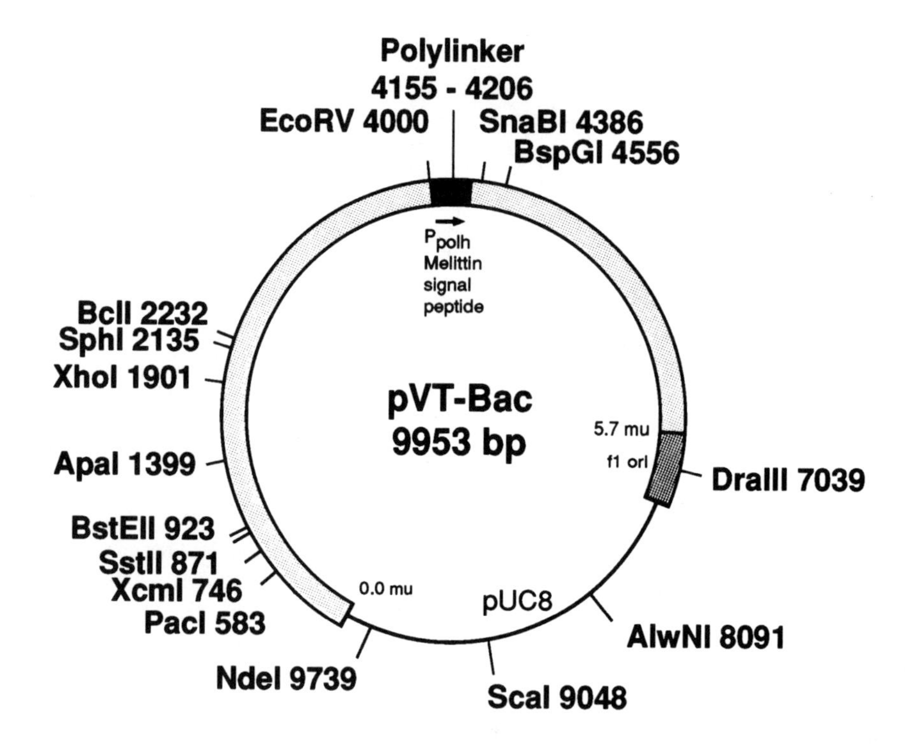

Plasmid	pVT-Bac
Promoter	P_{polh}
Parent virus	wt AcMNPV
Plaque phenotype	occ^-
Selection in *E. coli*	Amp^r
Source	T. Vernet or D.C. Tessier

pVT-Bac

```
    -90         -80         -70         -60         -50              ,     -30
EcoRV|           |           |           |           * polh mRNA 5'end → |
GATATCATGGAGATAATTAAAATGATAACCATCTCGCAAATAAATAAGTATTTTACTGTTTTCGTAACAG

    -20         -10       -1
     |           |         |          Melittin signal peptide →
TTTTGTAATAAAAAAACCTATAAAT ATG aaa ttc tta gtc aac gtt gcc ctt gtt ttt
                          Met Lys Phe Leu Val Asn Val Ala Leu Val Phe

                                                  BamHI          PstI
atg gtc gtg tac att tct tac atc tat gcg gatccaagcccgggctgcaggagctc
Met Val Val Tyr Ile Ser Tyr Ile Tyr Ala Asp...    SmaI          SstI
                                        ^

                   +631      +640
  NotI          EcoRI |         |
gcggccgcgctagcgaattcGGTACCGACT
         NheI         KpnI
```

The transfer plasmid pVT-Bac (Tessier et al., 1991) provides a signal peptide to facilitate the secretion of expressed gene products. The signal peptide used is the honeybee melittin signal. It is inserted immediately downstream from the *polh* ATG, and the heterologous gene is inserted at a polylinker downstream of the signal peptide. The signal peptide is shaded in the sequence above and the site of cleavage is indicated by the caret. Heterologous gene expression is under the control of the *polh* promoter, and the recombinant viruses are occ$^-$. An additional feature of this plasmid is the presence of a bacteriophage f1 origin of replication at the AcMNPV-pUC8 junction downstream of *polh*. Infection of transformed *E. coli* cells containing this plasmid with f1 helper phage results in the generation of single-stranded pVT-Bac DNA, which facilitates sequence analysis, site-directed mutagenesis, and the generation of strand-specific probes. The rescued strand is in the antisense orientation relative to the direction of *polh* transcription.

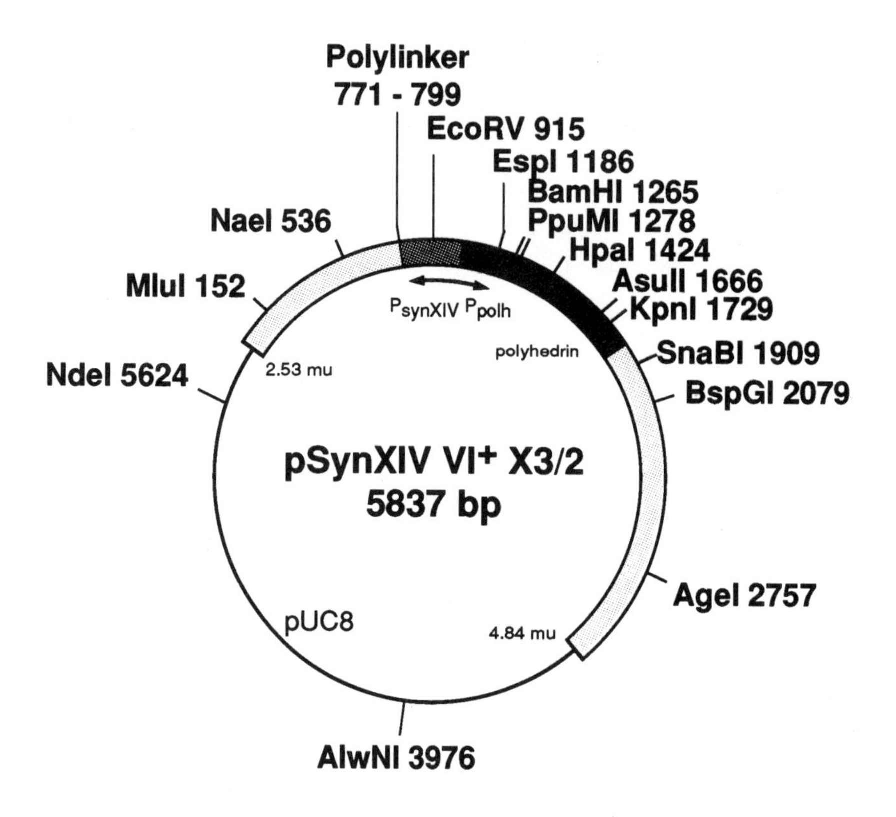

Plasmid	pSynXIV VI$^+$X3/2, pSynXIV VI$^+$X3/3, pSynXIV VI$^+$X3/4
Promoter	P_{synXIV}
Parent virus	vSynVI$^-$gal
Plaque phenotype	occ$^+$, white
Related plasmids	pLSXIV VI$^+$X3/2, pLSXIV VI$^+$X3/3, pLSXIV VI$^+$X3/4
Selection in *E. coli*	Amp^r
Source	L.K. Miller

pSynXIV VI+X3 (unfused)

```
     -10        +1                                                            -93
      |          |      EcoRI        PstI       SalI       poly(A)             |
ATAAAAAAACCTATAAATAgatcatcgaattctcgagctgcagatctgtcgacccgggaataaagagctccaATC
                                    XhoI       BglII        SmaI         SstI
```

pSynXIV VI+X3/2

```
 (+1)                 MET Glu Phe Cys Arg Ser Val Asp Pro Gly ...          -93
  |                       EcoRI       BglII          SmaI          SstI     |
AATAgatcatcgaattaat atg gaa ttc tgc aga tct gtc gac ccg gga ataaagagctccaAT
                                PstI          SalI          poly(A)
```

pSynXIV VI+X3/3

```
 (+1)                 MET Gly Ile Leu Gln Ile Cys Arg Pro Gly ...          -93
  |                        EcoRI       BglII          SmaI          SstI    |
AATAgatcatcgaattaat atg gga att ctg cag atc tgt cga ccc ggg aataaagagctccaA
                                 PstI          SalI           poly(A)
```

pSynXIV VI+X3/4

```
 (+1)                 MET Ala Asn Ser Ala Asp Leu Ser Thr ...              -93
  |                         EcoRI       BglII          SmaI         SstI    |
ATTAgatcatcgaattaat atg gcg aat tct gca gat ctg tcg acc cgggaataaagagctccaA
                                  PstI          SalI          poly(A)
```

The plasmids pSynXIV VI$^+$X3/2, X3/3, and X3/4 (Wang et al., 1991) allow the expression of heterologous genes lacking their own ATG in an occ$^+$ recombinant virus. They are based on pSynXIV VI$^+$X3, differing from it only in the polylinkers, which supply an ATG to the heterologous gene in all three reading frames. In the sequence above, the polylinker of pSynXIV VI$^+$X3 is shown for comparison, followed by the X3/2, X3/3, and X3/4 polylinkers. The sequences are presented in the direction of transcription of the heterologous gene, that is, anticlockwise relative to the figure opposite. See the description of pSynXIV VI$^+$X3 for further details of the structure of these plasmids. These plasmids are used with the occ$^-$, *lacZ*-positive parent virus vSynVI$^-$gal, and recombinants are occ$^+$ and white, in a blue occ$^-$ background.

The plasmids pLSXIV VI$^+$X3/2, X3/3, and X3/4 differ from the pSynXIV VI$^+$X3 series in that only P_{XIV}, rather than P_{synXIV}, is used to drive heterologous gene expression (Wang et al., 1991). In the nomenclature of these plasmids, XIV refers to the P_{XIV} promoter; VI$^+$ or $^-$ refers to the presence or absence of polyhedrin sequences in the HindIII-V/EcoRI-I region; and X3 refers to the polylinker used.

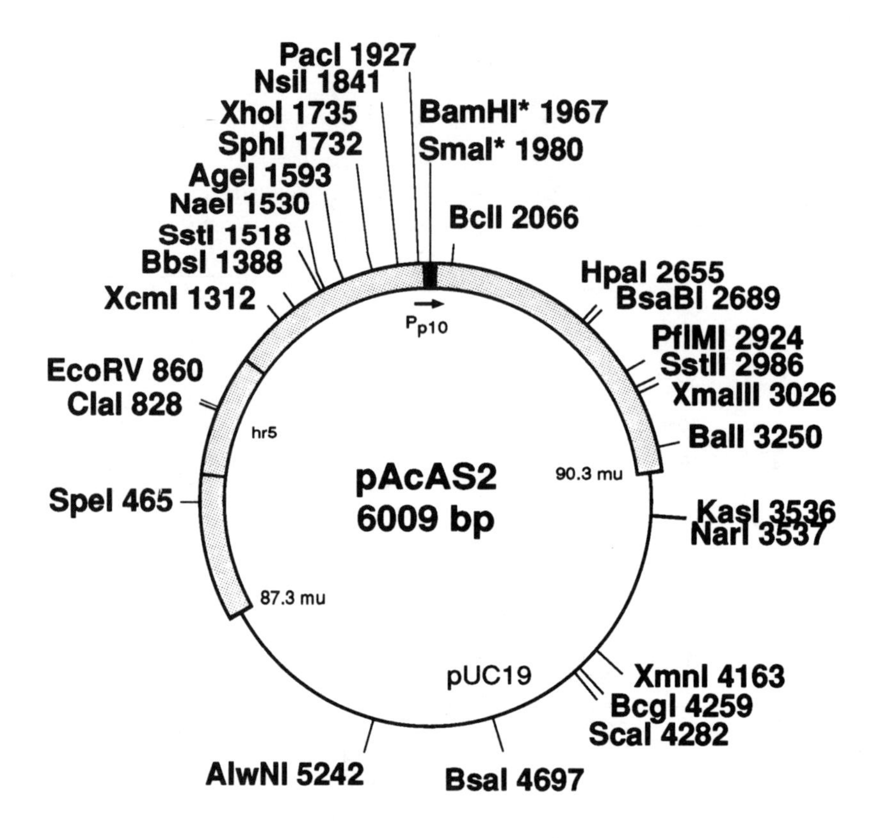

Plasmid	pAcAS2
Promoter	P_{p10}
Parent virus	vAcAS3
Plaque phenotype	occ^+, white
Related plasmids	pAcAS3
Selection in *E. coli*	Amp^r
Source	J.M. Vlak

pAcAS2

```
-90        -80        -70                    -50        -40        -30        -20
 |          |          * p10 mRNA 5′end→      |          |          |          |
ACAATATATTATAGTTAAATAAGAATTATTATCAAATCATTTGTATATTAATTAAAATACTATACTGTAAATTA
                                                        PacI

     -10        +1                     +283
      |          |BamHI      SmaI       |
CATTTTATTTACAATCAggatccctgaacccgggaagcttTAAATGAATCGTTT
                                   HindIII
```

The transfer plasmid pAcAS2 (Vlak et al., 1990) facilitates the insertion of heterologous genes into the *p10* locus of AcMNPV. It is best used with the parent virus, vAcAS3, in which the *lacZ* gene, under the control of the *Drosophila hsp70* promoter, is inserted at the *p10* locus. Recombinants are identified by a white, occ$^+$ plaque phenotype in a background of blue, occ$^+$ plaques.

The plasmid pAcAS3 is a derivative of pAcAS2 in which *lacZ*, under the control of the *Drosophila hsp70* promoter, is included in the transfer plasmid (Vlak et al., 1990). The *lacZ* gene is downstream of, and in the opposite orientation to, the *p10* promoter. Heterologous genes may be inserted at a unique BamHI site. (This plasmid lacks the SmaI site present in pAcAS2.) This plasmid is used in combination with wt AcMNPV, and recombinants are identified by screening for blue, occ$^+$ plaques in a white, occ$^+$ background.

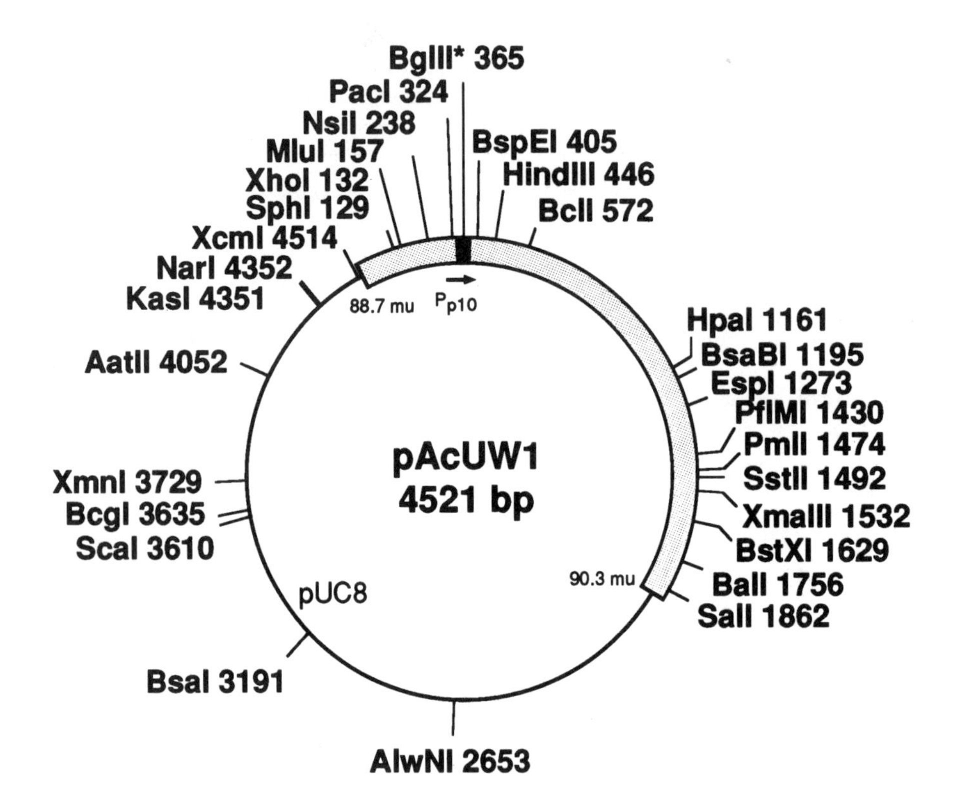

Plasmid	pAcUW1
Promoter	P_{p10}
Parent virus	AcUW1-*lacZ* or AcUW1-PH
Plaque phenotype	occ^+ or occ^-, white
Selection in *E. coli*	Amp^r
Source	R.D. Possee

pAcUW1

```
-90             -80             -70             ,       -50             -40             -30             -20
 |               |               * p10 mRNA 5 end →      |               |               |               |
ACAATATATTATAGTTAAATAAGAATTATTATCAAATCATTTGTATATTAATTAAAATACTATACTGTAAATTA
                                                                         PacI

        -10             +1+152
         |               | |
CATTTTATTTACAATCAcAGATCTCAG
                  BglII
```

The transfer plasmid pAcUW1 (Weyer et al., 1990) allows the insertion of heterologous genes into the *p10* locus of AcMNPV. Weyer et al. (1990) recommend that this plasmid be used either with AcUW1-*lacZ* or AcUW1-PH as the parent virus. In the former case, *lacZ* is inserted into the *p10* locus of the parent virus. Recombinant viruses may be identified as white, occ^+ plaques against a blue, occ^+ background due to replacement of *lacZ* with the heterologous gene. The large area of blue around the parent virus plaques, due to the high levels of expression from P_{p10}, may interfere with the identification of recombinant viruses with this screen. The parent virus AcUW1-PH expresses *polh* under the control of the *p10* promoter. This virus lacks an intact *polh* gene at its normal location. Thus, recombinants obtained when this parent virus is used are occ^- in an occ^+ background.

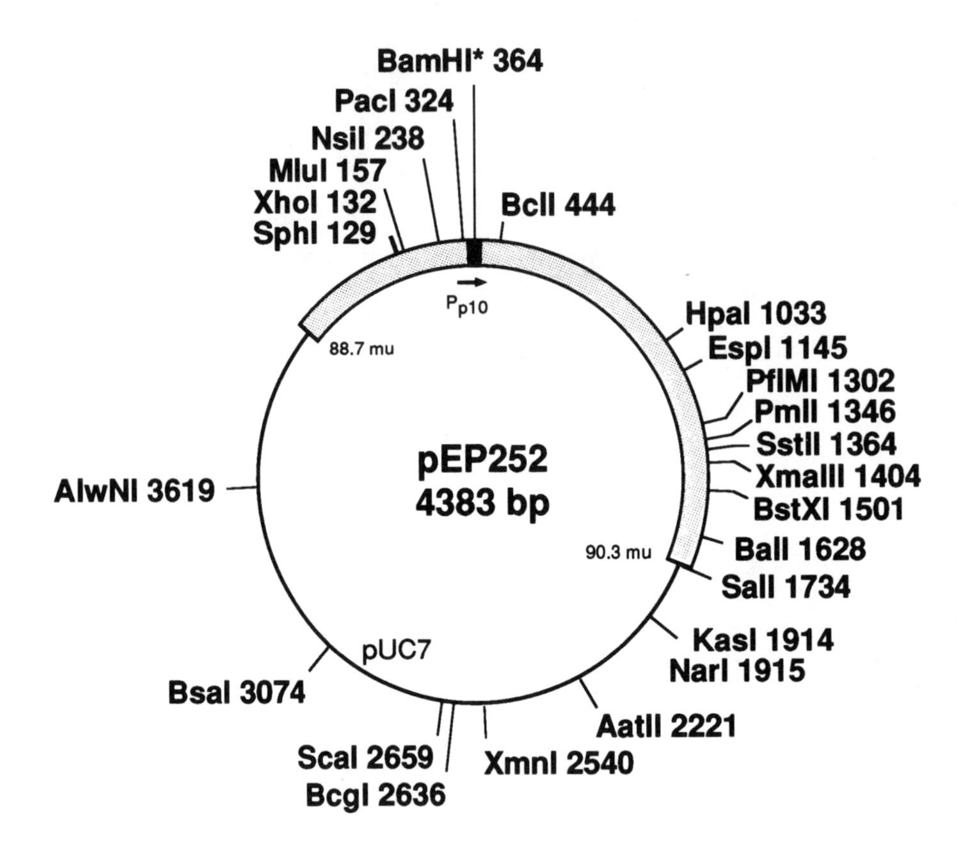

Plasmid	pEP252
Promoter	P_{p10}
Parent virus	Ac228z
Plaque phenotype	occ^+, white
Selection in *E. coli*	Amp^r
Source	P. Faulkner

pEP252

```
-90        -80        -70                   -50        -40        -30        -20
 |          |          * p10 mRNA 5' end→    |          |          |          |
ACAATATATTATAGTTAAATAAGAATTATTATCAAATCATTTGTATATTAATTAAAATACTATACTGTAAATTA
                                                       PacI

      -10        +1       +287
       |          |BamHI    |
CATTTTATTTACAATCAggatccggTGAATCGTTT
```

The plasmid pEP252 (J. Kuzio & P. Faulkner, personal communication) is a transfer plasmid designed for the insertion of heterologous genes into the *p10* locus of AcMNPV. It is best used with the parent virus, Ac228z, in which the *lacZ* gene is inserted at the *p10* locus (Kuzio et al., 1989). The *lacZ* gene is fused out-of-frame with the p10 N-terminus. However, low levels of β-galactosidase are produced so that Ac228z plaques are blue following staining with X-gal, but not excessively so. Recombinants are identified by a white, occ^+ plaque phenotype in a background of blue, occ^+ plaques. Note that the *lacZ* insertion disrupts part of the *p74* gene in Ac228z so that this virus is avirulent in insects (in the absence of wt helper virus). However, recombinants obtained with the plasmid and virus system should have a functional p74.

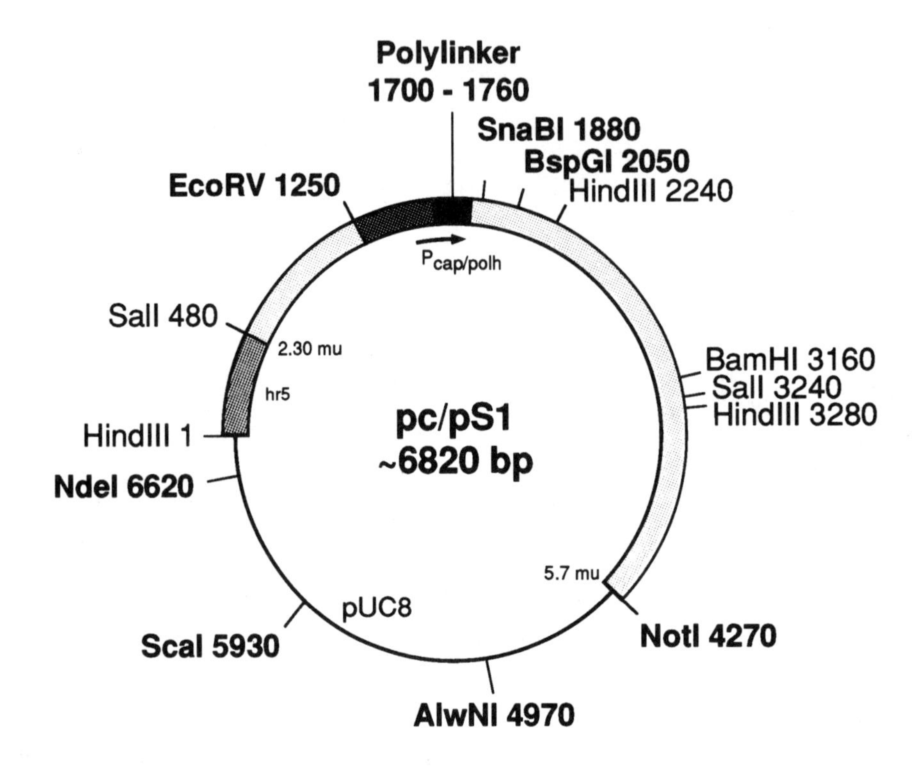

Plasmid	pc/pS1
Promoter	$P_{cap/polh}$
Parent virus	wt AcMNPV
Plaque phenotype	occ^-
Selection in *E. coli*	Amp^r
Source	L.K. Miller

pc/pS1

```
    -90                                    -60         -50                -30
     |      * cap mRNA 5′end #2→            |           * polh mRNA 5′end →|
GGAACAATATAAGAATTTAAAATTccaagcttggCGCAAATAAATAAGTATTTTACTGTTTTCGTAACA
                                  HindIII

     -20         -10          +1
      |           |           |BglII          SmaI           Bsu36I     EcoRI
GTTTTGTAATAAAAAAACCTATAAATAgatctcctaaggcccgggaccgcggccttaggtctagaattc
                                  Bsu36I           SstII           XbaI

                 +637
       PstI        |
tcgaggatcctgcaGGTACCGACT
XhoI BamHI      KpnI
```

The transfer plasmid pc/pS1 (Sihler, O'Reilly, and Miller, unpublished) is designed to allow the expression of heterologous genes under the control of a hybrid capsid/polyhedrin promoter, $P_{cap/polh}$. The late capsid promoter P_{cap} contains three transcriptional start sites. In $P_{cap/polh}$, P_{cap} sequences that include the two distal sites are fused to the modified polyhedrin promoter P_{XIV} at the HindIII linker inserted in P_{XIV} (Thiem and Miller, 1990). This arrangement allows gene expression to initiate in the late phase of infection from the two capsid start sites, with higher levels of expression very late in infection when P_{XIV} becomes active. In the sequence presented above, only one of the P_{cap} start sites is shown. The other is located 216 nt further upstream. The $P_{cap/polh}$ hybrid promoter and the polylinker are located within the polyhedrin locus so that recombinant viruses are occ⁻.

pc/pS1 was derived from the plasmid pCappolhcat (Thiem and Miller, 1990) by replacement of the *cat* gene with the polylinker above. In addition, a SmaI site near the *hr*5 sequence in pCappolhcat was removed, so that the SmaI site in the polylinker is unique. Note that the Bsu36I sites in the polylinker are the only Bsu36I sites in the plasmid; they effectively represent an additional, unique cloning site.

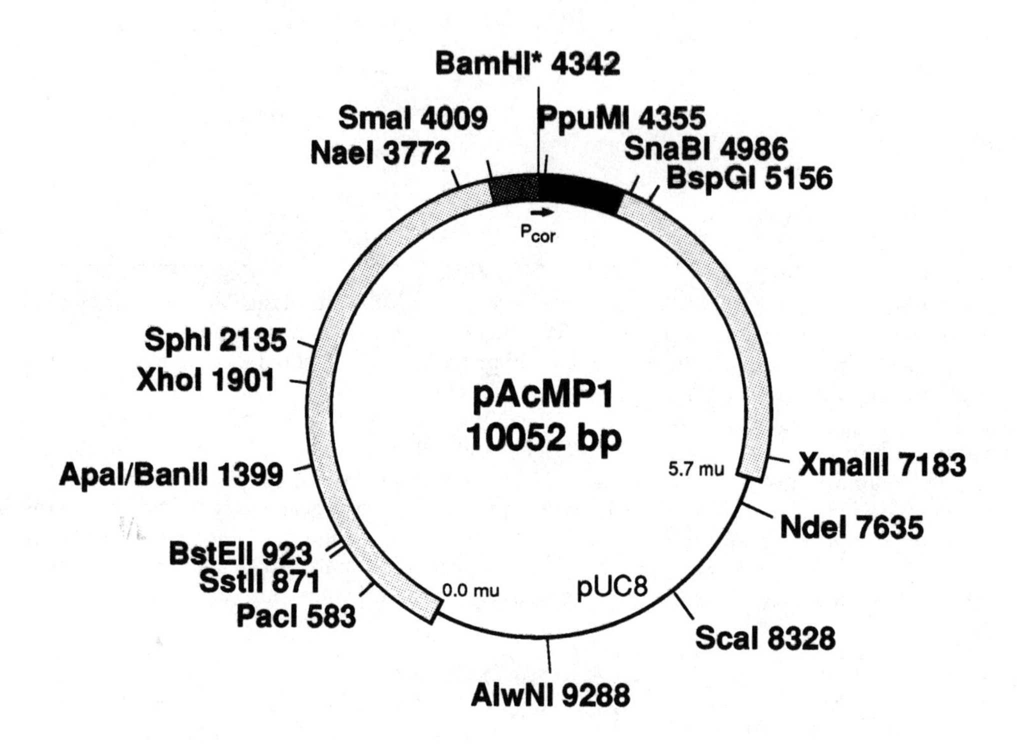

Plasmid	pAcMP1
Promoter	P_{cor}
Parent virus	wt AcMNPV
Plaque phenotype	occ^{-}
Selection in *E. coli*	Amp^{r}
Source	R.D. Possee

pAcMP1

```
-90        -80        -70        -60        -50        -40
 |          |          |          |          |          |* cor mRNA 5′end →
ACATATATTTGGGAGTTCAGTCGTCGAATGCAAAGCGTAAAAAATATTAATAAGGTAAAAATTACAGCTA

-20        -10        -1   +177
 |          |          |BamHI |
CATAAATTACACAATTTAAACggatccTTTCCT
                DraI
```

The plasmid pAcMP1 (Hill-Perkins and Possee, 1990) is a transfer plasmid that facilitates the insertion of heterologous genes into the *polh* locus of AcMNPV under the control of the P_{cor} promoter rather than under the control of P_{polh}. P_{cor} is a late promoter, so transcription of the heterologous gene should initiate earlier after infection than if a very late promoter, such as P_{polh} or P_{p10}, were used. However, the level of expression may not be as high very late in infection. Since the heterologous gene is introduced into the *polh* locus, recombinant viruses generated with this plasmid are occ^-. In the sequence above, numbers in bold (before the BamHI cloning site) refer to P_{cor}, whereas the regular script (after the BamHI site) refers to P_{polh}.

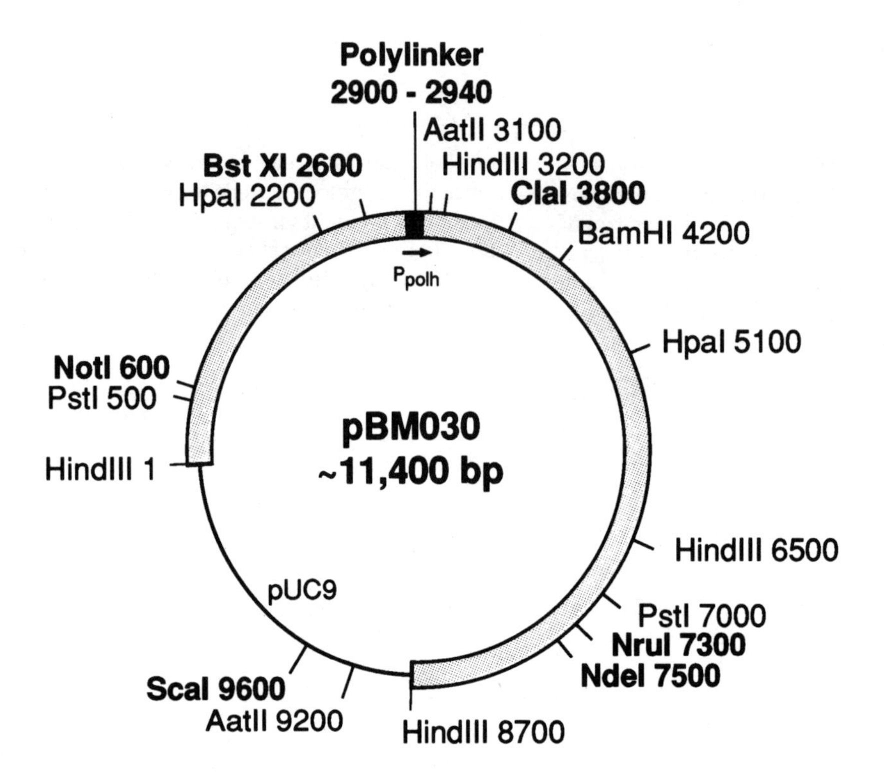

Plasmid	pBM030
Promoter	P_{polh}
Parent virus	wt BmNPV
Plaque phenotype	occ^-
Related plasmids	pBE274, pBE284, pBE520, pBK273, pBK283
Selection in *E. coli*	Amp^r
Source	S. Maeda

pBM030

```
-70       -60       -50              -30       -20       -10
 |         |         * polh mRNA 5'end →|        |         |
TAACCATCTCGCAAATAAATAAGTATTTTACTGTTTTCGTAACAGTTTTGTAATAAAAAAACCTA

-4                                           +736
 |         SstI     EcoRI     EcoRV      AatI  |
TAgatctaagagctcccgggaattccatggatatctagataggcctTAA
 BglII           SmaI      NcoI      XbaI
```

The plasmid pBM030 (Maeda, 1989b) is one of a series of transfer plasmids available for the expression of a heterologous gene under the control of the polyhedrin promoter of BmNPV. Recombinants are identified based on their occ⁻ phenotype. The transfer plasmids pBE274, pBE284, pBE520, pBK273, and pBK283, which differ only slightly from pBM030 and supply different cloning sites, are also available (Maeda, 1989b).

A series of plasmids (pBF series plasmids and p89BX40) is also available for the expression of a heterologous gene as a fusion with the N-terminal part of polyhedrin. Several different plasmids that differ in the amount of polyhedrin sequences present in the gene fusion are available. These plasmids are discussed in more detail in Maeda (1989b).

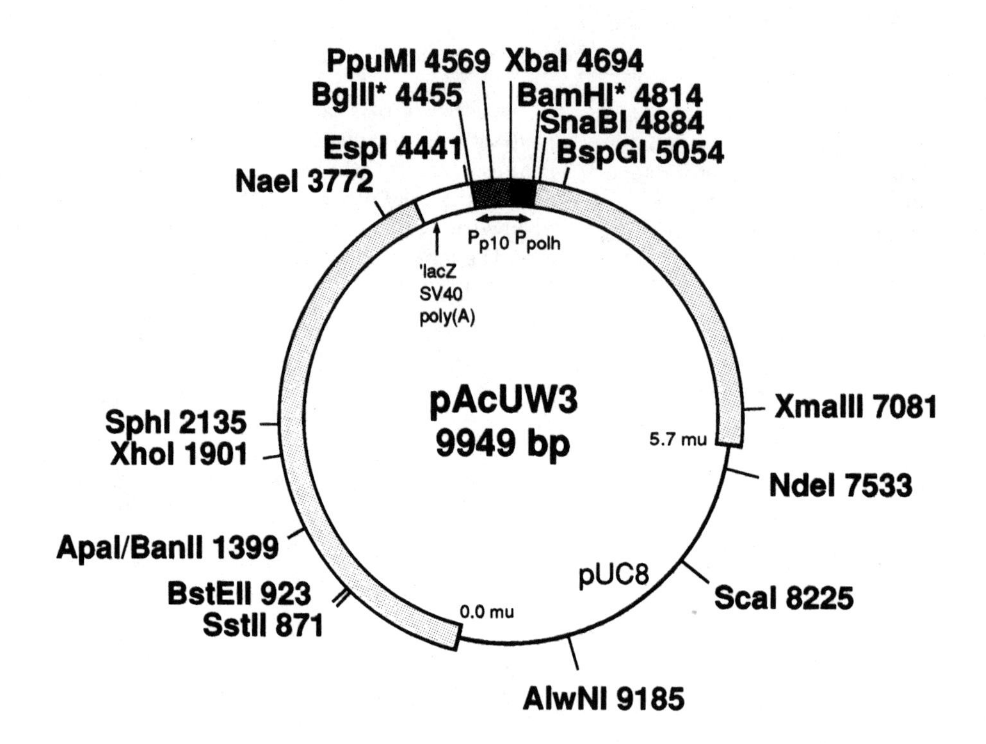

Plasmid	pAcUW3
Promoter	P_{p10} and P_{polh}
Parent virus	wt AcMNPV
Plaque phenotype	occ^-
Selection in *E. coli*	Amp^r
Source	R.D. Possee

pAcUW3

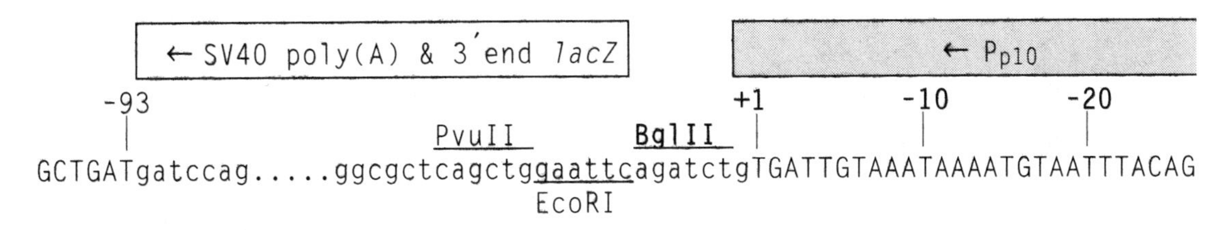

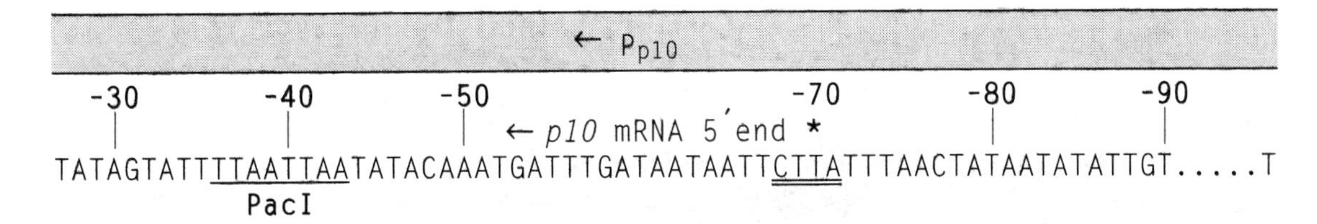

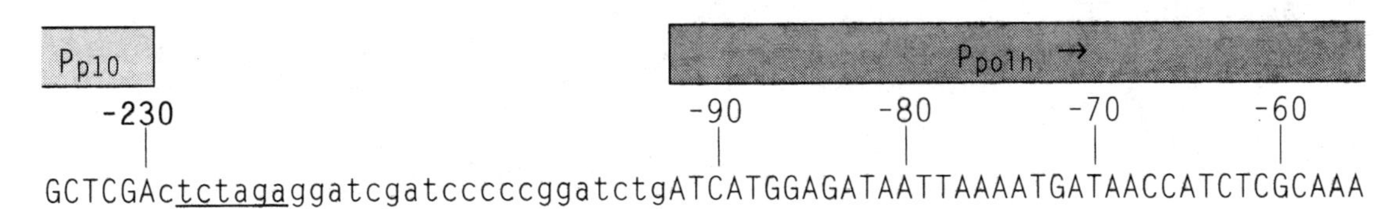

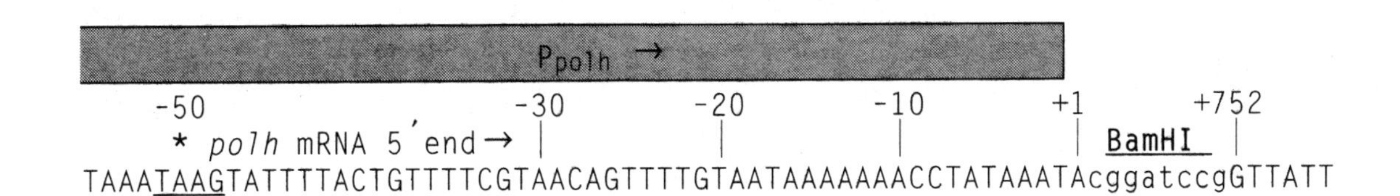

The plasmid pAcUW3 (Weyer and Possee, 1991) facilitates the introduction of two heterologous genes into a single recombinant virus. One gene is expressed under the control of the *p10* promoter, P_{p10}, while the other is expressed under the control of P_{polh}. The promoters are arranged back-to-back within the *polh* locus so that recombinant viruses are occ$^-$. In the sequence above, numbers in bold refer to P_{p10} while numbers in regular script refer to P_{polh}. The additional sequences between the promoters derive from the cloning procedure.

The plasmid pAcUW3 is a derivative of pAcUW2B (Weyer et al., 1990) in which the wt *polh* promoter and gene have been replaced by the *polh* promoter and cloning site from pAcYM1 (Matsuura et al., 1987). As in pAcUW2B, a sequence comprising 322 bp from the 3′ end of the *lacZ* gene, followed by 132 bp of SV40-derived sequences, is located immediately downstream of the BglII cloning site. The SV40 sequence provides polyadenylation signals for the P_{p10}-expressed heterologous gene. (Endogenous *polh* polyadenylation signals present in the recombinant virus will provide this function for the P_{polh}-expressed gene.)

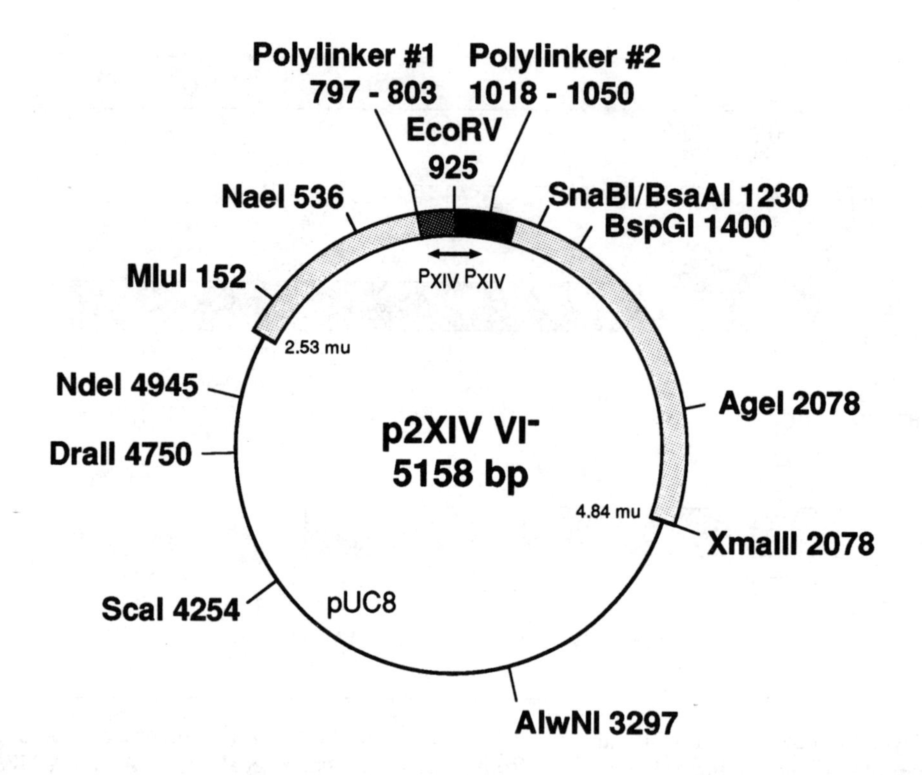

Plasmid	p2XIV VI$^-$
Promoter	Two P_{XIV}s
Parent virus	wt AcMNPV
Plaque phenotype	occ$^-$
Related plasmids	p2XIV VI$^-$X3, pSynwtVI$^-$, pSynXIVwVI$^-$
Selection in *E. coli*	Amp^r
Source	L.K. Miller

p2XIV VI$^-$

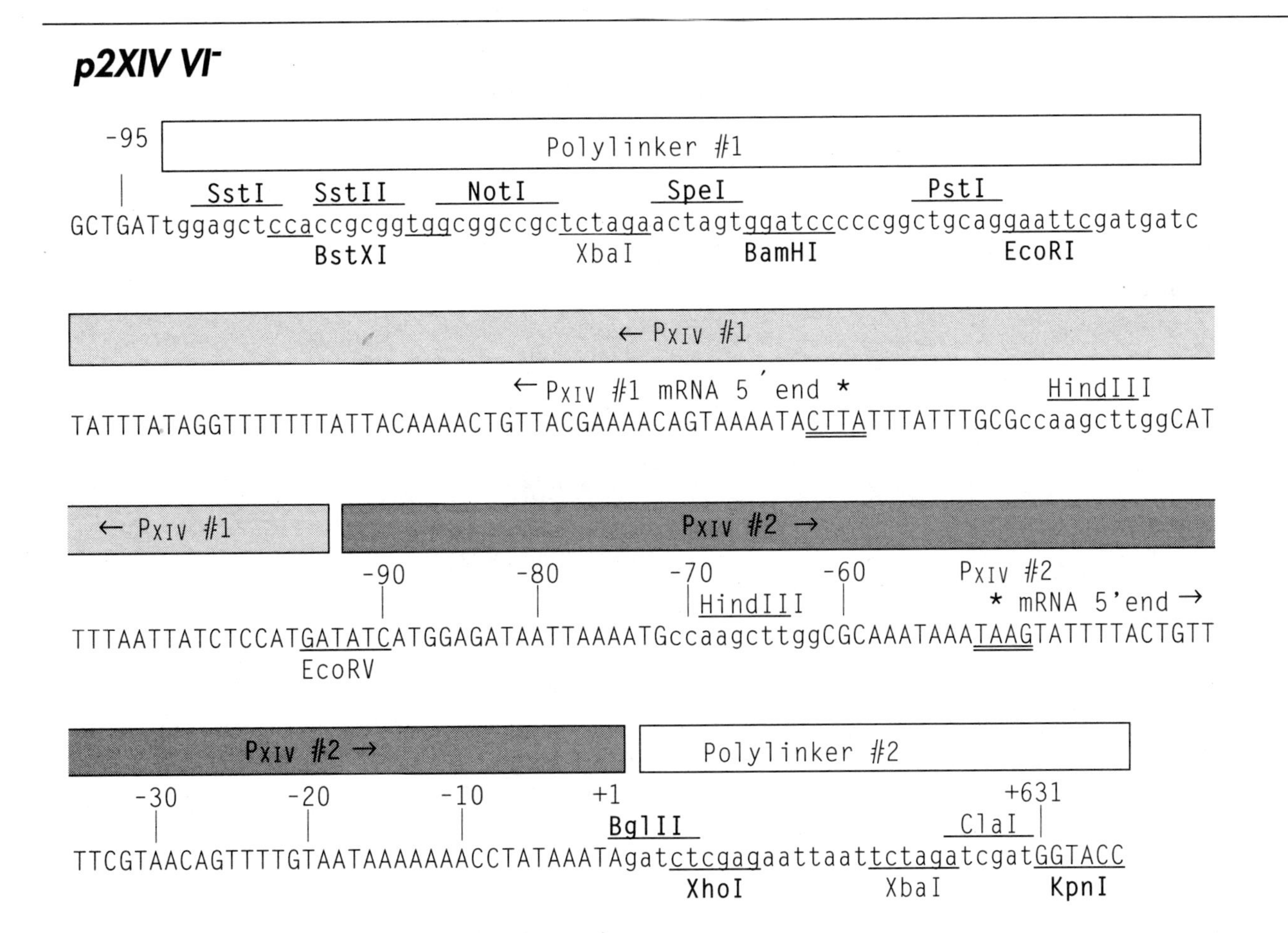

The plasmid p2XIV VI$^-$ (Wang et al., 1991) is designed to allow the expression of two heterologous genes by an occ$^-$ recombinant virus. It includes two P_{XIV} promoters in a back-to-back arrangement within the *polh* locus. The two promoters abut at an EcoRV site. P_{XIV} is a linker-scan mutant of P_{polh}, which has been reported to drive transcription ≈50% more efficiently than P_{polh} (Ooi et al., 1989). The HindIII linker is located upstream of the transcriptional start site. For the sake of clarity, we have only numbered the second P_{XIV} promoter (i.e., the one in the same orientation as P_{polh}) in the sequence above. Note that the XbaI site in polylinker #2 overlaps a site for *dam* methylation. This site might not be cleaved efficiently in DNA isolated from *dam*$^+$ *E. coli*.

The plasmids p2XIV VI$^-$X3, pSynwtVI$^-$, and pSynXIVwVI$^-$ are related transfer plasmids for the expression of two heterologous genes (Wang et al., 1991). They differ from p2XIV VI$^-$ in the promoters and polylinkers used. In some of these plasmids, the back-to-back promoters differ in strength. Interested users should consult Wang et al. (1991) for further information. In the nomenclature of these plasmids, XIV refers to the P_{XIV} promoter; VI$^+$ or $^-$ refers to the presence or absence of *polh* sequences in the HindIII-V/EcoRI-I region; and X3 refers to the polylinker used.

SOURCES OF TRANSFER PLASMIDS

R. Clark:

Molecular Biology Department
Chiron Corporation
Emeryville, California 94608
U.S.A.
Tel: (510) 420-3486
Fax: (510) 420-4414

P. Faulkner:

Department of Microbiology and
Immunology
Queen's University
Kingston, Ontario
Canada K7L 3N6
Tel: (613) 545-2464
Fax: (613) 545-6612

Invitrogen:

3985 B Sorrento Valley Boulevard
San Diego, California 92121
U.S.A.
Tel: 1-800-955-6288
(619) 597-6200
Fax: (619) 597-6201

S. Maeda:

Department of Entomology
University of California
Davis, California 95616
U.S.A.
Tel: (916) 752-0483
Fax: (916) 752-1537

L.K. Miller:

Departments of Entomology and
Genetics
University of Georgia
Athens, Georgia 30602
U.S.A.
Tel: (404) 542-2294
Fax: (404) 542-2279

R.D. Possee:

NERC Institute of Virology and
Environmental Microbiology
Mansfield Road
Oxford OX1 3SR
United Kingdom
Tel: (0)865 512361
Fax: (0)865 59962

C.D. Richardson:

Genetic Engineering Section
Biotechnology Research Institute
National Research Council of Canada
Montreal, Quebec
Canada H4P 2R2
Tel: (514) 496-6130/2
Fax: (514) 496-6232

D.C. Tessier:	Genetic Engineering Section Biotechnology Research Institute National Research Council of Canada Montreal, Quebec Canada H4P 2R2 Tel: (514) 496-6100 Fax: (514) 496-6232
T. Vernet:	Genetic Engineering Section Biotechnology Research Institute National Research Council of Canada Montreal, Quebec Canada H4P 2R2 Tel: (514) 496-6154 Fax: (514) 496-6213
J.M. Vlak:	Department of Virology Wageningen Agricultural University P.O.B. 8045 6700 EM Wageningen The Netherlands Tel: (0)8370 83090 Fax: (0)8370 84820

8

CHOOSING A PARENT VIRUS FOR USE IN VECTOR CONSTRUCTION

Chapter 6 discussed the basic options available in designing a baculovirus vector system. It was noted that it is particularly important to choose a transfer plasmid and parent virus combination that allows the rapid identification of double crossover recombinant viruses. This chapter describes several available AcMNPV recombinants that can be used in conjunction with specific transfer plasmids (described in Chapter 7) to facilitate recombinant vector identification. The viruses are divided into two basic groups depending on whether they are used with *polh*-based or *p10*-based transfer plasmids. (You can find information concerning the differences between the L-1, E2, C-6, and HR3 variants of AcMNPV from which these recombinants are derived in Chapter 5.)

PARENT VIRUSES FOR USE WITH polh-BASED TRANSFER PLASMIDS

vDA26Z

The L-1-derived occ$^+$ virus vDA26Z (O'Reilly et al., 1990) can be used to advantage with any of the *polh*-based transfer plasmids in which *polh* is to be replaced with a heterologous gene (i.e., in constructions that involve identifying occ$^-$ viruses in a background of occ$^+$ parental viruses). Thus, the virus can be helpful when used in conjunction with pAcYM1, pEVmXIV, pVL1393, pAcC4, pAc360, pc/pS1, pAcMP1, pAcUW3, and p2XIV VI$^-$. Although wt virus may be used as the parent virus with these transfer plasmids, many baculovirologists prefer to use parent viruses such as vDA26Z in which *lacZ* is located at a distance from the *polh* gene. Both the parent virus and recombinant vector will have a blue-plaque phenotype. The advantage of such a *lacZ*-containing parent virus over wild-type virus is that it ensures that occ$^-$ plaques are not overlooked during examination of the monolayer. Also, regions of the cell monolayer that appear to be occ$^-$ plaques can be distinguished clearly from aberrations of the monolayer by their blue color. Once located, such blue regions can be more thoroughly examined for polyhedra production under the light microscope.

The virus vDA26Z is particularly suitable for this purpose because the fusion of *lacZ* with *da26* has no detectable effect on the kinetics of protein synthesis during the baculovirus infection, and vDA26Z is as infectious as wild-type virus in larval bioassays (O'Reilly et al., 1990). Blue color produced from the *da26-lacZ* fusion is ample for plaque coloration but not excessive. Not all early gene/*lacZ* fusions work as well. For example, *egt*/*lacZ* fusions (O'Reilly and Miller, 1989) produce a fainter blue color and are not as easily distinguishable from wt virus as vDA26Z. In another example, *etl*/*lacZ* fusions (Crawford and Miller, 1988) are dark blue, but disruption of *etl* (*pcna*) results in the delay of late virus gene expression by four to six hours.

The vDA26Z virus may be obtained from L.K. Miller or D.R. O'Reilly:

L.K. Miller: Departments of Entomology and Genetics
University of Georgia
Athens, Georgia 30602, USA
Tel: (404) 542-2294
Fax: (404) 542-2279

D.R. O'Reilly: Department of Biology
Imperial College
Prince Consort Road
London, SW7 2BB, UK
Tel: (0)71 225-8245
Fax: (0)71 225-8238

Linearized DNA

As noted in Chapter 6, linearization of the parental viral DNA prior to transfection increases the proportion of recombinant viruses obtained. Also, use of linearized viral DNA should reduce the prevalence of single-crossover recombinants in a virus stock and thus facilitate the use of screens that rely on the acquisition of a marker. A C-6-derived occ$^-$ virus, AcRP6-SC (Kitts et al., 1990), carries a unique Bsu36I site in a polylinker that replaces a portion of *polh*. Readers should note that because this virus is occ$^-$ it is not possible to use an occ$^+$ to occ$^-$ screen when AcRP6-SC is used as the parent. However, the virus may be used with an occ$^+$ transfer plasmid such as pAcUW2B. In this case, an occ$^-$ to occ$^+$ screen may be used. The problem of obtaining predominantly single-crossover viruses by this screen should be minimized due to the use of linearized DNA. AcRP6-SC may be obtained from:

R.D. Possee: NERC Institute of Virology and Environmental Microbiology
Mansfield Road
Oxford OX1 3SR, UK
Tel: (0)865 512361
Fax: (0)865 59962

Many of the *lacZ*-containing viruses may also be linearized because *lacZ* contains a single Bsu36I site. For example, vSynVI$^-$gal (see the next section) can be linearized by Bsu36I cleavage of *lacZ* within the *polh* locus. Following the cotransfection, parent virus plaques will be blue, whereas recombinant virus plaques will be white. Although improperly recircularized parent virus will also be white, it will not be occ$^+$, and so can be distinguished from the recombinants.

Invitrogen sells linearized AcMNPV DNA which should be useful with *polh*-replacement transfer plasmids that yield occ$^-$ recombinants. The Invitrogen DNA (presumably derived from the E2 variant) is linearized at an introduced Sse8387I site upstream of *polh* so as not to disrupt *polh*. The occ$^-$ viruses resulting from cotransfection should be enriched for recombinants carrying double crossovers. In addition, improperly recircularized parent virus will be occ$^+$ (unless linearization and/or recircularization interferes with the *polh* promoter), and so will be easily distinguishable from the desired recombinants. As before, this viral DNA will also be useful in combination with plasmids that use a screen for the acquisition of a marker. For example, use of this DNA should increase the proportion of double-crossover recombinants identified by screening for β-galactosidase-producing plaques following cotransfection with pBlueBac2. This linearized DNA may be purchased from:*

Invitrogen: 3985 B Sorrento Valley Boulevard
San Diego, California 92121, USA
Tel: 1-800-955-6288; (619) 597-6200
Fax: (619) 597-6201

vSynVI$^-$gal

This L-1-derived occ$^-$ virus can be used in conjunction with plasmids that transfer *polh* into the *polh* locus, thereby converting the parent occ$^-$ virus to an occ$^+$ phenotype (Wang et al., 1991). Transfer plasmids that can be used in conjunction with vSynVI$^-$gal include pSynXIV VI$^+$, pSynXIV VI$^+$X3 and derivatives, and pAcUW2B. The vSynVI$^-$gal virus carries *lacZ* under the control of a synthetic promoter (P_{syn}), in place of *polh* and in the opposite orientation to normal *polh* transcription. Upon recombination with the appropriate transfer plasmid (containing both *polh* and the heterologous gene), *lacZ* is replaced with both the heterologous gene and *polh*, resulting in white, occ$^+$ recombinants. The advantage of using a parent virus that has *lacZ* at the *polh* locus is that double-crossover recombinants can be identified by the loss of *lacZ*. If recombinants are selected solely based on an occ$^+$ phenotype, most recombinants will be single crossovers and inherently unstable. The vSynVI$^-$gal virus may be obtained from:

L.K. Miller: Departments of Entomology and Genetics
University of Georgia
Athens, Georgia 30602, USA
Tel: (404) 542-2294
Fax: (404) 542-2279

PARENT VIRUSES FOR USE WITH p10-BASED TRANSFER PLASMIDS

AcAs3

This is a blue, occ$^+$ E2-derived virus carrying *lacZ* under the control of a *Drosophila hsp70* promoter at the *p10* locus (Vlak et al., 1990). It can be used in conjunction with transfer plasmids that replace *p10* with a heterologous gene such as pAcAs2, pAcUW1, and pEP252. The advantage of using a virus carrying *lacZ* in the *p10* locus is that double-crossover recombinant viruses can be identified by their white, occ$^+$ phenotype. The *hsp70* promoter provides enough β-galactosidase activity to produce a dark-blue plaque (without heat shock) but not excessive levels of coloration that would obscure other regions of the monolayer. Linearizing this virus with Bsu36I prior to cotransfection is not recommended because of the difficulty in

*PharMingen recently introduced linearized AcMNPV DNA (presumably derived from the C-6 variant) that includes a deletion around *polh* extending into the 1629 ORF. The 1629 ORF is essential for virus viability, so that the linearized DNA itself should be non-infectious. Use of this DNA should minimize the proportion of non-recombinant plaques obtained, because only DNA that has recombined with the transfer plasmid will be infectious. This system should work with any *polh*-based transfer plasmid, provided the plasmid includes sufficient downstream sequences to complement the 1629 deletion. However, we do not know the precise endpoints of the deletion in this DNA, and so we cannot provide more specific recommendations at this time. The linearized DNA, which is marketed under the name "BaculoGold DNA," is available from PharMingen: 11555 Sorrento Valley Road, Suite E, San Diego, CA 92121, USA; Tel: 1-800-848-6227, (619) 792-5730; Fax: (619) 792-5238.

distinguishing recombinants from improperly recircularized parent virus. AcAS3 may be obtained from:

J.M. Vlak: Department of Virology
Wageningen Agricultural University
P.O.B. 8045
6700 EM Wageningen, The Netherlands
Tel: (0)8370 83090
Fax: (0)8370 84820

AcUW1-lacZ

This C-6-derived virus is another blue, occ^+ virus carrying *lacZ* in the *p10* locus (Weyer et al., 1990). In this case, *lacZ* expression is under *p10* promoter control. It can be used with *p10*-based transfer vectors in the same way as AcAS3, but the excessive amount of β-galactosidase activity from this promoter may interfere with the screening process. The AcUW1-*lacZ* virus may be obtained from R.D. Possee (see following section).

AcUW1-PH

This C-6-derived virus carries *polh* under *p10* promoter control in the *p10* locus. It can be used with *p10*-based transfer plasmids such as pAcAS2, pEP252, and pAcUW1. The screen for recombinants using this parent virus and transfer plasmid combination is the conversion from occ^+ to occ^- plaques. AcUW1-PH and AcUW1-*lacZ* (preceding) may be obtained from:

R.D. Possee: NERC Institute of Virology and
Environmental Microbiology
Mansfield Road
Oxford OX1 3SR, UK
Tel: (0)865 512361
Fax: (0)865 59962

Ac228z

This HR3-derived virus is also a blue, occ^+ virus carrying *lacZ* at the *p10* locus (J. Kuzio and P. Faulkner, personal communication). It is used in the same manner as AcAS3 and AcAS2-*lacZ*. The *lacZ* gene of Ac228z is under *p10* promoter control but it is out-of-frame from *p10*, so the amount of β-galactosidase is not excessive. The *lacZ* gene of Ac228z replaces a portion of the C-terminus of the downstream gene *p74*, so the parent virus is non-infectious for insect larvae by the oral route of infection. However, recombinant viruses obtained following allelic replacement with any of the *p10* transfer plasmids described in this manual will be infectious by the oral route. Ac228z may be obtained from:

P. Faulkner: Department of Microbiology and
Immunology
Queen's University
Kingston, Ontario
Canada K7L 3N6
Tel: (613) 545-2464
Fax: (613) 545-6612

9

OPTIMIZING EXPRESSION: TAILORING THE HETEROLOGOUS GENE TO THE TRANSFER PLASMID AND THE BACULOVIRUS EXPRESSION SYSTEM

Many researchers are interested in maximizing heterologous gene expression and wish to know what factors will influence expression in the system. Chapter 6 considered features of the transfer plasmids that will optimize gene expression and post-translational modification. This chapter considers features of the heterologous gene itself that may influence the level of gene expression.

If you are simply trying out the system to see whether it might be useful for your purposes, you should not feel compelled to modify your gene extensively, because it is likely that many of the aspects considered here will influence expression by only two- to fivefold, if at all. The efficiency of heterologous gene expression in the baculovirus system can differ from gene to gene by approximately 1000-fold, so "tailoring" your gene extensively may have little influence on the overall level. There is little you can easily do to influence the intrinsic nature of the gene and gene product and, hence, its overall response in the system. However, there are some aspects worth considering in your initial constructions—if you can modify them without too much time or trouble. Those interested in optimizing their system for very large-scale (e.g., industrial) production purposes will probably devote considerable attention to tailoring their genes. Many of the factors considered in this chapter will also be useful to those constructing synthetic genes.

The basic conclusions reached in this chapter are summarized here:

- It is important to use genes without introns (e.g., cDNAs).
- Remove heterologous sequences in the 5′ noncoding region of the mRNA if they are likely to interfere with translational initiation at the correct AUG.
- The translational initiation codon (AUG) may need to be in a good context; an A at −3 (relative to AUG at +1,+2,+3) may be important.
- Remove unnecessary sequences in the 3′ untranslated mRNA sequences if they are likely to destabilize the mRNA.
- Polyadenylation signals are present in the transfer plasmids described in Chapter 7 and need not be supplied with heterologous genes.
- TAAG and CTTA sequences in your gene may specify a late or very late transcriptional initiation site, in the sense or antisense direction respectively. TAAGs could lead to shorter proteins due to internal initiations, whereas CTTA sequences could reduce overall levels of expression through antisense RNAs. However, before removing such sequences from a gene, we recommend that you first express your wt gene and then determine whether aberrant transcripts are indeed initiating from these sequences and/or interfering with expression.

- Codon usage will probably have very little influence on overall translational efficiency unless large numbers of rare codons are found in the reading frame.
- Consider removing the signal peptide and/or "pro" portion of a gene specifying a *preprotein* or *proprotein* if the signal or "pro" portions are not efficiently cleaved by the baculovirus expression system and are not required for proper folding or cellular localization. Transfer plasmids that can supply a signal or appropriate signals can be specifically designed (see Chapter 7).
- Consider changing the N-terminus of your protein (e.g., by using a fusion transfer plasmid) if the protein appears to be highly unstable in the system.

mRNA CONSIDERATIONS: TRANSCRIPTION INITIATION, SPLICING, 3′ END PROCESSING, AND STABILITY

Very few laboratories have studied the effects of heterologous gene sequences on the level of heterologous mRNA in the baculovirus expression system. With regard to the nature of the heterologous gene used, it is strongly recommended that it not contain introns. Although low levels of splicing have been reported for heterologous genes in the baculovirus expression system (Jeang et al., 1987b; Iatrou et al., 1989), strong expression of proteins from spliced mRNAs has not been observed.

As noted in Part I, late and very late promoters have essential TAAG sequences from which mRNAs initiate; TAAG sequences are otherwise rare in the AcMNPV genome. The context of the TAAG sequence influences the temporal and quantitative activity of the promoter. TAAG sequences within heterologous genes, either in the sense or antisense direction, may adversely influence the level of heterologous gene expression. TAAG sequences in the sense direction would be expected to initiate internally in the gene. If translated, such truncated transcripts would give rise to short, C-terminal segments of the protein of interest. TAAG sequences in the antisense direction could promote the formation of antisense transcripts, which might interfere with sense transcription, or form duplex RNAs with the sense strand thus leading to a translational block or destabilization of the transcripts. Antisense transcription influencing RNA levels has been reported in wt baculovirus infection (Ooi and Miller, 1990). Clearly, the impact of such TAAG sequences would depend on the level of transcriptional initiation from the TAAG; this, in turn, depends on the context of the TAAG. Since we do not yet know what sequence elements influence the rate of transcriptional initiation from TAAG sites, it is not possible to predict whether a particular heterologous TAAG sequence will strongly influence heterologous gene expression. Because altering TAAG sites within a heterologous gene will be timeconsuming and possibly unnecessary, it is recommended that expression be tested using the wt gene. If expression is low, the levels of both sense and antisense RNA can be determined and/or the 5′ start sites mapped to see if any of the heterologous TAAG sites are highly active transcriptionally. TAAG sites are certainly a feature that should be considered in building synthetic genes to be expressed using the baculovirus expression system.

Very little is known about transcriptional termination in the baculovirus system but most, if not all, viral transcripts are processed at their 3′ end at polyadenylation signals. Polyadenylation signals may not be used particularly efficiently during infection. For example, *polh* is transcribed into at least three sense mRNAs (Friesen and Miller, 1985; Ooi and Miller, 1990), which suggests that the polyadenylation signal immediately downstream of *polh* is not used efficiently. This polyadenylation signal, which is actually located in the downstream 1629 ORF, appears to be a standard type in which an A_2UA_3 sequence is found approximately 20 nucleotides upstream of the cleavage site (Possee et al., 1991).

Because the 1629 ORF appears to be essential for virus replication (Possee et al., 1991), *polh* deletions in polyhedrin-based transfer plasmids do not extend into the 1629 ORF. Thus, the *polh* polyadenylation site is maintained so that it is not necessary to introduce a polyadenylation site. For transfer plasmids supplying back-to-back promoters in the *polh* region, counterclockwise transcripts (i.e., those extending in the opposite direction to wt *polh* gene transcription) are expected to terminate at a polyadenylation signal at the 3′ end of the flanking 603 ORF (Gearing and Possee, 1990; Ooi and Miller, 1990). Some of these transfer plasmids (e.g., pSynXIV VI$^+$X3 series) also supply a polyadenylation signal at the 3′ side of the polylinker, although the efficiency of using this signal has not been tested or reported. For the *p10*-based vectors, an A_2UA_3 signal is located approximately 75 nucleotides downstream of the *p10* termination codon and is found in the transfer plasmids described in Chapter 7. Experiments demonstrating the use of this polyadenylation signal, however, have not been described in the literature yet. Thus, addition of polyadenylation signals is probably not necessary for any of the transfer plasmids described, although it is not known how the use of existing signals or the addition of new signals might affect the efficiency of RNA processing and/or RNA stability. It is generally not recommended that you make any specific effort to add polyadenylation signals from genes to be expressed in these transfer plasmids.

Because no one has systematically addressed the question of whether the 3′ untranslated tails of the *p10* and *polh* mRNAs may have a role in mRNA stability, it is not known whether the addition of another polyadenylation signal or putative transcriptional termination signal immediately 3′ to the heterologous gene open-reading frame would have an effect on mRNA levels. It is known that *cat* mRNA, having the same 5′ and 3′ ends as *polh* mRNA, has a half-life of approximately 12 hours (Ooi and Miller, 1991); *cat* is expressed relatively well in the baculovirus expression system. (The half-life of *polh* mRNA has not been determined.) The question of whether to specifically delete any heterologous 3′ processing or termination signals in the heterologous gene sequence prior to insertion into a transfer plasmid is thus difficult to answer. Because this type of gene tailoring might mean additional unnecessary work, this possibility should be investigated only if there is reason to suspect that this region has a destabilizing effect, or when optimal expression conditions are being systematically studied.

You may want to consider eliminating certain AU-rich residues in the 3′ untranslated tails of heterologous mRNAs in expression work. Specific AU-rich sequences in 3′ untranslated regions of mRNAs have been demonstrated in other systems, including insects (Yost et al., 1990), to govern mRNA instability. If it is simple to remove such sequences in the heterologous gene, it might be worthwhile to do so even in initial expression work. However, we do not recommend doing difficult heterologous gene tailoring work before using the expression system, because once an initial vector is made, that vector can be used to determine whether mRNA levels are low and whether RNA stability is a controlling influence on heterologous gene expression.

TRANSLATION: 5′ LEADER REGIONS, INITIATION CODON CONTEXT, AND CODON USAGE

Clones of many heterologous genes contain nucleotides upstream of the ATG initiation codon of interest. Generally, it is recommended that these nucleotides be removed if they contain additional ATGs that are not intended for initiation purposes. If no ATGs are present, the gene can probably be inserted into transfer plasmids and expression obtained. There is no guarantee, however, that the upstream sequences will not interfere with optimum gene expression. Consideration should be given to removing the leader region if it is long (e.g., over 30 nucleotides), if it has a high G+C content, or if it can form stable secondary structures (e.g., stem-loop structures). These suggestions should serve as guidelines, not as absolute rules. For example, although the leader sequences of abundant baculovirus mRNAs are usually short (20 to 60 nucleotides), some genes with long leaders (e.g., 150 nucleotides) have been expressed well using the baculovirus expression system. The leader regions are also generally low in G+C content, and there is one report that removal of G+C-rich 5′ noncoding sequences significantly improved the level of expression of the heterologous gene in the baculovirus expression system (Pendergast et al., 1989).

The importance of having the DNA sequence specifying the entire *polh* untranslated RNA leader region within the transfer plasmid was emphasized previously. It was noted that this sequence, including sequences up to the original polyhedrin translational initiation codon, constitutes the actual *polh* promoter. The existing evidence suggests that these sequences influence steady-state RNA levels rather than translational initiation (Ooi et al., 1989). Some researchers previously suggested that a "putative ribosome-binding site" was located in this region (Hu and Kang, 1991).

The immediate context of the translational initiation codon for the heterologous protein may be an important consideration in optimizing gene expression in the baculovirus expression system (Luckow and Summers, 1988a). Kozak systematically determined AUG contexts that correlate with efficient translational initiation in mammalian systems (Kozak, 1987). However, it is not certain that these studies/ contexts are applicable to the baculovirus gene expression system, nor is it clear what additional features might influence translational initiation in this system. Differences in the use of AUGs in different contexts have been reported in the baculovirus expression system (Price et al., 1988; Beames et al., 1991).

In lieu of a systematic study of immediate AUG contexts such as those conducted by Kozak, the AUG contexts of highly expressed AcMNPV proteins can be compared for some indication of features that might be important for translational initiation. (A caveat to this approach is that these genes are also abundantly transcribed, and translational efficiency is not certain.)

Gene		*AUG context*
polh	(Matsuura et al., 1987)	CCUAUAAAU**AUG**CCGG
p10	(Kuzio et al., 1984)	UUUACAAUC**AUG**UCAA
p6.9	(Wilson et al., 1987)	AAUUUAAAC**AUG**GUUU
vp39	(Thiem and Miller, 1989a)	GGCAACAAU**AUG**GCGC
	Consensus	A Y**AUG** Y

The consensus sequence shows that A is found at the −3 position relative to the AUG (at +1,+2,+3), and pyrimidines are found at −1 and +5. There also appears to be a strong preference *not* to have a G in the 7 nucleotides immediately preceding the AUG. It must be emphasized, however, that there is no evidence demonstrating that any of these observations are relevant to heterologous gene expression levels in the baculovirus expression system. The observations are only relevant in providing some guidance on whether the existing heterologous gene AUG context is similar to or distinct from what is known to be used by the baculovirus vector itself.

If there is sufficient reason to remove the 5′ leader region and/or change the context of the AUG of the heterologous gene, existing site-directed mutagenesis techniques can be used to tailor the gene. Transfer plasmids with an NcoI site in place of the *polh* ATG (e.g., transfer plasmid pAcC4) may prove to be useful in some constructions. The NcoI recognition sequence contains an ATG (i.e., it recognizes C′CATGG) so that, if cloned appropriately, the ATG can replace precisely the polyhedrin ATG. The context of the NcoI ATG appears to be compatible with the consensus sequence for ATGs noted above. If the N-terminal sequence of the heterologous protein is not important for biological activity, one may also consider using fusion vectors that supply an ATG in the appropriate context.

Although there is a bias in codon usage by highly expressed baculovirus genes such as *polh* and *p10*, it is not clear that the codon usage of the heterologous gene will have a major influence on the level of expression. For example, *E. coli lacZ* is well expressed by the baculovirus expression system (Pennock et al., 1984), and yet this gene uses many relatively poor codons. Although even higher levels might be achieved by altering codon usage of this gene, it is clearly not worth the effort to do so unless maximum production levels are of the utmost importance. It is also worth noting that the UAA codon is used at the termination codon for most AcMNPV genes defined currently. However, the significance of this observation is not known, and it is most likely that the baculovirus system will recognize the other two termination codons. For synthetic genes, a UAA codon is clearly preferable.

PROTEINS: PROCESSING AND STABILITY

Readers concerned with the processing of signal peptides and proteolytic cleavage of proproteins in baculovirus-infected cells are referred to the introduction to post-translational modification in Chapter 15. In the author's experience, converting a gene encoding a preproprotein to one encoding a preprotein (i.e., fusing sequences encoding an insect signal peptide to the sequences specifying the mature protein, thereby eliminating the need to cleave the "pro" sequence) provided significant improvement to expression levels (O'Reilly and Miller, unpublished observations).

It is quite likely that the inherent stability of the heterologous gene product itself has a major influence on the level of expression that can be achieved in the baculovirus expression system. There is at least one example in the literature showing that steady-state levels of heterologous mRNA of a poorly produced protein were equivalent to those of *polh* mRNA (Carbonell et al., 1988). Inherent protein instability probably accounted for the low expression level observed in this case although the protein turnover rate was not determined. Generally, proteins that are inherently stable, such as virus structural proteins, are expressed to high levels. Here, we discuss aspects that might influence protein stability.

The sequences that target proteins to a degradation pathway are not known for the baculovirus expression system. In other systems, selective protein degradation is known to involve the ATP-dependent coupling of ubiquitin, a highly conserved, small (ca. 76 residues) protein, to the target protein [reviewed by Finley and Varshavsky (1984)]. An important component of the degradation signal in short-lived proteins is the amino terminal residue. Recognition of this fact has led to the establishment of what are referred to as N-end rules (Bachmair and Varshavsky, 1989; Tobias et al., 1991). These N-end rules differ depending on the organism (Tobias et al., 1991). In yeast, N-terminal amino acids that provide stability (half-life > 20 hours) are Val, Met, Gly Pro, Ala, Ser, Thr, and Cys; those residues with intermediate stability (10-30 minutes) are Ile, Glu, His, Tyr, and Gln; those residues that result in rapid degradation (2-3 minutes) are Asp, Asn, Phe, Leu, Trp, Lys, and Arg. A second determinant of the degradation signal is a conformationally mobile lysine residue, usually near the N-terminus, that can be ubiquitinated (Bachmair and Varshavsky, 1989).

It is far from clear how valid these rules will be in the baculovirus system. Baculoviruses carry a gene encoding a unique ubiquitin (Guarino, 1990). The possible function of this ubiquitin in protein degradation or stability remains to be established. It is therefore difficult to recommend that the N-terminus of the heterologous protein be modified if there is evidence to indicate that the protein is inherently unstable in the expression system. However, specific modifications of the residue in question can be made to test this possibility if necessary. Alternatively, one may consider using one of the available N-terminal fusion vectors.

Baculovirus-infected cells and their extracellular fluids contain proteolytic enzymes that may be active in nonubiquitin-mediated pathways of protein turnover. Baculoviruses contain a gene known as *v-cath,* which encodes a homolog of cysteine protease (J. Kuzio and P. Faulkner, personal communication). No enzymatic activity has been ascribed to the product of *v-cath,* but a cysteine protease has been found in extracellular fluids of baculovirus-infected cells at very late times pi (Vernet et al., 1990; Yamada et al., 1990). It is possible that this is the viral-encoded protease, but it is also possible that it is a cellular lysosomal cysteine protease that has been released upon lysis at very late times pi. One group has reported that p-chloromercuribenzene sulphonic acid (pCMBS), an inhibitor of cysteine proteases, is effective in blocking the action of the protease found in the extracellular fluid of Bm-N cells (Yamada et al., 1990). This reagent might be generally useful in baculovirus expression systems for extracellular protein stabilization where required.

III

METHODS FOR VECTOR CONSTRUCTION AND GENE EXPRESSION

Part III of this manual is devoted to the methods used in the generation and characterization of a recombinant baculovirus expression vector. It begins with a discussion of some safety aspects that should be considered before undertaking the construction of a recombinant baculovirus. Following this, the methods are organized in the order in which they will be used, although readers will need to refer to certain sections, such as cell culture and virus methods, many times during the course of constructing a virus. Figure III-1 provides an overview of the different

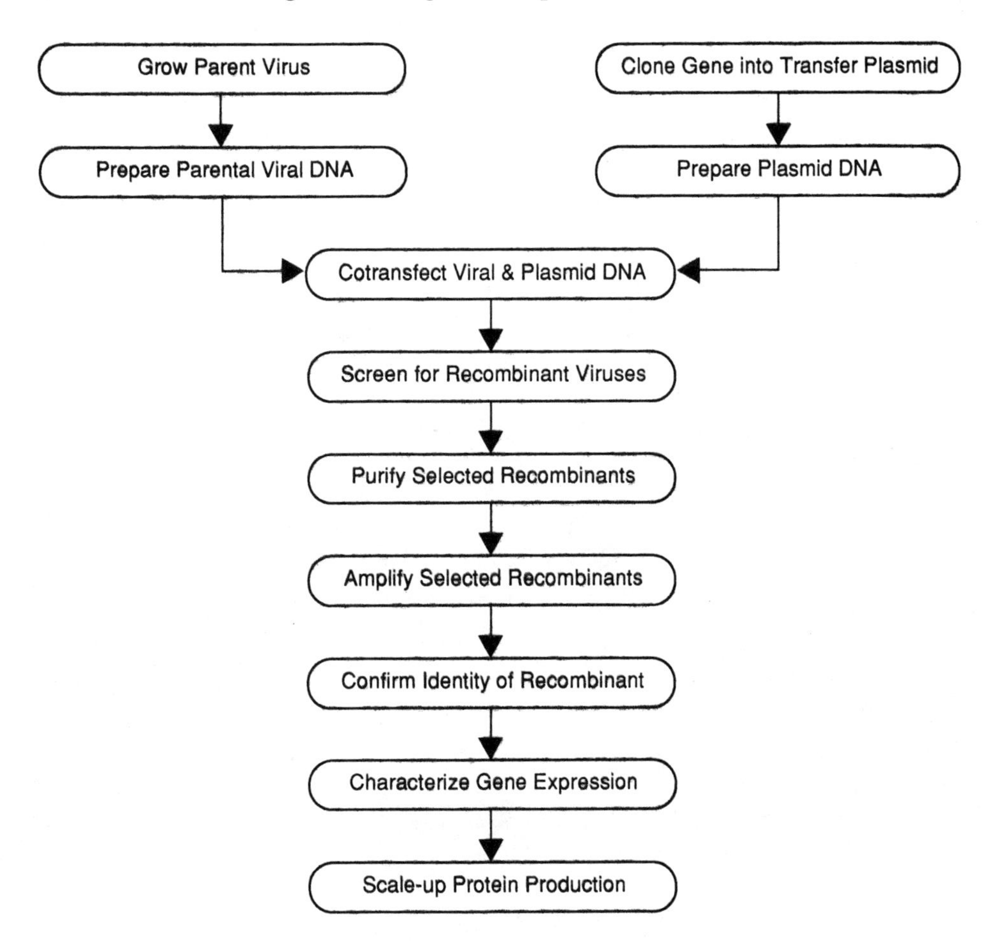

FIGURE III-1 *Flow chart of stages in the construction of a recombinant baculovirus expression vector.*

stages in the generation of a baculovirus expression vector. Protocols are provided for all the stages listed in Figure III-1 except for protocols for the insertion of a heterologous gene into the transfer plasmid or for the preparation of transfer plasmid DNA. These are common molecular biology procedures, and the reader may consult any good cloning manual, such as Sambrook et al. (1989) or Berger and Kimmel (1987) for descriptions of the methodology involved. Methods for the large-scale culture of insect cells and for the scale-up of recombinant protein production are provided in Part IV.

For each method, there is a list of the materials required, including the names of suppliers and catalog numbers for most items. Note that the suppliers' names are only provided as suggestions. Clearly, there are perfectly acceptable alternative sources for many of the materials described. A list of suppliers' addresses, with telephone and fax numbers, is in Appendix 4.

The compositions of commonly used solutions are only listed the first time each solution is used. However, recipes for common stock solutions are provided in Appendix 5.

10

SAFETY CONSIDERATIONS

Baculoviruses have a number of features that make them comparatively safe viral expression vectors (see host-range discussion in Chapter 3). This section discusses safety considerations in the laboratory and notes features of the NIH *Guidelines for Research on Recombinant DNA Molecules* that relate to biological containment practices for research with recombinant baculoviruses. As with other recombinant organisms, recombinant baculoviruses should be treated as if they were potential biohazards, and precautions should be taken to limit their realm to the laboratory. For those working in the United States, the NIH *Guidelines for Research on Recombinant DNA Molecules* should be consulted for laboratory safety procedures. Those working in other countries should consult the guidelines or regulations concerning recombinant DNA research in their own country. Some specific suggestions for work with baculoviruses in the United States follow.

The NIH guidelines, which are in effect for research involving recombinant DNA molecules in the United States, appeared in the May 7, 1989 issue of the U.S. *Federal Register.* A copy of these guidelines and additional major actions since May 1989 can be obtained upon request from:

Office of Recombinant DNA Activities
National Institutes of Health
Building 31, Room 4B11
Bethesda, MD 20892

This section provides some suggestions for finding key sections in the NIH guidelines. However, readers are responsible for ensuring that their specific research meets the appropriate guidelines for research in their country, state, and institution.

Because baculoviruses are not classified as Class 2, 3, 4, or 5 human or animal pathogens, the relevant NIH guidelines for insect baculovirus expression work are under Sections III-B-3, III-B-4, and III-B-5 *unless* covered under another more restrictive section dealing specifically with the nature of the gene(s) being expressed. For example, potent toxin genes would be covered in Section III-A, and experiments involving major portions of the genome of an animal pathogen would be covered under such sections as III-B-1 or III-B-2. Research performed under Section III-B must be approved by the local Institutional Biosafety Committee (IBC) before it is initiated. Section III-B-3-e covers most baculovirus vector research in cultured cells as baculoviruses are not considered Class 2 to Class 5 pathogens. The guidelines recommend BL1 containment for this work; readers should familiarize themselves with the requirements of BL1 containment and, for comparison, BL2 containment. A special cautionary statement at the beginning of Section III-B-3 emphasizes that serious consideration should be given to using BL2 containment if the intended experiments will extend the host range of viruses or increase their pathogenicity. Readers should consider the nature of their gene as well as baculovirus host-range limitations when choosing appropriate containment for the vectors they construct. Occlusion-negative (occ$^-$) recombinants are already partially disabled for survival in the environment.

Consideration should be given to the potential of baculoviruses being carried from the laboratory by invertebrates—either experimental organisms or pests. An effective program of insect pest control should be implemented. Occ$^-$ viruses are less able to survive outside the insect hemolymph or culture media than occ$^+$ viruses. The latter, however, are reported to survive for 20 or more years in soil. Treatment of waste liquids with alkali to approximately pH 10.5 should solubilize the occlusion matrix. All materials that come in contact with recombinant viruses should be sterilized before disposal.

Section III-B-4-b covers most experiments involving recombinant baculovirus-infected insect larvae. Containment for such experiments is to be determined by the IBC prior to initiation of the experiment. The potential user should propose to the IBC an effective method for containing infected insects. The primary consideration here will be whether both the virus and the insects can be contained. Most baculovirus-based research involving live insects will use larvae because it is the larval stage that is susceptible to infection. The lepidopteran larval hosts of AcMNPV and BmNPV can be contained easily, although some (e.g., *Heliothis virescens*) may be voracious and able to eat through a variety of container materials.

You should do preliminary studies and gain experience in effective control methods for uninfected larvae before approaching the IBC. If experimental larvae are secured in escape-proof containers during the infection and destroyed by the pupal stage, the insects will not escape by flying. Thus, you should design containment methods that confine test larvae and keep other invertebrates out. You must also be conscientious about ensuring that insects do not reach the adult stage. If, for some reason, adult insects or highly mobile insects are used in an experiment with recombinant virus, you will need to design a more sophisticated containment system. One possibility is a mobile double-containment system in which the outer container is clear; any insects that have escaped from the first container can be seen within the second container. Additional consideration should be given to how highly mobile insects can be immobilized while they are being studied. Methods used will depend on the insect and may include cooling the containers to 5° C before opening the containers, or treating with carbon dioxide, or simply killing the experimental insects by some other means before opening the containers. All containers should be sterilized, preferably autoclaved, before reuse or disposal.

Section III-B-5 covers experiments involving more than 10 liters of culture. Appendix K-II specifies the containment required for large-scale (greater than 10 liter) cell cultures using most baculovirus vectors. If higher laboratory containment is required due to the nature of the inserted genes (e.g., toxins or large proportions of viral genomes), Appendix K-III or K-IV may also be relevant.

The reader is reminded that environmental release of recombinant organisms is covered in separate sections of the guidelines and is not considered further in this manual.

11

INSECT CELL CULTURE

This chapter describes the propagation of cultured insect cells. First, some of the major pieces of equipment required in a laboratory doing insect cell culture work are listed and discussed. Second, the preparation and use of serum-supplemented and serum-free insect media is described. Finally, routine procedures for maintaining cells in tissue-culture flasks and spinner flasks are described. Strategies for adapting cells from one medium to another and for long-term storage of cells are also discussed. The chapter focuses on methods for the small-scale culture of insect cells. Protocols for large-scale cell culture are provided in Chapters 16 and 17.

EQUIPMENT REQUIRED

The maintenance of cultured insect cells for the propagation of baculoviruses requires a number of major items of equipment. However, not all of the items on the following list will be required in every case. Items at the end of this list, such as high-capacity filtration equipment, will not be needed if only commercially prepared liquid media are used.

1. Laminar flow hood
2. Temperature- and humidity-controlled incubators
3. Stereo dissecting microscope
4. Inverted tissue-culture microscope
5. 4° C storage space
6. –20° C and –80° C freezers
7. Liquid nitrogen refrigerator
8. Several water baths
9. Autoclave
10. Sterilizing oven
11. Water-purification equipment. Use of triple-distilled deionized water (ddH_2O), milliQ water, or water of equivalent quality is recommended for all cell culture work.
12. High-capacity filtration equipment for media sterilization. (See materials list in Insect Cell Culture Media section following.)
13. Osmometer

In general, standard equipment used for mammalian cell culture is suitable. If two laminar flow hoods are available, one should be reserved for virus-free work, particularly passaging uninfected stock cultures. The hoods should be equipped with an ultraviolet light to maintain sterility when the hood is not in use.

The incubators should be capable of maintaining a temperature of 27° C to 28° C. Apply caution when selecting an incubator because the lowest temperature maintained by many incubators is 5° C above ambient temperature. If ambient temperatures exceed 23° C, such incubators will not be able to maintain a temperature of 28° C. Thus, in particularly warm environments, it may be necessary to use an incubator with a built-in cooling system. A CO_2 supply to the incubator is not required (see discussion later in this chapter). However, water should be kept in the pan at the bottom of the incubator to maintain a humid environment. If two incubators are available, it is helpful to keep uninfected and infected cell cultures separate. It may also be helpful to reserve one refrigerator for the storage of unused media and media components and to store virus stocks and media bottles that are currently in use in a separate refrigerator.

It is essential to keep these items of equipment clean at all times. The work area of the hood and the incubator trays should be wiped down with 70% ethanol before and after use. Spills of media or agarose should be immediately wiped up, cleaned with disinfectant solution [0.5% Amphyl (Scientific Products # D1306-1A), 1.5% isopropanol], rinsed well with water, and rinsed once with 70% ethanol. Periodically (about once a month), it will be necessary to clean out and disinfect the chambers of humidified incubators. The moist environment of these chambers will promote the growth of fungi if the chambers are not kept spotlessly clean. Clean the stainless steel walls of the incubator with a disinfectant solution, rinse well with distilled water, rinse once with 70% ethanol, and wipe dry with a tissue. Clear Bath (10–15 drops/gal; Spectrum # 105540) can be added to the water kept in a stainless steel pan on the bottom of the chamber to inhibit growth of bacteria or mold. The refrigerators used for storing media, and so on, should also be kept clean to avoid the growth of mold.

INSECT CELL CULTURE MEDIA

Serum-Supplemented Media

Three basic formulations of media are most commonly used to support the growth of lepidopteran cell lines that are used for the propagation of baculoviruses. These are Grace's medium (Grace, 1962), IPL-41 (Weiss et al., 1981; Weiss and Vaughn, 1986), and TC-100 (Gardiner and Stockdale, 1975). TNM-FH medium (Hink, 1970), which is also widely used, is Grace's basal medium supplemented with lactalbumin hydrolysate and yeastolate. All of these supply basic nutrients (salts, sugars, amino acids, organic acids, vitamins, and trace elements) and differ primarily in the relative amounts of these components (Table 11-1). These media have a pH of approximately 6.2 and are buffered with sodium phosphate. This contrasts with mammalian media, which generally use a CO_2-bicarbonate buffer system, and explains why CO_2 is not required for insect cell culture. Phenol red, a pH indicator commonly used in mammalian media, is not used, so most insect media are clear to yellowish in color.

If desired, an antibiotic-antimycotic solution may be added to the tissue-culture medium to minimize bacterial or fungal contamination. Some workers prefer not to add antibiotics to medium for routine subculturing of stock cultures, reasoning that contaminated cultures can be more rapidly identified and discarded if antibiotics are not used. However, including antibiotics in medium for larger-scale cultures is recommended. It is sometimes possible to "cure" valuable cultures that have become contaminated by short-term treatment with the antibiotic-antimycotic

TABLE 11-1 Composition of insect cell culture media (mg/l)

Component	*Grace's*	*IPL-41*	*TC-100*
Inorganic Salts			
$CaCl_2$ (anhyd)	750.00	500.00	996.49
$CoCl_2 \cdot 6H_2O$		0.05	
$CuCl_2 \cdot 2H_2O$		0.20	
$FeSO_4 \cdot 7H_2O$		0.55	
KCl	4100.00	1200.00	2870.00
$MgCl_2 \cdot 6H_2O$	2280.00		
$MgCl_2$			1067.95
$MgSO_4 \cdot 7H_2O$	2780.00		
$MgSO_4$		918.00	1357.67
$MnCl_2 \cdot 4H_2O$		0.02	
NaCl		2850.00	500.00
$NaHCO_3$	350.00	350.00	
$NaH_2PO_4 \cdot H_2O$	1013.00	1160.00	1008.36
$(NH_4)(Mo_7O_{24} \cdot 4H_2O)$		0.04	
$ZnCl_2$		0.04	
Sugars, Other			
D(−)-Fructose	400.00		
Fumaric acid, free acid	55.00	4.40	
D(+)-Glucose	700.00	2500.00	1100.00
α-Ketoglutaric acid	370.00	29.60	
L(−)-Malic acid	670.00	53.60	
Maltose		1000.00	
Succinic acid	60.00	4.80	
Sucrose	26,680.00	1650.00	
Amino Acids			
β-Alanine	200.00	300.00	
L-Alanine	225.00		225.00
L-Arginine·HCl	700.00	800.00	665.11
L-Asparagine	350.00	1300.00	350.00
L-Aspartic acid	330.00	1300.00	350.00
L-Cystine	22.00		
L-Cysteine·2Na		119.40	28.68
L-Glutamic acid	600.00	1500.00	600.00
L-Glutamine	600.00	1000.00	600.00
Glycine, free base	650.00	200.00	650.00
L-Histidine	2500.00	200.00	
L-Histidine·HCl·H_2O			3697.89
L-Isoleucine	50.00	750.00	50.00
L-Leucine	75.00	250.00	75.00
L-Lysine·HCl	625.00	700.00	625.00
L-Methionine	50.00	1000.00	50.00
L-Phenylalanine	150.00	1000.00	150.00
L-Proline	350.00	500.00	350.00
Hydroxy-L-Proline		800.00	
DL-Serine	1100.00	400.00	
L-Serine			550.00
L-Threonine	175.00	200.00	175.00
L-Tryptophan	100.00	100.00	100.00
L-Tyrosine	50.00		
L-Tyrosine 2Na·$2H_2O$		360.40	72.08

TABLE 11-1 Composition of insect cell culture media (mg/l) *(continued)*

Component	*Grace's*	*IPL-41*	*TC-100*
L-Valine	100.00	500.00	100.00
Vitamins			
Para-aminobenzoic acid	0.02	0.32	0.02
Biotin	0.01	0.16	0.01
D-Ca pantothenate	0.02	0.008	0.11
Choline chloride	0.20	20.00	
Folic acid	0.02	0.08	0.02
iso-Inositol	0.02	0.40	0.02
Niacin	0.02	0.16	0.02
Pyridoxine·HCl	0.02	0.40	0.02
Riboflavin	0.02	0.08	0.02
Thiamine·HCl	0.02	0.08	0.52
Vitamin B_{12}		0.24	0.01
Undefined Ingredients			
Tryptose			2000.00

solution. However, the most reliable method of ensuring the survival of a cell line is to maintain adequate frozen stocks.

The media are commercially available from a variety of sources in liquid or powdered form. Before using liquid medium, simply add supplements and antibiotics as required and complete with fetal bovine serum (FBS). Powdered medium must first be dissolved in the appropriate quantity of water and filter-sterilized. Available basal media and supplements and other required materials are listed here. Next, a table indicating which supplements are used with which basal media is presented. Finally, the methods for preparing powdered medium and completing medium with 10% FBS are discussed.

Materials

1. Cell culture media
   a. Grace's Insect Cell Culture Medium (1X)—liquid (GIBCO # 350-1590AJ) or powder (GIBCO # 440-1300EB)
   b. Grace's Insect Medium—Supplemented (1X), liquid (GIBCO # 350-1605AJ)
   c. IPL-41 Insect Medium (1X)—liquid (GIBCO # 350-1405PK) or powder (GIBCO # 440-1400EB)
   d. TC-100 Insect Medium—powder (GIBCO # 440-1600EB)
2. Supplements
   a. LH: Lactalbumin Hydrolysate Solution (50X)—liquid (GIBCO # 670-8080AG)
   b. YL: Yeastolate Solution (50X)—liquid (GIBCO # 670-8190AG)
   c. TPB: Tryptose Phosphate Broth (50X)—liquid (GIBCO # 670-8060AG)
   d. TB: Tryptose Broth (Difco # 0062-01-4)
3. Antibiotic-antimycotic solution (100X; GIBCO # 600-5240PE)
4. 10 N KOH
5. 15% NaCl
6. High-capacity filtration equipment for media sterilization
   a. Wide-diameter tripod filter holder (e.g., Fisher Type 304 142 mm stainless-steel tripod holder, # 09-753-17E)
   b. Prefilters and filters to fit tripod holder (e.g., Millipore AP prefilters, # AP25 127 50; 1.2 μm RA filters, # RAWP 142 50; 0.45 μm HA filters, # HAWP 142 50; 0.22 μm GS filters, # GSWP 142 50; and mesh spacer # AP32 124 50)
   c. High-volume peristaltic pump (e.g., Millipore # XX80 000 00)
   d. Filling bell (e.g., Millipak 60; Millipore # MPGL 065 H2)

7. 500-ml glass medium bottles (Wheaton # 219119). It is preferable to reserve these bottles for medium use only. Sterilize by autoclaving.
8. FBS: Fetal bovine serum, low endotoxin levels (J.R.H. Biosciences # 12-10378). Store at –20° C until required.

Supplements

Medium	*Basal Medium*	*Supplements*
TNM-FH	Grace's[a]	LH
		YL
IPL-41	IPL-41	TPB
TC-100[b]	TC-100	TB

[a]Alternatively, Grace's medium can be purchased with supplements added, so that complete TNM-FH can be made simply by completing with serum.

[b]Although most commercial formulations of powdered TC-100 contain TB, adding additional TB at a concentration of 2.6 g/l is recommended.

Preparation of Powdered Medium (10 l)

Depending on the source, powdered media often lack certain components (e.g., $NaHCO_3$, glutamine, $CaCl_2$), which must be added at the time of preparation. Read the manufacturer's instructions carefully.

Method

1. Dissolve the contents of one 10-liter packet of powdered medium in 9.5 liters of ddH_2O.
2. Add extra components (e.g., $NaHCO_3$, glutamine, etc.; see manufacturer's instructions) and supplements (see preceding table) as required.
3. Add antibiotics if desired. Mix on a magnetic stir-plate for one to two hours to ensure thorough mixing.
4. While the medium is mixing, assemble the filters in the tripod stand. Start with the 0.22 μm filter and place a mesh spacer between each filter. We do not sterilize the tripod assembly and filters because the filling bell contains an in-line 0.22 μm filter.
5. Adjust the pH of the medium to 6.2 with 10N KOH.
6. Check the osmolality and adjust to 350 mOsm/kg with 15% NaCl if necessary.
7. Filter-sterilize and dispense into sterile 500 ml media bottles. With the preceding setup, you can filter the medium at 200 to 300 mls/minute.
8. Store at 4° C for up to three months. The glutamine is labile, so the shelf life of tissue-culture medium is limited. However, glutamine may be readded to medium over three months old.

Completing with FBS

These media are generally completed with 10%, which provides additional nutrients and also protects the cells from hydrodynamic stresses encountered in suspension culture. The quality of the serum is of vital importance for optimal cell growth and viral replication. Serum from different vendors or even different lots of serum from the same vendor can vary in their ability to support optimal cell growth. We strongly recommend that you test a small aliquot of a new lot of serum for cell growth and viral replication before purchasing it in large quantities.

Method

1. Thaw the required amount of FBS. Once thawed, FBS should not be refrozen, as repeated freezing and thawing is detrimental.
2. Heat-inactivate the serum by heating in a water bath at 56° C for 30 minutes. Swirl frequently during this period to ensure even heat distribution.
3. Aseptically add serum to the medium to a final concentration of 10% (v/v).
4. Store completed medium at 4° C and use within one to two months.

Preparation of Medium from Individual Components

The preparation of tissue-culture medium from the individual components is occasionally necessary for the preparation of specialized media such as deficient media lacking particular amino acids, sugars, phosphate, and so on. Some media of this type are commercially available (e.g., methionine-deficient). The preparation of TC-100 is described here. Other media may be prepared in an analogous fashion. Generally, deficient media are prepared simply by omitting the relevant component. However, to prepare phosphate-free medium, it is necessary to include 7.5 mM Hepes, pH 6.2, to replace the sodium phosphate as a buffer. More details are provided in the description of methods that require these specialized media.

Materials (from Sigma unless noted otherwise)

1. Amino acids

L-alanine	A 3534	L-leucine	L 1512
L-arginine	A 3909	L-lysine HCl	L 1262
L-asparagine	A 4159	L-methionine	M 2893
L-aspartic acid	A 4534	L-phenylalanine	P 5030
L-cystine	C 7777	L-proline	P 4655
L-glutamic acid	G 5638	L-serine	S 5511
L-glutamine	G 5763	L-threonine	T 1645
Glycine	G 6388	L-tryptophan	T 0271
L-histidine HCl	H 8125	L-tyrosine	T 1145
L-isoleucine	I 7383	L-valine	V 6504

2. Vitamins

p-Aminobenzoic acid	A 3659	Nicotinic acid	N 0761
Biotin	B 4693	Pyridoxine HCl	P 6280
Calcium pantothenate	P 5155	Riboflavin	R 9504
Folic acid	F 8758	Thiamine HCl	T 1270
i-Inositol	I 7508	Vitamin B_{12}	V 6629

3. Salts and other components

$CaCl_2 \cdot 2H_2O$	C 7902	$MgSO_4 \cdot 7H_2O$	M 1880
Glucose	G 7021	$NaHCO_3$	S 5761
KCl	P 5405	$NaH_2PO_4 \cdot H_2O$	S 9638
10 N KOH	P 1767	15% NaCl	S 5886
$MgCl_2 \cdot 6H_2O$	M 2393	Tryptose broth	Difco # 0062-01-4

Method

1. L-leucine stock
   Dissolve 0.75 g L-leucine in 50 ml ddH_2O. This may be stored at –20° C.
2. Insoluble amino acids
   a. Weigh the following reagents:
      0.22 g L-cystine
      0.50 g L-tyrosine
      1.00 g L-tryptophan

b. Mix in 800 ml ddH_2O and 50 ml 2N HCl.
c. Heat to 80° C and stir well.
d. If undissolved reagents remain, add concentrated HCl and mix well until dissolved.
e. Bring volume to 1 l. Store at –20° C if required.

3. Vitamin Stock
   a. Weigh the following:
      0.02 g p-Aminobenzoic acid
      0.11 g Calcium pantothenate
      0.02 g Folic acid
      0.02 g i-Inositol
      0.02 g Nicotinic acid
      0.02 g Pyridoxine HCl
      0.02 g Thiamine HCl
      0.01 g Vitamin B_{12}
   b. Dissolve in 800 mls ddH_2O.
   c. Dissolve 0.01 g Biotin and 0.02 g Riboflavin in a small volume of 1 N KOH. Add to the solution prepared in step b.
   d. Bring to 1 l. Store at –20° C in 10 ml aliquots.
4. Weigh the following in a 10 l flask and dissolve in 2.5 l ddH_2O:
   2.25 g L-alanine
   5.50 g L-arginine
   3.50 g L-asparagine
   3.50 g L-aspartic acid
   6.00 g L-glutamine
   6.50 g Glycine
   33.8 g L-histidine HCl
   0.50 g L-isoleucine
   6.25 g L-lysine HCl
   0.50 g L-methionine
   1.50 g L-phenylalanine
   3.50 g L-proline
   5.50 g L-serine
   1.75 g L-threonine
   1.00 g L-valine
   26.0 g Tryptose broth
   28.7 g KCl
   13.2 g $CaCl_2 \cdot 2H_2O$
   22.8 g $MgCl_2 \cdot 6H_2O$
   27.8 g $MgSO_4 \cdot 7H_2O$
   10.0 g Glucose
5. Dissolve 11.4 g $NaH_2PO_4 \cdot H_2O$ and 3.5 g $NaHCO_3$ separately in 250 ml each ddH_2O. Add to 10 l flask and stir well.
6. Add the following to 10 l flask, stirring constantly:
   50 ml L-leucine stock (from step 1)
   1 l insoluble amino acids (from step 2)
   10 ml vitamin stock (from step 3)
7. Bring volume to 9.95 l with ddH_2O.
8. Adjust pH to 6.2 with 10 N KOH.
9. Adjust osmolarity to ≈350 mOsm/kg with 15% NaCl.
10. Bring to 10 l and filter-sterilize. Store at 4° C.

Serum-Free Media

Recently, a variety of serum-free and low-serum media have been developed to propagate cultured lepidopteran insect cells (Cho et al., 1989; Inlow et al., 1989; Mitsuhashi and Goodwin, 1989; Léry and Fedière, 1990; DiSorbo et al., 1991; Hink, 1991). These have several potential advantages over serum-supplemented media. First, fetal bovine serum is expensive and can be difficult to obtain. Second, the composition of serum can vary from lot to lot and can affect cell growth and viral replication. Third, some of these media have very low protein contents, which greatly facilitates the purification of secreted proteins.

ISFM (Inlow et al., 1989) is a simple serum-free medium prepared by supplementing IPL-41 medium with ultrafiltered yeastolate and a lipid suspension containing cholesterol, alpha tocopherol acetate, cod liver oil fatty acid methyl esters, Tween 80, and Pluronic F-68 as an emulsifying agent. This lipid emulsion and ultrafiltered YL are available from a variety of vendors. Alternatively, Inlow et al. (1989) provide detailed protocols for their preparation. Other lipid emulsions, such as Inserum I (J.R.H. Biosciences) or Chemically Defined Lipid Concentrate (GIBCO), can be used to prepare serum-free media and may offer slight nutritional advantages over the original formulation for particular cell lines.

Ex-Cell 400, Ex-Cell 401, Sf900 (DiSorbo et al., 1991), and Sf900-II are protein-free or low-protein media, which are also based on IPL-41. Currently, these are available only as liquids and are generally much more expensive than powdered media. Liquid media are very convenient to use, however, and unlikely to have significant lot-to-lot variations that can affect the growth of cells. ISFM, Ex-Cell 400, and Sf900-based media are all commonly used with SF cells, and we suggest choosing one of these if you are trying serum-free medium for the first time.

SFM-LP (Hink, 1991) is a serum-free medium prepared by supplementing Grace's medium containing lactalbumin hydrolysate and yeastolate with Pluronic F-68 and a low-protein lipid supplement (Ex-Cyte VLE). The amount of Ex-Cyte present in SFM-LP is adjusted to 100 μg cholesterol/ml.

The level of foreign proteins produced by recombinant baculoviruses propagated on cell lines maintained in different media can vary significantly. In some cases, serum-free media outperform serum-supplemented media. In other cases, the situation is reversed, or there is no difference in expression levels. Even if expression levels are lower for a secreted protein in serum-free media, more rapid purification and higher yields from serum-free media might offset any advantage of higher expression in serum-supplemented media. To date, the examination of the expression of many genes in baculovirus-infected cells in a variety of media has failed to reveal a single best combination of cell line and medium that works well with all recombinant viruses.

The following materials list includes various serum-free media and supplements that are available. This list is followed by methods for preparing a selection of different serum-free media.

Materials

1. Cell culture media
   a. IPL-41 Insect Medium (1X)—liquid (GIBCO # 350-1405PK)
   b. Grace's Insect Medium—Supplemented (1X) (GIBCO # 350-1605AJ)
   c. Ex-Cell 400—liquid (J.R.H. Biosciences # 14-40078)
   d. Ex-Cell 401—liquid (J.R.H. Biosciences # 14-40150)
   e. Sf-900 Serum-Free Insect Cell Culture Medium (1X)—liquid (GIBCO # 350-0900PK)
   f. Sf-900-II Optimized Serum-Free Medium for Insect Cell Culture—Complete (1X), liquid (GIBCO # 350-0902PK)
2. Supplements
   a. UFY: Yeastolate Ultrafiltrate (50X)—liquid (GIBCO # 670-8200AG)
   b. LC: Lipid Concentrate (100X)—liquid (GIBCO # 680-1900AG)
   c. CDLC: Chemically Defined Lipid Concentrate (100X)—liquid (GIBCO # 680-1905AG)
   d. Inserum I—liquid (J.R.H. Biosciences # 58-90477)
   e. Ex-Cyte VLE (Miles # 81-129-1)
   f. 10% Pluronic F-68 in H_20 (GIBCO # 670-4040AG)

Method

Ex-Cell 400 and 401, and Sf-900 and 900-II are purchased ready to use. Antibiotics may be added if required. Preparation of ISFM (with different lipid supplements) and SFM-LP is detailed here.

Medium	Basal Medium	Supplements (per liter medium)
ISFM (IUL)	IPL-41	20 ml 50X UFY 10 ml 100X LC
IUCDL	IPL-41	20 ml 50X UFY 10 ml 100X CDLC
I-Ins	IPL-41	30 ml Inserum I
SFM-LP	Grace's supplemented	10 ml 10% Pluronic F68 ≈10 ml Ex-Cyte VLE (100 μg/ml cholesterol)

Low-Serum Media

A number of vendors market serum extenders that are intended to replace fetal bovine serum in complete medium. These extenders can be considerably cheaper than fetal bovine serum. Although they are prepared from serum, the manufacturing processes yield products that are more consistent from lot to lot than natural products like serum. The protein concentration tends to be much lower in extenders than in serum, which may facilitate the purification of proteins from cells maintained in these media.

For the most part, the currently available serum extenders have not been tested with cultured insect cells, and it is not clear whether they will be of any benefit. Two types of extender we have used are CPSR-3 (Sigma # C 9155) and NuSerum IV (Collaborative Research # 50004). They were used at a concentration of 10% (v/v) in basal IPL-41 supplemented with ultrafiltered yeastolate (UFY). CPSR-3 (Controlled process serum replacement-type 3) is a serum replacement that contains bovine embryonic fluid and plasma, and has been used to support the growth of SF-21AE cells in the absence of serum (Vaughn and Fan, 1989). NuSerum IV is a serum extender containing 25% fetal bovine serum. In our hands, SF-9 cells appear to adapt much better to medium containing NuSerum IV than to medium containing CPSR-3 (V. Luckow, unpublished).

INSECT CELL CULTURE TECHNIQUES

This section describes the routine maintenance of insect cell cultures. The discussion is limited to the propagation of SF cells. Researchers using other cell lines should be aware that they may differ dramatically in their behavior in cell culture, so cell culture protocols will have to be optimized for each cell line. The discussion that follows is subdivided into sections dealing with routine subculturing procedures for cells in serum-supplemented medium, modifications and notes for the propagation of SF cells in serum-free medium, and protocols for the long-term storage of SF cells. Only the small-scale culture (up to 1 l) of cells is discussed. Scale-up to larger cultures is described in detail in Chapters 16 and 17.

Routine SF Cell Culture Procedures Using Serum-Supplemented Medium

SF cells grow well both in suspension and as monolayer cultures, and they can be transferred from one to the other without any adaptation. We have found that continuous propagation of SF cells in suspension for more than a few passages can result in decreased cell viability and growth rate. Thus, we generally maintain and passage our stock cultures in tissue-culture flasks and use these to seed suspension cultures when required. Some workers report that continuous propagation in suspension is not problematic if the cells are maintained in medium containing 0.1% Pluronic F-68 (V.A. Luckow, unpublished observations).

As discussed in the section on insect cell culture media, it is up to each researcher to decide whether to include antibiotics in the tissue-culture medium for stock cultures. We recommend including antibiotics in medium for larger-scale cell culture, including growth in spinner bottles, and so on.

SF cells are routinely grown at 27° C. However, they tolerate temperatures from 25° to 28° C well, and because they have no requirement for CO_2, they may be cultured successfully on the lab bench if the room temperature does not fluctuate excessively. Cell growth rate will be decreased when the cells are grown at lower temperatures. Under ideal conditions, SF cell populations double every 18 to 24 hours. The cells are normally subcultured twice a week.

Passaging Monolayer Cultures

Materials

1. Insect cell culture medium (described earlier in this chapter). Warm the medium to 27° C before use.
2. Antibiotic-antimycotic (100X)—liquid (GIBCO # 600-5240PE)
3. Tissue-culture flasks, dishes, and plates. Like mammalian cell culture, SF cell culture requires the use of plasticware that has been treated to permit cell attachment. *Bacterial Petri dishes are not suitable for SF cell culture.* In general, tissue-culture ware designed for mammalian cells can be used successfully. We commonly use plasticware from Falcon or Corning. Catalog numbers for some commonly used items follow. Note that while either of these brands can be used successfully, you should be consistent in the brand you use. The cells may require an adaptation period to adjust from one surface type to another.

Size	*Falcon #*	*Corning #*
25 cm^2 flask	3013	25102
75 cm^2 flask	3024	25110
35 mm dish	3001	25000
60 mm dish	3002	25010
100 mm dish	3003	25020
96 well plate	3072	25860
24 well plate	3047	25820
6 well plate	3046	25810

4. Sterile pipets. Standard glass pipets can be used. They should be plugged with cotton and sterilized by baking at 180° C for at least two hours. Alternatively, presterilized disposable pipets can be used.
5. Trypan blue (Sigma # T 6146). Prepare a 0.4 % (w/v) solution in incomplete tissue-culture fluid or in PBS. Alternatively, the solution can be purchased ready-made (e.g., GIBCO # 630-5250AG).
6. Hemocytometer (Fisher # 0267110)

Methods

1. Generally, SF cells should be subcultured when they are 80% to 90% confluent. Unlike most mammalian cell lines, SF cells are not susceptible to contact inhibition and will begin to float if the cell density is too high. Before subculturing, examine the cells under a tissue-culture microscope to evaluate the confluence and general health of the cells. Healthy cells will have distinct cell boundaries, and the nucleus and smaller organelles may be visible under high magnification (320X–400X). The cells should be nicely rounded and should not appear granular.

Examine the cultures closely for signs of contamination. The tissue-culture medium of cultures contaminated with bacteria will become cloudy very rapidly, and bacterial contamination can be confirmed by examination at high magnification. Yeast contamination will also be manifested by cloudiness in the medium. Yeasts are easily visible under high magnification as refractive, round to oval-shaped bodies. With fungal contamination, the mycelium will often be macroscopically visible. Low-level contamination may become evident by the microscopic detection of mycelial fragments in the tissue-culture fluid. Viral contamination can be detected by the presence of occlusion bodies in the nuclei of infected cells if the contamination is due to an occ^+ virus. Contamination by an occ^- virus is harder to detect and is generally not clearly manifested until the cells begin to lyse. The morphology of virus-infected cells is described in more detail in Chapter 14. *Do not* open any culture flask that you suspect is contaminated. Contaminated cultures should be disposed of in biohazard bags and autoclaved.

Occasionally, a certain amount of particulate material may be present in serum-supplemented tissue-culture medium. This material is serum proteins that have precipitated during storage. This precipitate will appear amorphous under the microscope and should not be confused with medium contamination.

2. Wash the cells off the surface of the flask by gentle pipeting. It is not necessary to use trypsin or EDTA to detach SF cells from the substrate.
3. Take a small aliquot of cells and add 0.1 volumes of trypan blue solution. Examine immediately on a hemocytometer under a tissue-culture microscope. Exclusion of the trypan blue stain is a working definition of cell viability. Determine the concentration of viable cells in the suspension and also the proportion of viable cells. Ideally, viability, as determined by trypan blue staining, should be >98%.
4. The cells should be diluted 1:8 at each subculture so that they will be ready to pass again in three to four days. A confluent 25 cm^2 flask of SF cells contains $\approx 1\text{-}2 \times 10^7$ cells. Thus, each 25 cm^2 flask should be seeded with $\approx 2 \times 10^6$ cells. The final volume of tissue-culture fluid should be 5 mls to 7 mls. Flasks of 75 cm^2 should be inoculated with $\approx 5 \times 10^6$ cells in a final volume of 10 mls to 12 mls. It is better to add the medium to the flasks before the cells. This helps ensure that the cells are evenly dispersed in the flask following inoculation. Label each tissue-culture flask with the cell type, cell media, splitting ratio, passage number, and the current date.
5. It is a good idea to seed one backup flask at half the above density so that it becomes confluent the day after the stock cultures. The day after the stock cultures are subcultured, they should be examined closely for signs of contamination (described earlier in this section). If the cultures are contaminated, the backup flask from the previous passage can be used to continue the culture.

   The old flask from which the cells were taken can also be kept and observed for signs of contamination. Low-level contamination may become evident here before it appears in the fresh cultures. If more than 30% of the volume of media in the old flask is removed, replace it with an equivalent volume of fresh media.
6. Incubate the flasks in a humidified incubator at 27° C. The caps should be left about one-quarter of a turn open to ensure adequate aeration.

Growing SF Cells in Suspension Cultures

SF cells may be grown in suspension in a variety of different vessels. Most simply, they can be cultured in glass bottles containing a magnetic stir bar or in flasks that are agitated in incubators containing a shaking platform. Specially designed spinner flasks are also commonly used to maintain small cultures in suspension. These contain a teflon-coated magnetic stir bar suspended from a glass or stainless steel rod that is driven from below by a magnetic stirrer. Microcarrier spinner flasks have a large plastic paddle that is perpendicular to the stir bar. In these vessels, more media is stirred at a given speed than in vessels without the paddle. The increased aeration may favor cell growth, especially when larger volumes are used.

The use of a good quality magnetic stir plate is essential for the propagation of SF cells in suspension. The most critical features are: (1) the stir plate should have a strong magnetic field that enables it to operate smoothly at slow stirring speeds; (2) it should not generate excess heat; and (3) it should be very reliable. A stir plate that meets these requirements will be adequate for most general cell culture work. However, laboratories growing large numbers of cells in suspension (e.g., for use as inoculum for large-scale bioreactors) may want to consider stir plates with more advanced features. For example, some stir plates have digital tachometers to indicate stirring speed (in RPM) and can accelerate slowly to the desired speed after a setting change, or from a restart. Several vendors provide optional backup power supplies for the speed controller to ensure continuous operation in the event of a power failure (e.g., Sci/ERA Quad Drive Master Controller and SciERA Backup Power Pack; Bellco # 7766-00110 and 7765-88110).

Materials

1. Insect cell culture medium (described earlier in this chapter). Warm the medium to 27° C before use.
2. Antibiotic-antimycotic (solution 100X; GIBCO # 600-5240PE)
3. Sterile pipets. Standard glass pipets can be used. They should be plugged with cotton and sterilized by baking at 180° C for at least two hours. Alternatively, presterilized disposable pipets can be used.
4. Trypan blue (Sigma # T 6146). Prepare a 0.4% (w/v) solution in incomplete tissue-culture fluid or in PBS. Alternatively, the solution can be purchased ready-made (e.g., GIBCO # 630-5250AG).
5. Hemocytometer (Fisher # 0267110)
6. Spinner bottles or flasks. A selection of different vessels that are suitable for the suspension culture of SF cells are listed here. Instructions follow for the preparation of these vessels for use.
   a. Sterile 500 ml medium bottle (Wheaton # 219759) with a stir bar on the bottom
   b. Sterile adjustable hanging-bar spinner flask, 50 ml (Bellco # 1967-00050)
   c. Sterile microcarrier spinner flasks (Bellco: 100 ml # 1965-00100; 250 ml # 1965-00250; 500 ml # 1965-00500; 1000 ml # 1965-01000)
7. Slow-speed magnetic stirrer [e.g., Cellgro deluxe programmable; Thermolyne, # S45725; or Bellco # 7760-06005 (4 place)]
8. Linbro tissue-culture glassware soap (Flow Laboratories # 76-670-94)

Methods

1. Add the appropriate volume of prewarmed tissue-culture medium to the spinner flask. Inoculate with cells so that the starting density is 2 to 4×10^5 cells/ml. A 200 ml suspension culture can thus be conveniently inoculated with cells from a single 75 cm^2 flask.
2. Incubate at 27° C with constant stirring at 60 RPM to 80 RPM. Leave the lid of the vessel open one-quarter of a turn to ensure adequate aeration.
3. If you are growing these cells as inoculum for monolayer cultures (e.g., for plaque assays or time-course experiments), they should be used when they reach a density of $\approx 1 \times 10^6$ cells/ml (three to four days after inoculation). If

convenient, you may dilute the cells in half at this stage and incubate in suspension for a further day (by which stage the cells will again be at a density of $\approx 1 \times 10^6$ cells/ml). If the cells are grown under conditions of very efficient aeration, it may be acceptable to let them grow to a density of 1.5×10^6 cells/ml before use. In general, we do not recommend letting the cells grow to densities greater than this in small-scale cultures.

4. After use, clean and prepare the spinner flasks for reuse as follows:
   a. Discard media and cells into waste beaker containing Amphyl or 2% clorox bleach, and rinse the flask once with water. Do *not* put Amphyl disinfectant or bleach into spinner flasks or any other reusable tissue culture glassware!
   b. Add enough water to cover the scum line of dead cells, and autoclave the spinner flask for more than 20 minutes.
   c. Discard the waste water and clean the spinner using a gentle liquid or powdered soap. Ivory liquid dishwashing soap or Linbro are suitable because they do not leave soap residues. Be sure to scrub the flask with a test tube brush to remove the scum line of dead cells. Rinse well with deionized water.
   d. Scrub the spinner bar or spinner paddle and suspension rod. Remove any magnetic particles that are attracted to the magnet. Rinse well with deionized water.
   e. Examine the flask for any broken parts (e.g., cracked caps, tops, spinner bars, or glass rods; split paddles; pinched neoprene gaskets) and replace if necessary.
   f. If using hanging bar or microcarrier spinner flasks, assemble by screwing the top containing the spinner bar into the glass spinner flask. The bottom of the spinner bar should be adjusted, if necessary, so that it does not touch the bottom of the flask, nor should it be suspended above the uppermost volume mark indicating the capacity of the flask. If you are only using a medium bottle with a stir bar as a spinner bottle, simply place the stir bar in the bottle.
   g. Check to make sure that flexible paddles, if present, are perpendicular to the magnetic spinner bar.
   h. Screw the top on tight and verify that the spinner bar spins freely on a magnetic stirrer or stirrer platform and that the glass or stainless steel rod holding the spinner bar is correctly aligned above the center of the bottom of the flask.
   i. Rinse the flask again with high-quality water. Screw the caps onto the side arms, but leave open about one-quarter of a turn. Cover the side arm caps, if present, with aluminum foil.
   j. Autoclave the spinner flasks for 30 minutes with 45 minutes of drying time.

Propagation of SF Cells in Serum-Free Medium

We recommend that you adapt cells gradually for growth in serum-free medium as follows. Subculture the cells from complete serum-supplemented medium to medium composed of a 1:1 mix of serum-containing and serum-free medium. Allow the cells to grow until nearly confluent. Subculture these cells into a 1:3 mix of serum-containing and serum-free medium. Repeat this procedure using a 1:9 mix. Then, subculture the cells into serum-free medium. Several passages in the serum-free medium may be required before the cells are fully adapted to the medium. During this period, the cells may display longer lag times, reduced growth rates, and decreased viability. The cytoplasm may become somewhat granular, and the cells will appear swollen. If you experience difficulty adapting the cells, you may wish to try changing the serum concentration in smaller increments, or maintaining the cells at an intermediate serum concentration for more than one passage.

It may also be necessary to adapt cells from one serum-free medium to another, but this process is generally quicker.

Cells are propagated in serum-free medium using essentially the same protocols described above for cells in serum-containing medium. One significant difference is that trypan blue cannot be used to estimate viability of cells cultured in medium containing Pluronic F-68 (most serum-free media; see Murhammer and Goochee, 1988). Thus, viability must be estimated by direct visual observation. Dead cells will appear shriveled and more refractive than live cells.

Cells grown in serum-free medium may differ in a number of other ways from cells grown in serum-containing medium. They are less able to grow at low-starting densities. At densities less than 2×10^5 cells/ml, they display longer lag times before they begin to grow exponentially. On the other hand, they can achieve higher final densities in suspension in spinner flasks than cells grown in serum-containing medium. (In both types of media, the actual final density achieved is very dependent on the aeration.) Cells adapted to serum-free medium seem to stick tighter to plasticware than those maintained in complete medium. More vigorous pipeting may be required to resuspend the cells when they are being subcultured. Alternatively, it may be necessary to gently scrape the cells to detach them from the plastic. Be careful to avoid excessive cell damage.

Similarly, SF cells show a greater propensity to clump together when grown in serum-free medium. Because of this, it is necessary to stir suspension cultures at 100 RPM rather than 60 RPM to 80 RPM for cells in medium with serum. Cells grown in serum-free medium may be more sensitive to antibiotics added to combat bacterial or fungal contamination than cells grown in medium containing serum (Mitsuhashi, 1989). Similarly, cells in serum-free medium may be more easily damaged by freezing for long-term storage, and appear to be more sensitive to damage during centrifugation. Finally, some workers have found that cells that have been passaged in serum-free medium for prolonged periods of time (>50 passages) begin to show decreased susceptibility to infection and reduced expression of foreign genes (see Chapter 16).

Long-Term Cell Storage

SF cells can be stored for long periods of time by freezing at –80° C or in liquid nitrogen. We recommend that aliquots of stock cultures be frozen down periodically to provide a backup in case of loss of the culture due to contamination or other reasons. The most critical factor for the successful freezing and recovery of cells is the condition of the cells at the time of freezing. They should be healthy (>98% viable), free from contamination, and in exponential growth.

Materials

1. Insect cell culture medium (described earlier in this chapter). Warm the medium to 27° C before use.
2. Sterile pipets. Standard glass pipets can be used. They should be plugged with cotton and sterilized by baking at 180° C for at least two hours. Alternatively, presterilized disposable pipets can be used.
3. Trypan blue (Sigma # T 6146). Prepare a 0.4 % (w/v) solution in incomplete tissue-culture fluid or in PBS. Alternatively, the solution can be purchased ready-made (e.g., GIBCO # 630-5250AG).
4. Hemocytometer (Fisher # 0267110)
5. Sterile, disposable polypropylene centrifuge tubes with plug-seal cap, 15 ml (Corning # 25319-15)

6. Sterile cryovials, 2.0 ml (Nalgene # 5000-0020)
7. Freezing medium: 85% complete tissue-culture medium, 15% DMSO (Sigma # D 2650). Prepare fresh and sterilize by filtration through a 0.22 μm filter. The DMSO should be reserved for this purpose. Do not use DMSO that is more than one year old.

Methods

Freezing

1. Examine the cells for health, viability, and signs of contamination as described already. Cells should be from cultures that are >98% viable and in exponential growth ($\approx 1 \times 10^6$ cells/ml).
2. Pellet the cells by gentle centrifugation (1000 × g for five minutes), discard the old media, and resuspend in fresh media to a density of 1 to 2×10^7 cells/ml.
3. Mix equal volumes of cell suspension and freezing medium and place on ice. Aliquot 1.0 ml of diluted cells into each cryovial and screw the cap on tightly.
4. Place vials into an insulated box and freeze slowly. Transfer to a –20° C freezer for about an hour, and then place at –70° C for one to two days. Alternatively, put the vials in a Nalgene Cryo 1° C Freezing Container (# 5100-0001) and place directly at –80° C. Freeze in liquid nitrogen for long-term storage.
5. One to two weeks after the cells are placed in liquid nitrogen, thaw one vial to check for viability and contamination.

Thawing Frozen Cells

1. Remove vials from liquid nitrogen storage and thaw rapidly (within 40 to 60 seconds) at 37° C.
2. Decontaminate the outside of the vial by immersing in 70% ethanol at room temperature. Carry out subsequent steps under aseptic conditions in a laminar flow hood.
3. Transfer the contents of one vial to a 25 cm^2 tissue-culture flask containing 10 mls fresh insect cell culture medium. It may be helpful to use medium with twice the normal concentration of serum.
4. Incubate at 27° C for two to three hours and examine to see whether the cells have attached. If not, incubate for a further two to three hours and check again. If none of the cells have attached after this time, discard and thaw another vial.
5. When the cells have attached, remove the medium and replace with 10 mls fresh, complete medium.
6. Incubate at 27° C for 24 hours. Refeed with 7 mls complete medium.
7. Incubate at 27° C until the cells are confluent. The cells may take several days to recover before they can be subcultured routinely. It may help to partially or completely replace the medium with fresh complete medium every 24 to 48 hours.

12

VIRUS METHODS

This chapter includes basic procedures you will use to produce and maintain baculovirus stocks, starting with a description of how to prepare a pure (cloned) stock of the virus. This is followed by protocols for the preparation of large-scale virus stocks. Procedures to determine the titer of the stocks are described. Then, approaches to purifying either budded virus or occluded virus particles are discussed. Finally, both short- and long-term storage of baculovirus stocks are discussed.

If you have received a small stock of virus from a reliable source, and you are confident the virus has not been mishandled, it will probably not be necessary to clone this virus again. You should first determine its titer, and then prepare a large stock of the virus by one or two low-multiplicity infections as described later in this chapter.

GENERATING PURE VIRUS STOCKS

The generation of a pure virus stock involves the preparation of a stock starting from a single infectious unit. The procedures given next are for producing a clonal inoculum for the stock either by plaque purification or by end-point dilution. The section that follows then describes how to scale up from this clonal inoculum to large virus stocks.

Plaque Purification

Plaque formation by an animal virus in cell culture was originally described by Dulbecco (1952), and the first baculovirus plaque assay was described by Hink and Vail (1973). The following plaque purification protocol is a modification of that described by Lee and Miller (1978). The principle of the method is to infect cells with extremely low numbers of infectious virus particles so that only isolated cells become infected. The cells are then overlaid with a solid medium, which limits the spread of virus particles. When the original infected cells release virus, only adjacent cells become infected. After several cycles of infection, a focus of infected cells, known as a plaque, is formed. All the virus particles in a plaque derive from a single infectious particle and so represent a clonal population. This infectious virus can therefore be used as the inoculum for generation of a pure virus stock. By counting the plaques obtained with different dilutions of a virus stock, you can also determine the number of infectious virus particles (i.e., the titer) in that stock (see the virus titration protocol later in this chapter).

The most important factor for obtaining good plaques is the condition of the cells. They must be healthy and actively growing at the time of infection. The cell density is of critical importance. Overconfluent cells do not support good occlusion-body formation or plaque development. If the cells are too sparse, the plaques will be difficult to see or will not form at all. Hence, it is very important that the correct number of cells are distributed evenly on the plate. SF cells attach to the substrate rapidly, and each plate should be rocked gently to disperse the cells *immediately* after their addition to the plate.

To ensure good purification, it is best that plaques are picked from plates with *no more than 10 plaques.* This minimizes problems due to diffusion of the virus or perturbation of the overlay during the pick. Thus, it is important that the virus is diluted so that only a small number of plaques are obtained on each plate.

> CONTROLS: It is useful to include one or two uninfected plates to monitor the condition of the cell monolayers.

Materials

1. Complete tissue-culture medium. (See Chapter 11 for details concerning different media.)
2. SF cells grown to approximately 1×10^6 cells/ml
3. 60 mm tissue-culture dishes. (See Chapter 11.)
4. Virus stock to be purified
5. Agarose (Seakem ME (see note 1); FMC # 50014)
6. Stain: 1 mg/ml stock of neutral red (Sigma # N 7005) in incomplete tissue-culture medium. Filter-sterilize and store at 4° C in the dark.

Method

1. Seed 60 mm tissue-culture dishes with 2×10^6 cells per plate (see note 2). The final volume of tissue-culture medium should be 4 mls per plate. To help disperse the cells evenly, it is best to add some tissue-culture medium to the plates before adding the cells. Alternatively, dilute the cells to their final volume before adding them to the plate. Rock each plate gently *immediately* after addition of the cells to ensure an even monolayer. Incubate at 27° C for 30 minutes to 1 hour to allow the cells to attach. They will attach more rapidly if seeded in medium lacking serum. Alternatively, the cells may be plated out at a density of 1×10^6 cells per dish (in complete tissue-culture medium) and incubated at 27° C overnight.
2. Dilute the virus stock in tissue-culture medium. Dilutions of 10^{-5}, 10^{-6}, 10^{-7}, and 10^{-8} will usually be satisfactory.
3. Aspirate the tissue-culture medium from the cell monolayers and infect with 0.5 ml of each virus dilution. *Be careful not to let the plates dry out before addition of the virus.* For the control, add 0.5 ml tissue-culture fluid instead of a virus dilution. Incubate at room temperature for one hour, with gentle rocking.
4. While the cells are rocking, prepare the agarose overlay.
   a. Prepare a solution of 5% agarose in triple-distilled, deionized (or similar quality) water (ddH_2O) and then autoclave. Make up 0.5 mls for each 60 mm tissue-culture dish to be overlaid. (The agarose may be prepared in advance and melted by microwaving when needed.) This solution of agarose will subsequently be diluted tenfold, so prepare it in a container that is large enough to accommodate 10 times the starting volume.
   b. Heat complete tissue-culture medium (5 mls for each 60 mm plate) to 60° C in a water bath.

c. Cool the melted agarose to 60° C.
d. Dilute the agarose to a concentration of 0.5% with the heated tissue-culture medium. For example, for ten 60 mm tissue-culture dishes, prepare 5 mls of 5% agarose and dilute with 45 mls tissue-culture medium. Mix well by swirling.
e. Cool the 0.5% agarose to 40° to 42° C.

5. Remove the virus inoculum from the cells and add 4 mls of agarose overlay to each plate. Try to remove the viral inoculum thoroughly; excess liquid underneath the agarose will lead to smeared plaques. However, be careful not to let the monolayer dry out. It is important that the agarose has cooled sufficiently before overlaying the cells. Do not let the agarose drop on top of the cells. Instead, let it run gently down the side of the plate to avoid disturbing the cell monolayer.
6. Leave the cells at room temperature for 15 to 20 minutes until the agarose has hardened. (If you experience problems with plaques that appear smeared, it may help to dry the plates by leaving them open in a laminar flow hood for several hours after overlaying. Also, as mentioned above, it is important to remove as much of the viral inoculum as possible before overlaying.) Incubate at 27° C.
7. Four to five days later, overlay with stain as follows (see note 3).
   a. Prepare a 0.5% agarose solution in tissue-culture medium as described in Step 4.
   b. Cool the agarose to 40° to 42° C and add the stain to a final concentration of 50 μg/ml.
   c. Add 3 mls to each plate. Let harden and incubate at 27° C overnight.
8. Plaques will appear as clear, circular areas approximately 0.5 mm to 3 mm in diameter against a red background. Comparison with the control plate should help identify the plaques. If there are too many plaques on the plate, or if the cells have died, the whole plate will appear uniformly light red and clear. Select a well-isolated plaque on a plate that has *10 plaques or less.* If you are plaque-purifying wild-type (wt) virus [or any occlusion-positive (occ^+) virus], examine the plaque under the microscope to ensure that many occlusion bodies are found in the nucleus of each infected cell (i.e., that the plaque does not have an FP phenotype; see Chapter 2). Mark the precise position of the plaque with a dark pen on the bottom of the plate.
9. Place the tip of a sterilized Pasteur pipet or glass micropipet directly onto the plaque. Carefully apply *gentle* suction until a small plug of agarose is drawn into the pipet. It helps to lift the pipet very slightly while applying suction.
10. Place the agarose plug in 1 ml complete tissue-culture medium. Vortex well to release the budded virus particles from the plug.
11. This 1 ml stock should be used to plaque-purify the virus at least once, and preferably twice more, to ensure that the virus is pure. Proceed essentially as described above using 10^{-1}, 10^{-2}, and 10^{-3} dilutions of the 1 ml stock. The final plaque-purified stock may be scaled up, as described in the protocols for amplification of virus stocks provided later.

Notes:

1. We recommend SeaKem ME agarose for overlays. We do not recommend low-gelling termperature agarose such as SeaPlaque because the lower gel strength necessitates a higher concentration of agarose, which tends to obscure the plaques.

2. Proper cell density is critically important to the success of a plaque assay. Generally, 2×10^6 cells/60 mm tissue-culture dish is the best cell density to start with if you are unfamiliar with the properties of a particular cell line and their growth in different insect cell culture media. If the plaques are too small after five days, the cell density used was too high. If the plaques are large and diffuse, the density used was too low. Repeat the plaque assay using a series of cell densities to find the optimal cell density for your particular cell line and plaque overlay medium.
3. Instead of preparing the neutral red stain in 0.5% agarose made up with tissue-culture fluid, you may prepare it in 0.5% agarose in water. Omitting the tissue-culture fluid from this preparation does not adversely affect the cells.

End-point Dilution

In generating a pure viral stock by end-point dilution, the aim is to dilute the virus such that, if multiple cultures are exposed to the diluted inoculum, any cultures that become infected will have received only a single infectious unit. To be 95% confident that any infected culture received only one infectious unit, you should dilute the virus to the point where only 10% of the cultures become infected (see note).

CONTROLS: We recommend including several uninfected wells on each plate.

Materials

1. Complete tissue-culture medium. (See Chapter 11 for details concerning different media.)
2. SF cells grown to approximately 1×10^6 cells/ml
3. 96 well tissue-culture plates. (See Chapter 11.)
4. Virus stock to be purified

Method

1. Dilute the cells to a concentration of 1×10^5 cells/ml with complete tissue-culture medium.
2. Prepare tenfold serial dilutions of the virus stock in tissue-culture medium. Dilutions of 10^{-6} and 10^{-7} will probably be appropriate for most stocks.
3. Mix 10 µl aliquots of each dilution with 100 µl aliquots of the cell suspension and seed into a 96 well plate. It is convenient to use four rows for each dilution. If four wells are seeded with 100 µl of cell suspension only as the uninfected controls, one plate can accommodate both dilutions, with 46 replicates per dilution.
4. Incubate the plate at 27° C. To avoid dehydration, you should incubate the plate in a humidified environment. It may be sealed in a plastic bag or other container lined with damp paper towels. We recommend incubating the plate for four to five days for occ^+ viruses and seven days for occ^- viruses.
5. Examine each well for virus replication. All wells with signs of infection should be scored as positive.
6. Harvest the tissue-culture medium from an infected well *provided only 10% or less of the wells were infected at that dilution.* This inoculum should be purified twice more before generating a large-scale stock as described following.

Note:

Virus particles in solution are distributed according to the Poisson distribution. The proportion (p) of cultures receiving any particular number of infectious units (r) is given by the equation

$$p = \mu^r e^{-\mu}/r!$$

where μ is the mean concentration of infectious units in the dilution. Hence, the proportion of cultures receiving no infectious units (r = 0) is $p = e^{-\mu}$, and the proportion receiving one or more infectious particles is $1 - e^{-\mu}$. The fraction of infected cultures that received only one infectious unit is $\mu e^{-\mu}$. The ratio of pure cultures to the total number of cultures in which the virus replicated is

$$\mu e^{-\mu}/(1 - e^{-\mu})$$

If we require that 95% of all infected cultures are initiated by a single infectious unit, then $\mu e^{-\mu}/(1 - e^{-\mu}) = 0.95$. Solving this for μ, we get $\mu = 0.094$. The proportion of wells remaining uninfected is $e^{-\mu} = 0.91$. Thus, to be 95% confident that any infected culture received only one infectious unit, the virus should be diluted until only 10% of the cultures are infected.

AMPLIFICATION OF VIRUS STOCKS

Large stocks of budded virus may be prepared simply by infecting cell cultures and harvesting the tissue-culture medium four to five days later. However, it is necessary to exercise caution to avoid two potential problems. First, serial passage of AcMNPV in cell culture is known to give rise to mutant viruses that can become the predominant species in a viral stock. This has been well-documented for mutants displaying the *few polyhedra* (FP) phenotype. FP mutants form plaques with very few PIBs per infected cell. They accumulate within serially passaged stocks because they produce higher titers of budded virus than wt virus, apparently at the expense of occluded virus.

The second concern is the accumulation of defective virus particles in the stock. This is a frequently observed phenomenon with many viruses and has been demonstrated to occur with AcMNPV (Kool et al., 1991). These defective particles have extensive mutations in their genome and are helper-dependent (i.e., they require coinfection of the same cell by wt virus to supply essential functions for their replication). This is most likely to occur when the cells are infected at a high multiplicity of infection (MOI), and many cells are infected by both wt and defective viruses. The replication of these defectives has been shown to interfere with the replication of the wt virus, resulting in a substantial reduction of infectious virus yield.

Both problems can be avoided by taking a few simple precautions. First, all virus stocks should be initiated with only a thoroughly purified inoculum (see protocols for generation of pure virus stocks earlier in this chapter). Avoid FP mutants by choosing plaques with many occlusion bodies per infected cell. [In general, if you are working with an occ$^-$ recombinant expression vector, you need not worry about FP mutants because the FP mutation affects the formation of PIBs but not the expression of the *polyhedrin* (*polh*) gene. However, if you are preparing occ$^-$ virus as parent virus for cotransfection with a transfer plasmid carrying *polh,* it is best to try to verify that the virus does not carry an FP mutation. Most FP mutations involve an insertion within the *fp25* gene, and so can be identified by analyzing the HindIII restriction fragment profile of the viral DNA and screening for insertions in the HindIII-I fragment. Protocols for the restriction analysis of viral DNA are provided in Chapter 13.]

Second, the number of times the virus is passaged in cell culture should be kept to a minimum. A large low-passage stock should be prepared, and small aliquots of this used as the inocula for working stocks. Finally, care should be taken to infect at low MOI when preparing large-scale stocks.

The procedures outlined here describe the scale-up and production of a working stock of virus beginning with a clonal inoculum. If a virus stock becomes contaminated at any stage, it can be sterilized by filtration through a 0.22 μm filter. The stock should be retitered after filtration.

Materials

1. Complete tissue-culture medium. (See Chapter 11 for details concerning different media.)
2. SF cells grown to approximately 1×10^6 cells/ml
3. 35 mm and 100 mm tissue-culture plates. (See Chapter 11.)
4. Purified virus stock to be amplified

Method

1. Seed a 35 mm tissue-culture dish with 1×10^6 cells in a final volume of 2 mls of tissue-culture medium. To help disperse the cells evenly, it is best to add some tissue-culture medium to the plates before the cells. Rock each plate gently *immediately* after addition of the cells to ensure an even monolayer. Let attach for 30 minutes to 1 hour at 27° C. Alternatively, the cells may be seeded at 5×10^5 cells per 35 mm dish and used the following day.
2. If the inoculum was purified by end-point dilution, bring the volume of the virus stock to 1 ml with complete tissue-culture medium. If plaque purification was used, then the agarose plug will already be in 1 ml tissue-culture medium.
3. Aspirate the tissue-culture medium from the plate. Infect with 0.5 ml purified inoculum and incubate at room temperature for one hour with gentle rocking.
4. Add 1.5 ml complete tissue-culture medium (there is no need to remove the virus inoculum) and incubate at 27° C for four days. Since this is a very low MOI infection, it will be difficult to see any cytopathic effects early after infection. However, with an occ^+ virus it should be possible to see some cells with occlusion bodies by three days postinfection (pi).
5. Collect the tissue-culture fluid and centrifuge at $1000 \times g$ for five minutes at 4° C to remove cells and debris. Be careful to keep the tissue-culture fluid sterile. Store the supernatant at 4° C. This is designated the passage one stock.
6. Seed four 100 mm tissue-culture dishes with 5×10^6 cells each. The final volume in each plate should be 10 mls.
7. Dilute 1 ml of the passage one stock (from Step 5) with 3 mls of complete tissue medium. Save the other 1 ml of passage one stock at 4° C as an emergency back-up.
8. Aspirate the tissue-culture medium from the plates and infect with 1 ml each of the diluted passage one stock. Incubate at room temperature for one hour with gentle rocking.
9. Feed each with 9 mls complete tissue-culture medium. Incubate at 27° C for four days.
10. Harvest the tissue-culture fluid, and centrifuge it to remove debris. Determine the titer of this passage two stock and take several 1 ml aliquots for storage at –80° C as long-term back-ups (see recommendations later in this chapter). The remainder of the stock should be kept at 4° C and used only as inoculum

for working stocks. However, for the first scale-up of a novel recombinant virus, 10 mls of this passage two stock may be used as a source of DNA to verify the structure of the recombinant.

11. A large-scale working stock of virus may be prepared by repeating Steps 6 through 10, using as many tissue-culture dishes as required and infecting with an MOI of 0.1 plaque-forming units (pfu)/cell (see virus titration protocol in the next section). Alternatively, cells may be infected in suspension as described here.
    a. Prepare a 200 ml spinner culture (Chapter 11) and grow until the cells reach a density of 1×10^6 cells/ml.
    b. Collect the cells by centrifugation at $1000 \times g$ for five minutes at room temperature and resuspend *very gently* in 20 mls of complete tissue-culture medium.
    c. Add passage two virus (from Step 10) to give an MOI of 0.1 (in other words, add 1×10^7 pfu of virus). Incubate at room temperature for one hour with gentle rocking. Swirl periodically to mix.
    d. Return the cells and virus to a spinner flask and feed with 180 mls complete tissue-culture medium. Incubate at 27° C with stirring for three to four days. Harvest by centrifugation to remove cells and debris, and titer the supernatant as described following. Store at 4° C and use as a working stock.

TITERING VIRUS STOCKS

The titer of a virus stock is the concentration of infectious virus particles in that stock. Note that the titer of a stock is not equivalent to the total concentration of particles in the stock. In general, the total concentration of particles is much greater than the concentration of infectious units (probably about 100-fold greater for most budded baculovirus stocks). The titer of a stock can be determined either by plaque assay or by end-point dilution. Methods for both procedures are described following.

Plaque Assay

The plaque purification of a virus stock has already been described. Plaque assay of a virus stock involves essentially the same protocol. Because each plaque derives from a single infectious unit, counting the number of plaques formed by different dilutions of a virus stock allows you to determine the concentration of infectious units in the stock. Viral titers determined in this manner are expressed in plaque-forming units (pfu)/ml. For most purposes, an adequate estimation of the titer of a stock will be obtained using one plate of cells per dilution. However, for precise determination of the titer, the dilutions should be plated multiple times, at least in triplicate. For a more thorough treatment of the analysis and significance of plaque assay data, consult the reviews by Cooper (1967) and Kleczkowski (1968).

As discussed earlier for plaque purification, the condition of the cells is the most critical parameter when doing plaque assays. Specifically, you should ensure that the cells are healthy and actively growing, and are evenly spread at the correct density on the plates.

CONTROLS: It is useful to include one or two uninfected plates to monitor the condition of the cell monolayers.

Materials

1. Complete tissue-culture medium. (See Chapter 11 for details concerning different media.)
2. SF cells grown to approximately 1×10^6 cells/ml
3. 60 mm tissue-culture dishes. (See Chapter 11.)
4. Virus stock to be titered
5. Agarose [Seakem ME (see note 1); FMC # 50014]
6. Stain: 1 mg/ml stock of neutral red (Sigma # N7005) in incomplete tissue-culture medium. Filter-sterilize and store at 4° C in the dark.

Method

1. Seed 60 mm tissue-culture dishes with 2×10^6 cells per plate (see note 2). The final volume of tissue-culture medium should be 4 mls per plate. To help disperse the cells, it is best to add some tissue-culture medium to the plates before the cells. Alternatively, dilute the cells to their final volume before adding them to the plate. Rock each plate gently *immediately* after addition of the cells to ensure an even monolayer. Incubate at 27° C for 30 minutes to 1 hour to allow the cells to attach. They will attach more rapidly if seeded in medium lacking serum. Alternatively, the cells may be plated out at a density of 1×10^6 cells per dish (in complete tissue-culture medium) and incubated at 27° C overnight.
2. Dilute the virus stock in tissue-culture medium. Dilutions of 10^{-4}, 10^{-5}, 10^{-6}, and 10^{-7} will be adequate for most stocks.
3. Aspirate the tissue-culture medium from the cell monolayers and infect with 0.5 ml of each virus dilution. *Be careful not to let the plates dry out before addition of the virus.* For the beginner, it is a good idea to prepare an uninfected plate as well. Simply infect with 0.5 ml tissue-culture medium. Incubate at room temperature for one hour with gentle rocking.
4. While the cells are rocking, prepare the agarose overlay.
   a. Prepare a solution of 5% agarose in ddH_2O and autoclave. Make up 0.5 mls for each 60 mm tissue-culture dish to be overlaid. (The agarose may be prepared in advance and melted by microwaving when needed.) This solution of agarose will subsequently be diluted tenfold, so prepare it in a container that is large enough to accommodate 10 times the starting volume.
   b. Heat complete tissue-culture medium (5 mls for each 60 mm plate) to 60° C in a water bath.
   c. Cool the melted agarose to 60° C.
   d. Dilute the agarose to a concentration of 0.5% with the heated tissue-culture medium. For example, for ten 60 mm tissue-culture dishes, prepare 5 mls of 5% agarose and dilute with 45 mls tissue-culture medium. Mix well by swirling.
   e. Cool the 0.5% agarose to 40° to 42° C.
5. Remove the virus inoculum from the cells and add 4 mls of agarose overlay to each plate. Try to remove the viral inoculum thoroughly; excess liquid underneath the agarose will lead to smeared plaques. However, be careful not to let the monolayer dry out. It is important that the agarose has cooled sufficiently before overlaying. Do not let it drop on top of the cells. Instead, let it run gently down the side of the plate to avoid disturbing the cell monolayer.
6. Leave the cells at room temperature for 15 to 20 minutes until the agarose has hardened. (If you experience problems with plaques that appear smeared, it may help to dry the plates by leaving them open in a laminar flow hood for several hours after overlaying. Also, as mentioned previously,

it is important to remove as much of the viral inoculum as possible before overlaying.) Incubate at 27° C.

7. Occ^+ plaques should become visible after three days. However, occ^- plaques will be harder to see and may not be easily visible until six or seven days pi. However, in both cases, the plaques should be well enough developed for staining by four to five days pi. Note that the neutral red stain used is a vital stain (i.e., it only stains living cells). Thus, it is important to stain the plates while the monolayer is still healthy. We do not recommend waiting more than seven days before staining. Overlay with stain as follows (see note 3):
   a. Prepare a 0.5% agarose solution in tissue-culture medium as described in Step 4.
   b. Cool the agarose to 40° to 42° C and add the stain to a final concentration of 50 μg/ml.
   c. Add 3 mls to each plate. Let harden and incubate at 27° C overnight.
8. Plaques will appear as clear, circular areas approximately 0.5 mm to 3 mm in diameter against a red background. Comparison with an uninfected plate should help identify the plaques. If there are too many plaques on the plate, or if the cells have died, the whole plate will appear uniformly light red and clear. The titer of the virus stock can be determined by counting the number of plaques present on the plates. Generally, the most reliable counts will be obtained from plates with between 10 and 100 plaques. Counts from replicate plates should be averaged. The titer of the stock is calculated by multiplying by the appropriate dilution factor and by 2, as 0.5 mls were inoculated onto each plate. For example, if an average of 30 plaques are counted on the plates infected with a 10^{-6} dilution, the titer of the stock is $30 \times 10^6 \times 2 = 6 \times 10^7$ pfu/ml.

Notes:

1. We recommend SeaKem ME agarose for overlays. We do not recommend low-gelling temperature agarose such as SeaPlaque because the lower gel strength necessitates a higher concentration of agarose, which tends to obscure the plaques.
2. Proper cell density is critically important to the success of a plaque assay. Generally, 2×10^6 cells/60 mm tissue-culture dish is the best cell density to start with if you are unfamiliar with the properties of a particular cell line and their growth in different insect cell-culture media. If the plaques are too small after five days, the cell density used was too high. If the plaques are large and diffuse, the density used was too low. Repeat the plaque assay using a series of cell densities to find the optimal cell density for your particular cell line and plaque overlay medium.
3. Instead of preparing the neutral red stain in 0.5% agarose made up with tissue-culture fluid, you may prepare it in 0.5% agarose in water. Omitting the tissue-culture fluid from this preparation does not adversely affect the cells.

End-Point Dilution

The determination of a viral titer by end-point dilution involves the inoculation of multiple cultures with different dilutions of the virus, and estimation of the dilution of virus that would infect 50% of the cultures (the end-point dilution). This quantity of virus is known as the 50% tissue-culture infectious dose, or $TCID_{50}$. Virus titers determined in this manner may be expressed as $TCID_{50}$/ml or converted to pfu/ml. To get an accurate estimate of a virus titer, it is necessary to have several (at least two) dilutions that infect some but not all of the cultures. The method described here to derive the $TCID_{50}$ is that of Reed and Muench (1938). This has been the most widely used in animal virology and is quite adequate for most purposes. However, there are a number of other methods that may be preferable in certain circumstances (such as when an estimate of the error of the titer obtained is required). The methods available and their statistical basis have been reviewed in depth by Dougherty (1964) and Hughes and Wood (1986).

CONTROLS: We recommend including several uninfected wells on each plate.

Materials

1. Complete tissue-culture medium. (See Chapter 11 for details concerning different media.)
2. SF cells grown to approximately 1×10^6 cells/ml
3. 96 well tissue-culture plates. (See Chapter 11.)
4. Virus stock to be titered

Method

1. Prepare tenfold serial dilutions of the virus stock. Dilutions of 10^{-5}, 10^{-6}, 10^{-7}, and 10^{-8} should be appropriate in most cases.
2. Dilute the cells to a concentration of 1×10^5 cells/ml with complete tissue-culture medium.
3. Mix 10 μl aliquots of each virus dilution with 100 μl aliquots of the cell suspension, and seed into a 96 well plate. Seed 4 wells with 100 μl of cells, only as uninfected controls. Alternatively, the cells can be seeded into the microtiter plate before addition of the virus. If necessary, they can be seeded at half-density (5×10^3 cells/well) and used the following day. It is convenient to use one row for each dilution so that one plate can accommodate four dilutions each of two separate stocks.
4. Incubate the plate at 27° C. To avoid dehydration, incubate the plate in a humidified environment. It may be sealed in a plastic bag or other container lined with damp paper towels. Infection by an occ^+ virus will become evident after two to three days, but infection by an occ^- virus will be harder to detect. We recommend incubating the plate for four to five days for occ^+ viruses and seven days for occ^- viruses.
5. Examine each well for virus replication. All wells with signs of infection should be scored as positive.
6. The basic principle of the method of Reed and Muench (1938) is to assume that all cultures infected at a particular dilution would have been infected at all lower dilutions, and conversely, that all cultures uninfected at that dilution would have been uninfected at all higher dilutions. Consider the following example:

Dilution	*Infected Wells*	*Uninfected Wells*
10^{-5}	12	0
10^{-6}	8	4
10^{-7}	1	11
10^{-8}	0	12

At the 10^{-5} dilution, all 12 wells were infected. However, 8 wells were infected at the 10^{-6} dilution, and 1 at the 10^{-7} dilution. Assuming that these wells would also have been infected by the 10^{-5} dilution, the total number of infected wells at 10^{-5} is taken to be 21. In the same manner, while there are 12 wells uninfected at the 10^{-8} dilution, we assume that those uninfected at 10^{-7} and 10^{-6} would also have been uninfected at 10^{-8}, resulting in a total of 27 uninfected wells at 10^{-8}. This procedure is repeated for both infected and uninfected wells at all dilutions, and the numbers obtained are as follows:

Dilution	*Infected*	*Uninfected*	*% Infected*
10^{-5}	21	0	100.0
10^{-6}	9	4	69.2
10^{-7}	1	15	6.3
10^{-8}	0	27	0.0

In this case, the dilution that would have given 50% infection lies between 10^{-6} and 10^{-7}. This dilution is calculated by linear interpolation between the infection rates observed at these dilutions. First, the proportionate distance (PD) of a 50% response from the response above 50% is calculated using the following formula:

PD = (A–50)/(A–B)

where A is the % response above 50%, and B is the % response below 50%. In our example

PD = (69.2 – 50) /(69.2 – 6.3)
= 0.305

The dose that would have given a 50% response, the $TCID_{50}$, is then calculated using the following formula:

Log $TCID_{50}$ = log of the dilution giving a response greater than 50% – the PD of that response.

Thus, log $TCID_{50}$ = –6 – 0.305
= –6.305

Therefore, $TCID_{50} = 10^{-6.305}$

The titer of the virus is the reciprocal of this = 2.02×10^6 $TCID_{50}$/10 μl = 2.02×10^8 $TCID_{50}$/ml. This can be converted to pfu/ml using the relationship pfu = $TCID_{50} \times 0.69$ (see note). Thus, in our example the titer of the stock is 1.4×10^8 pfu/ml.

A sample Microsoft Excel spreadsheet for these calculations is provided in Appendix 6.

Note:

According to the Poisson distribution, the proportion (p) of cultures remaining uninfected (i.e., receiving no infectious units) at any given dose is $e^{-\mu}$, where μ is the mean concentration of infectious particles at that dose. The $TCID_{50}$ is the dose at which 50% of the cultures become infected, that is, p = 0.5. Thus, $0.5 = e^{-\mu}$, which implies that μ, the mean concentration of infectious units at that dose, = 0.69. Hence, $TCID_{50} \times 0.69$ = pfu.

VIRUS CONCENTRATION AND PURIFICATION

Budded Virus

For most purposes, there is no need to purify budded virus extensively. Virus stocks can be stored as tissue-culture medium supernatants for extended periods (see Virus Storage later in this chapter) and used as inocula without further purification. There is inevitably a loss of titer during the purification protocol, and we do not recommend purification of budded virus unless there is a specific need (e.g., purification away from expressed gene products secreted from infected cells).

While budded virus purification should not be necessary very frequently, you may occasionally need to concentrate budded virus stocks (e.g., to allow infection at high MOIs). In addition, concentration of the virus stock is one of the initial steps in the purification of viral DNA.

Many workers have described the concentration and purification of budded virus from tissue-culture medium. All of these protocols are very similar and differ only in detail. The protocol described here is a modification of the procedure used by Lee and Miller (1978). For virus purification, a large quantity of virus should be used to allow easy visualization of the virus band after gradient centrifugation. The protocol provided is for 200 mls of virus stock. Much smaller quantities of virus can be used if the stock is only going to be concentrated and the procedure can be scaled down easily.

Materials

1. Virus stock to be concentrated
2. Phosphate-buffered saline (PBS) pH 6.2: 1 mM $Na_2HPO_4 \cdot 7H_2O$; 10.5 mM KH_2PO_4; 140 mM NaCl; 40 mM KCl; pH 6.2
3. Sucrose Cushion: 25% sucrose (w/w) in 5 mM NaCl, 10 mM EDTA
4. Sucrose gradient solutions:
   a. 25% sucrose (w/w) in PBS
   b. 60% sucrose (w/w) in PBS
5. Ultracentrifuge tubes (e.g., for Beckman SW28 and SW28.1 rotors, # 326823 and # 337986)
6. 0.2 μm filter (Gelman Acrodisk, # 4192)

Method

Virus Concentration

1. Load 33 mls virus stock in each of six 38 ml polyallomer ultracentrifuge tubes (Beckman SW28 tubes or equivalent).
2. Underlay with 3 mls of sucrose cushion solution. (The sucrose cushion may be loaded in the tubes first, and then overlaid carefully with the virus. However, it is easier to avoid perturbation of the cushion by underlaying.) If the amount of virus used is less than 200 mls and the centrifuge tubes are only partially full, add PBS until there is at least 35 mls per tube.
3. Centrifuge at 80,000 × g (24,000 RPM in an SW28 rotor) for 75 minutes at 4° C.
4. Decant the supernatant and carefully remove traces of sucrose. The viral pellet should be translucent white with a faint, blue tinge around the edges when viewed against a dark background.
5. Resuspend the pellets in a total of 2 mls PBS. This represents a concentrated and partially purified virus stock that may be used without any further purification (see the introduction to this section). It should be filtered through a 0.2 μm filter and retitered before use. Store at 4° C. If further purification is necessary, proceed as follows.

Virus Purification

1. Pour a 14 ml 25% to 60% linear sucrose gradient (see note 1) in a 17 ml polyallomer ultracentrifuge tube (Beckman SW28.1 or equivalent). Carefully load the concentrated virus onto the gradient.
2. Centrifuge at 96,000 × g (26,500 RPM in SW28.1) for three hours at 4° C.
3. Budded virus forms a white band at 47% to 49% sucrose. Collect this band in a minimum volume.
4. Dilute the virus to 15 mls with PBS and load into a 17 ml ultracentrifuge tube.
5. Centrifuge at 80,000 × g (24,000 RPM in SW28.1) for 75 minutes at 4° C.
6. Decant the supernatant and carefully remove any remaining sucrose.
7. Resuspend the pellet in a small volume of PBS. Filter through a 0.2 μm filter and retiter before use. Store at 4° C.

Occluded Virus

Viral PIBs are most easily purified from insect cadavers. The purification involves both differential centrifugation and sucrose-gradient centrifugation procedures. As for budded virus, there is likely to be substantial loss of PIBs during the gradient-purification procedure. While the preparation of DNA from occluded virus is best performed with purified occlusion bodies, a partially purified occlusion body preparation (without gradient centrifugation) is adequate for most other purposes [e.g., infecting insects by feeding (*per os*)].

The protocol described here is a modification of the procedure of Miller and Dawes (1978) for the purification of PIBs from infected insects. The protocols for purification of PIBs from cell culture and from infected larvae are quite similar. In our experience, the yield of PIBs from infected cells can be variable, and insect cadavers are the preferred source. However, there may be occasions when you need to obtain PIBs from cell culture (e.g., for the co-occlusion of an occ^- virus). Thus, we have indicated appropriate modifications to the following protocol to allow the purification of PIBs from cell culture.

Materials

1. Insects infected with the appropriate virus. (See Chapter 18 for recommendations for infection and harvesting of insects.)
2. Tissue homogenizer. (Use any standard brand. We use a Janke and Kunkel Ultra-Turrox homogenizer.)
3. 110 μm PE macro filter (Spectra/Mesh # 146398)
4. 0.5% SDS
5. 0.5 M NaCl
6. Sucrose gradient solutions:
   a. 40% sucrose (w/w) in PBS
   b. 65% sucrose (w/w) in PBS
7. Ultracentrifuge tubes (e.g., Beckman SW28, # 326823)

Method

From Insects

1. Place insect cadavers in 1 mls to 2 mls ddH_2O/insect.
2. Homogenize well (one to two minutes) with any conventional tissue homogenizer.
3. Filter the homogenate through a 110 μm PE macro filter to remove pieces of cuticle, and so on. It may be necessary to filter the homogenate twice.
4. Pellet the PIBs by centrifugation at 5000 × g in a swinging-bucket rotor (e.g., 7000 RPM in a Beckman superspeed JS13.1 rotor or equivalent) for 10 minutes at room temperature.

5. Discard the supernatant and resuspend the pellet in 1 ml 0.5% SDS per insect equivalent. The pellet can be resuspended by brief homogenization with the tissue homogenizer.
6. Centrifuge at 5000 × g and resuspend the pellet in the same volume of 0.5 M NaCl.
7. Pellet the PIBs and resuspend in a small volume of ddH_2O (0.5 ml per insect equivalent). At this stage, the PIBS are sufficiently pure for most purposes. They should be stored frozen. If further purification is required, proceed as follows.
8. Pour 34 ml linear 40% to 65% sucrose gradients (see note 1) in 38 ml polyallomer centrifuge tubes (SW28 or equivalent). Carefully load 2 mls of the occlusion body preparation onto each gradient.
9. Centrifuge at 96,000 × g (26,500 RPM in an SW28) at room temperature for three hours.
10. The occluded virus will form a thick, white band at 54% to 56% sucrose. Collect this band and dilute up to 25 mls with ddH_2O.
11. Pellet the PIBs by centrifugation at 5000 × g for 10 minutes. Discard the supernatant and drain thoroughly. Resuspend the pellet (see note 2) in a small volume of ddH_2O and store at 4° C, frozen or freeze-dried (see discussion on long-term storage of virus following this section).

From Cell Culture

1. Harvest infected cells by centrifugation at 1000 × g for five minutes at room temperature.
2. Discard supernatant and resuspend in 0.5% SDS. Ten mls of 0.5% SDS per 2×10^8 cells (one spinner culture) is a convenient volume.
3. The remainder of the purification is the same as described for insect-derived occlusion bodies. Continue from Step 6.

Notes:

1. The sucrose gradient is best generated using a gradient former. However, if one is not available, a step gradient can be poured. For example, for a 25% to 60% gradient, prepare 25%, 30%, 35%, 40%, 45%, 50%, 55%, and 60% sucrose solutions. Pour 1.5 mls of the 25% solution into the tube and then underlay successively with 1.5 mls of each of the 30% to 55% solutions and 2.5 mls of 60% sucrose. The gradient can be used immediately, but it is preferable to let it stand for three hours at room temperature or 16 hours at 4° C to allow the solutions to diffuse and form a gradient.
2. After sucrose gradient centrifugation, purified PIBs tend to clump together and become very difficult to count. Brief sonication in a bath sonicator will help disperse the occlusion bodies.

VIRUS STORAGE

Budded Virus

Working stocks of budded virus can most conveniently be stored in tissue-culture medium at 4° C. It is best to supplement serum-free tissue-culture medium with 5% FBS for optimal virus stability. Infected cells and cell debris should be removed from the virus stock by centrifugation at 1000 × g for five minutes at 4° C. There is generally no appreciable loss in titer of virus stored in this manner for at least a year, and we have kept stocks for much longer periods with good recovery of activity. However, virus stocks should be retitered before use after prolonged storage. Purified budded virus can similarly be stored in tissue-culture medium or in PBS. Budded virus is in fact reasonably stable at ambient temperatures, and there is no need to keep the virus at 4° C when transporting it for short periods of time.

Naturally, sterile technique should be used at all times to avoid contamination of virus stocks. However, if budded virus stocks do become contaminated, they can be sterilized by filtration through a 0.2 μm filter. Filtered stocks should be retitered before use.

For long-term storage, aliquots of the virus stock in tissue-culture medium should be stored frozen at –80° C or in liquid nitrogen. The virus may be maintained for long periods of time in this manner. Hemolymph from infected insects, which contains high titers of budded virus, may also be stored for long periods of time at –80° C or in liquid nitrogen. However, in both cases, repeated freezing and thawing causes marked reduction in the virus titers, and frozen stocks should only be used as long-term back-ups.

Occluded Virus

Occluded virus stocks are generally very stable under laboratory conditions. They can be maintained as suspensions in ddH_2O or PBS and stored at 4° C. If the PIBs are not intended for feeding to insects, 0.01% sodium azide (Sigma # S 2002) may be added. For long-term storage, PIBs should be freeze-dried and stored at room temperature or 4° C. They may also be stored at –20° C, –80° C, or in liquid nitrogen. Another convenient method of storing occluded virus is to freeze the infected insects. Insect cadavers, which contain high numbers of occlusion bodies, may be stored indefinitely at –20° C, –80° C, or in liquid nitrogen, with little or no loss of infectivity.

Caution: Sodium azide is poisonous. Label all solutions clearly and handle with care.

13

COTRANSFECTION AND RECOMBINANT VIRUS IDENTIFICATION

This chapter deals with the actual construction and identification of your recombinant virus. First, you learn how to prepare DNA suitable for transfection purposes from the large virus stock you have already generated (Chapter 12). Then, various techniques for the cotransfection of this viral DNA and your transfer plasmid into insect cells are described. It is at this stage that the allelic replacement reaction, which gives rise to the formation of recombinant virus, takes place. Several procedures for the identification of recombinant virus among the background of parent virus are provided. The approach you use will depend, to a large extent, on the type of transfer plasmid and parent virus you have chosen. Finally, the chapter discusses how you will confirm that the virus you have isolated is in fact the required recombinant.

PREPARING VIRAL DNA

Preparation of viral DNA is necessary at two stages in the generation of a recombinant baculovirus expression vector. High-quality DNA is vital for the efficient transfection of insect cells. In addition, confirmation of the structure of a recombinant viral genome will require the preparation of viral DNA. A selection of useful protocols is provided in this section. Note that not all of these protocols are suitable for the preparation of DNA for transfection purposes.

The starting material for viral DNA preparation may be infected cells, budded virus, or purified PIBs. In general, we have had excellent success, both in transfections and in restriction enzyme analysis, with viral DNA that has not been purified by cesium-chloride gradient centrifugation. However, we have included a description of this technique if required. For all the techniques described here, it is important to remember that baculovirus genomes are quite large (e.g., 128 kbp for AcMNPV) and thus are sensitive to damage by shearing. To help avoid damage, *never vortex any viral DNA solution* and always use wide-mouth pipets. Disposable pipet tips can be cut 2 mm to 3 mm from the point to generate a wider mouth.

The first protocols described here encompass some basic techniques for the purification of DNA, which are required for all procedures involving the manipulation of viral DNA. These techniques are followed by detailed protocols for the preparation of viral DNA from a variety of different sources.

Extraction and Purification of DNA

The most common approach to the purification of DNA is extraction of the sample with phenol and/or chloroform, followed by ethanol precipitation or dialysis of the

DNA. The protocol outlined here involves the sequential extraction of the DNA sample with phenol, a phenol-chloroform mix, and chloroform. During extraction with phenol, most proteins are partitioned into the organic phase and/or denatured. The addition of chloroform to the phenol improves the efficiency of the extraction. Extraction with chloroform alone removes any residual phenol. In all extractions involving chloroform, isoamylalcohol is added to minimize frothing of the sample.

Following extraction, residual traces of organic solvent are removed either by ethanol precipitation or by dialysis. Ethanol precipitation also serves to concentrate the DNA sample. In this procedure, ethanol is added to the DNA solution to a final concentration of approximately 65%. In the presence of moderate concentrations of monovalent cations, the DNA forms an insoluble precipitate and can be collected by centrifugation. The pellet is resuspended in a small amount of the required buffer. An added advantage of this method is that the DNA is sterilized during the procedure. However, ethanol-precipitated viral DNA is often extremely difficult to resuspend. This disadvantage can be avoided if the DNA is dialyzed rather than precipitated. It is up to the individual researcher to decide which method to use.

Materials

1. 0.5 M EDTA: Prepare a 0.5 M solution of the disodium salt of EDTA (Sigma # E 5134). Adjust the pH to 8.0 with NaOH. The EDTA will not dissolve until the pH approaches 8.0.
2. TE: 10 mM Tris-HCl, pH 8.0; 1 mM EDTA
3. Phenol: Prepare liquefied phenol (Baker # JT2858-1) containing 0.1% (w/v) 8-hydroxyquinoline (Sigma # H 6878). Extract twice with an equal volume of 1 M Tris-HCl, pH 8.0 and once with an equal volume of TE. Verify that the pH of the aqueous phase is greater than 7.8. If not, repeat the extractions. Store under an equal volume of TE at 4° C in a dark bottle. Alternatively, buffer-saturated phenol is commercially available (e.g., BRL # 5513UA). **Caution:** Phenol is highly corrosive and causes severe burns. Wear protective clothing and safety glasses. In case of contact with skin, apply 80% glycerol to the affected area and wash extensively with water.
4. Phenol-chloroform-IAA; TE-saturated phenol:chloroform:isoamylalcohol (25:24:1). Store under TE at 4° C in a dark bottle.
5. Chloroform-IAA; Chloroform:isoamylalcohol (24:1)
6. 3 M sodium acetate, pH 5.2
7. 95% ethanol in triple-distilled, deionized water (ddH_2O)
8. 70% ethanol in ddH_2O
9. Dialysis tubing (e.g., Spectrapor cellulose tubing; Spectrum # 25223-226)

Method

1. Add an equal volume of phenol to the sample and mix well by inverting the tube repeatedly until an emulsion has formed. *Do not vortex to mix the phases.*
2. Centrifuge at 12,000 × g for five minutes at 4° C in a microfuge.
3. Collect the aqueous (upper) phase. Do not take any material at the interface. Add one-half volume TE to the organic phase and interface, and mix well. (This back-extraction helps reduce the loss of DNA that may become trapped at the interface in very crude samples. With purer samples, back-extraction is not necessary.)
4. Centrifuge at 12,000 × g for five minutes at 4° C. Combine the two aqueous phases, add an equal volume of phenol-chloroform-IAA, and mix well.
5. Centrifuge at 12,000 × g for three minutes at 4° C. Collect the aqueous phase and extract with an equal volume of chloroform-IAA. Centrifuge for one minute and collect the aqueous phase.

6. Dialyze the DNA at 4° C against 500 volumes of TE with two changes for at least four hours each time. Alternatively, ethanol-precipitate the DNA as follows:
   a. Add two volumes cold 95% ethanol and one-tenth volume 3 M sodium acetate pH 5.2. Mix well and centrifuge at 12,000 × g for 10 minutes at 4° C.
   b. Rinse the pellet carefully with cold 70% ethanol and air-dry. (Vacuum-drying the pellet will make it even more difficult to resuspend.) Resuspend gently in a small volume of TE.
7. Baculovirus DNA should be stored at 4° C. We do not recommend storing the DNA frozen.
8. To quantify the amount of DNA present, measure the OD_{260} of the solution and calculate the DNA concentration using the formula 1 OD unit = 50 µg/ml in a 1 cm path cell. The OD_{280} should be approximately half the OD_{260} (OD_{260}/OD_{280} for pure DNA is 1.8). If the OD_{280} is very close to the OD_{260}, the DNA is probably contaminated with protein or phenol and should be further purified.

Purification of DNA from Budded Virus

Budded virus is the most convenient starting material for the preparation of viral DNA. We have not found it necessary to use highly purified budded virus to obtain high-quality DNA preparations suitable for restriction digestion and transfection. The protocol presented here is based on that described by Miller et al. (1986).

In addition to the full-scale viral DNA preparation protocol, we have also provided a "miniprep" procedure that allows the preparation of DNA from small quantities of virus (Kumar and Miller, 1987). This procedure is particularly useful for a preliminary screen of multiple clones when full-scale preparations would be too tedious, or when a small amount of DNA for a single restriction digest is required. It is also useful for the preparation of small amounts of DNA for PCR amplification. The DNA obtained is not suitable for transfection purposes.

Materials

1. Passage two or three stock of virus (from Chapter 12)
2. Sucrose cushion: 25% sucrose (w/w) in 5 mM NaCl, 10 mM EDTA
3. Virus disruption buffer: 10 mM Tris-HCl, pH 7.6; 10 mM EDTA, 0.25% SDS
4. 10 mg/ml proteinase K (BRL # 5530UA) in ddH_2O. Store frozen.
5. TE
6. Phenol, phenol-chloroform-IAA, and chloroform-IAA
   Caution: Phenol is highly corrosive and causes severe burns. Wear protective clothing and safety glasses. In case of contact with skin, apply 80% glycerol to the affected area and wash extensively with water.
7. Dialysis tubing (e.g., Spectrapor cellulose tubing; Spectrum # 25223-226)
8. 3 M sodium acetate, pH 5.2
9. 95% ethanol in ddH_2O
10. 70% ethanol in ddH_2O

Method

Full-scale Method

1. Centrifuge some high-titer budded-virus stock through a sucrose cushion as described in Steps 1 through 4 of the budded-virus concentration protocol in Chapter 12. The quantities given below are for 10 mls of virus. This should yield up to 100 µg of viral DNA if the titer of the stock was at least 1×10^8 plaque forming units (pfu)/ml.
2. Add 1 ml disruption buffer and resuspend the pellet carefully by pipeting up and down gently using a pipet with a cut-off tip.

3. Add proteinase K to a final concentration of 500 μg/ml and digest at 37° C with gentle mixing. Allow the digestion to proceed from four hours to overnight. If the solution has not cleared, add more disruption buffer and proteinase.
4. When the solution is clear, purify the DNA by sequential phenol, phenol-chloroform-IAA, and chloroform extractions (see the preceding method: a TE back-extraction of the first interface is recommended).
5. Ethanol-precipitate or dialyze the DNA. Measure the OD_{260} and OD_{280} of the DNA solution and calculate the concentration, given that 1 OD_{260} = 50 μg/ml.

Miniprep Method

1. Centrifuge 1.5 mls of high-titer budded virus stock at 12,000 × g for 15 minutes at 4° C in a microfuge.
2. Discard the supernatant and resuspend the pellet *carefully* in 100 μl virus-disruption buffer. Add proteinase K to a final concentration of 500 μg/ml and incubate at 37° C with gentle mixing, until the solution is clear.
3. Extract and ethanol-precipitate or dialyze as described in Steps 4 and 5 of the full-scale method.
4. Use the entire sample for a single restriction endonuclease digest.

Purification of DNA from PIBs

To purify viral DNA from PIBs, it is first necessary to release the embedded virions from the occlusion bodies. You accomplish this by solubilizing the occlusion bodies in an alkaline solution. The remainder of the procedure is the same as the preparation of DNA from budded virus. The following protocol is based on that described by Miller and Dawes (1978):

Materials

1. Gradient-purified PIBs, prepared as described in Chapter 12
2. 1 M sodium carbonate. Prepare fresh.
3. 1 M Tris-HCl, pH 7.6

Method

1. As a rough guide, 10^9 PIBs should yield at least 10 μg of DNA.
2. Bring the occlusion-body solution (at a concentration of approximately 1×10^9 PIBs/ml) to 0.1 M sodium carbonate, and incubate at room temperature for at least 30 minutes. The solution should clear and darken somewhat.
3. Add 1 M Tris-HCl, pH 7.6 to a final concentration of 0.1 M to neutralize the solution.
4. Centrifuge at 7000 × g (e.g., 10,000 RPM in a Beckman superspeed JS13.1 rotor) for 10 minutes at room temperature to remove any undisrupted occlusion bodies. For maximum yield, the pellet may be resuspended in H_2O, and the solubilization procedure (Steps 2-4) repeated. However, this is not normally necessary.
5. The supernatant contains the released virions. DNA may be prepared from these, as described for budded virus (full-scale method).

Density-Gradient Purification of Viral DNA

If further purification of the DNA is required, it can be achieved by cesium chloride/ethidium bromide (CsCl/EtBr) density-gradient centrifugation of the sample. In this procedure, the DNA sample is mixed with a solution of CsCl and EtBr. This mixture is then centrifuged until a density gradient of CsCl is formed. The DNA will form a sharp band at a position corresponding to the density of the DNA-EtBr complex. It will be well separated from RNA (which will pellet or form a band along the side of the tube in a vertical rotor) and proteins (which float to the top of the tube).

Materials

1. Viral DNA prepared as described in the preceding protocols
2. TE
3. Cesium chloride (Sigma # C 4036)
4. Ultracentrifuge tubes (e.g., Beckman # 342413)
5. 10 mg/ml ethidium bromide (Sigma # E 7637) in ddH_2O. Store at room temperature in a dark bottle.
   Caution: Ethidium bromide is moderately toxic and a powerful mutagen. Wear gloves when handling reagents containing this dye.
6. Hand-held UV lamp
   Caution: UV light is harmful, particularly to the eyes. Wear protective goggles or a safety mask that efficiently blocks UV light.
7. 18-gauge syringe needles (Becton Dickinson # 5196)
8. Saturated isobutanol: Mix isobutanol vigorously with an equal volume of ddH_2O and let the phases separate. Use the upper phase.
9. Dialysis tubing (e.g., Spectrapor cellulose tubing, Spectrum # 25223-226)
10. Phenol-chloroform-IAA
    Caution: Phenol is highly corrosive and causes severe burns. Wear protective clothing and safety glasses. In case of contact with skin, apply 80% glycerol to the affected area and wash extensively with water.
11. 3 M sodium acetate, pH 5.2
12. 95% ethanol in ddH_2O
13. 70% ethanol in ddH_2O

Method

The following method is for 12.5 ml sealable ultracentrifuge tubes (Beckman TY65 rotor or equivalent).

1. Dilute the DNA sample to 10 mls with TE. Add 10 g CsCl and dissolve by gentle mixing. Heat at 37° C, if necessary, to dissolve the CsCl.
2. Transfer the CsCl solution containing the DNA to an ultracentrifuge tube, and add 500 µl 10 mg/ml ethidium bromide. Once the ethidium bromide has been added, the sample should be kept dark as much as possible.
3. Fill the tube with TE; seal and centrifuge at 200,000 × g (55,000 RPM in a TY65) for 21 hours at 20° C.
4. View with a hand-held UV lamp and collect the UV-fluorescent band. First, cut off the top of the tube, or pierce it with an 18-gauge needle. Then, insert a second 18-gauge needle into the centrifuge tube approximately 5 mm below the band. Insert the needle bevel-side up and angle it up so that it is just below the band. Collect the DNA by allowing the CsCl solution to drip out slowly. If the solution does not flow, withdraw the band by applying *very gentle* suction with a 3 ml syringe. If two bands are visible, collect the lower one. This is covalently closed circular DNA.
5. Remove the ethidium bromide by repeated extractions (mix well but *do not vortex*) with saturated isobutanol. Discard the upper (organic phase) after each extraction. Repeat the extractions (generally three to four times) until no pink color is detectable in the organic phase.
6. Dialyze the sample twice against 500 mls TE, for at least 30 minutes each time, to remove the CsCl.
7. Extract the sample with phenol-chloroform-IAA, and then ethanol-precipitate or dialyze. Store at 4° C. Quantify the DNA as described already.

Purification of DNA from Infected Cells

Total DNA, extracted late in infection, contains substantial amounts of viral DNA. Because of the large molar excess of viral DNA relative to genomic DNA in these preparations, virus-specific bands are generally visible against the genomic DNA background on agarose gels. Thus, DNA prepared in this manner may be used for restriction-endonuclease and Southern blotting analysis of the viral DNA and is convenient for confirmation of the structure of a recombinant virus. However, it is not suitable for transfection purposes.

The following protocol is a modification of that of Summers and Smith (1987). It involves the isolation of infected nuclei, followed by the extraction and purification of total DNA.

Materials

1. Recombinant virus-infected cells (e.g., the cells remaining after production of a passage two stock of recombinant virus)
2. Lysis Buffer: 30 mM Tris-HCl, pH 7.5; 10 mM magnesium acetate; 1.0% NP-40 (v/v)
3. 15-ml conical centrifuge tubes (e.g., Falcon # 2097)
4. PBS, pH 6.2
5. Extraction Buffer: 10 mM Tris-HCl, pH 8.0; 5 mM EDTA; 0.5% SDS
6. 10 mg/ml proteinase K (BRL # 5530UA) in ddH_2O. Store frozen.
7. Phenol, phenol-chloroform-IAA, and chloroform-IAA
   Caution: Phenol is highly corrosive and causes severe burns. Wear protective clothing and safety glasses. In case of contact with skin, apply 80% glycerol to the affected area and wash extensively with water.
8. Dialysis tubing (e.g., Spectrapor cellulose tubing, Spectrum # 25223-226)
9. 3 M sodium acetate, pH 5.2
10. 95% ethanol in ddH_2O
11. 70% ethanol in ddH_2O

Method

1. Collect and save the tissue-culture fluid from the preparation of your passage two recombinant virus stock. Add 5 mls lysis buffer (per 100 mm dish) to the cells remaining on the plate. Incubate on ice for 5 to 10 minutes, rocking periodically.
2. Transfer the lysate to a 15 ml conical tube, and vortex well. Spin at 1000 × g for five minutes at 4° C.
3. Discard the supernatant and resuspend the pellet gently in 5 mls cold PBS.
4. Centrifuge at 1000 × g for five minutes at 4° C.
5. Resuspend the pellet gently in 200 µl extraction buffer. *Do not vortex the preparation from this point onward.* The pellet may be resuspended by gentle pipeting with a cut-off tip.
6. Add proteinase K to a final concentration of 1 mg/ml, and incubate at 37° C with gentle mixing for at least four hours. The incubation may be continued overnight if convenient.
7. Extract the sample sequentially with phenol, phenol-chloroform-IAA, and chloroform-IAA as described preceding. A back-extraction of the initial interface with TE will probably be necessary.
8. Ethanol-precipitate or dialyze the DNA as described previously. Quantify the DNA in the sample and store at 4° C. To allow easy visualization of viral bands on an agarose gel, use 8 µg to 10 µg of this DNA preparation per digest.

TRANSFECTION METHODS

The insertion of a gene of interest into the baculovirus genome is usually achieved by homologous recombination between the transfer plasmid and viral DNA. This occurs when the plasmid and viral DNAs are simultaneously introduced (cotransfected) into host insect cells (see note). Some factors to consider when setting up the cotransfection experiment are addressed first. These are followed by descriptions of several protocols for the transfection of DNA into insect cells.

> **Note:**
>
> It has been reported that, instead of cotransfecting the viral and plasmid DNAs into the cells, the viral DNA can be introduced by infecting the cells prior to transfection (Goswami and Glazer, 1991). The advantage is that it is unnecessary to prepare viral DNA for the transfection procedure. However, the efficiency of this approach has not been documented. We would expect that the percentage of recombinant viruses in the resulting stock would be greatly reduced.

Before Transfection

Quality of DNA

The quality of both the plasmid and viral DNA is critical for successful cotransfections. The protocols described previously will allow the isolation of viral DNA suitable for transfection. It is important that the DNA is relatively fresh as the efficiency of transfection decreases with the age of the DNA preparation. We do not recommend using viral DNA that is more than one month old. The transfer plasmid DNA should be thoroughly purified. Plasmid DNA purified by CsCl-EtBr density gradient centrifugation will work well in the cotransfection protocol. Alternatively, the plasmid DNA may be purified by anion exchange chromatography (e.g., through a "QIAGEN"-type column). See Sambrook et al. (1989) or other molecular cloning manuals for procedures for the purification of plasmid DNA.

Using Linearized Viral DNA

With certain parent viruses, the background of nonrecombinant virus remaining after cotransfection may be reduced by linearizing the parental viral DNA within the target area for allelic replacement prior to transfection. This approach also reduces the proportion of single-crossover recombinants obtained. For the moment, this strategy is employed with parent viruses, which carry a unique Bsu36I or Sse8387I site at the appropriate location. The recombinant virus AcRP6-SC contains a single Bsu36I site downstream from the polyhedrin promoter (See Kitts et al., 1990; see also Chapter 8.) Alternatively, any parent viruses that carry the *lacZ* gene at the site of insertion may be linearized with Bsu36I, because a single site for this enzyme occurs within *lacZ*. Invitrogen has constructed a recombinant virus in which a unique Sse8387I site is positioned immediately upstream of the polyhedrin promoter. In this case, the engineered virus retains an intact polyhedrin gene, so that occlusion-positive (occ^+) to occ^- screening may be carried out. PharMingen markets linearized DNA, termed "BaculoGold DNA," which contains a lethal deletion around the *polh* gene. This deletion can be complemented by recombination with an appropriate transfer plasmid, so that only recombinant viruses should form plaques.

The following protocol describes the linearization of viral DNA with Bsu36I. Viral DNA, linearized with Sse8387I, may be purchased from Invitrogen (# B825-03). BaculoGold DNA may be purchased from PharMingen (# 21100D).

Materials

1. Viral DNA purified as described above
2. Bsu36I (New England Biolabs # 524. A 10X reaction buffer is supplied with the enzyme.)

Method

1. Calculate the quantities of the reaction mixture components that will be required. A 1 µg quantity of viral DNA will be cut with three to four units of Bsu36I three times, for two to three hours each time. The final volume of the reaction mix should be at least ten times greater than the total volume of restriction enzyme added. (Restriction enzymes are typically supplied in 50% glycerol, which must be diluted to 5% or less for correct cleavage.) The volume of the 10X buffer used will be one-tenth of the final reaction volume, and ddH_2O will be used to adjust the volume as necessary.
2. Mix the DNA, 10X buffer, and the appropriate volume of ddH_2O on ice. Add the enzyme last.
3. Incubate at 37° C for two to three hours.
4. Add the same amount of enzyme twice more, and incubate for two to three hours each time.
5. Inactivate the enzyme by heating the DNA at 70° C for 15 minutes. Store at 4° C. If you wish, you may analyze an aliquot by agarose-gel electrophoresis (see protocols later in this chapter) to confirm that the DNA has been linearized. It would be easiest to digest the aliquot with a second enzyme and look for loss of the fragment that contains the Bsu36I site (see Fig. 13-2 later in this chapter for a restriction map of the viral genome).

Setting Up the Cotransfection Reactions

The protocols in the next section describe various methods for the cotransfection of viral and plasmid DNA into SF cells. We strongly recommend preparing reagents for the cotransfection of at least two separate plates of cells. Selecting putative recombinants from both plates will then ensure that at least two truly independent isolates will be obtained. This is particularly important if you have used linearized parental DNA for cotransfection. In this case, the yield of progeny virus obtained after cotransfection is quite low, and there is a high probability that different isolates derived from transfection of the same plate of cells simply represent separate isolates of the same clone.

Calcium Phosphate Coprecipitation

By far, the most common method for the introduction of DNA into insect cells is the calcium phosphate coprecipitation method. This method was first devised for an animal cell line by Graham and Van Der Eb (1973) and adapted for use with insect cells by a number of workers (Burand et al., 1980; Carstens et al., 1980; Potter and Miller, 1980). The method involves the formation of a coprecipitate of calcium phosphate and DNA that is taken up by the cells. In the protocol outlined here, adapted from that of Summers and Smith (1987), the DNA, in a buffered solution containing $CaCl_2$, is added to cells that have been incubated in Grace's or IPL-41 medium. The sodium phosphate in the medium causes the formation of the calcium phosphate coprecipitate directly in the culture medium.

The animal cell-transfection literature is replete with methods to increase the efficiency of calcium phosphate transfection. We do not generally use such protocols because the efficiency of transfection routinely obtained is more than adequate. However, there are protocols for glycerol (Potter and Miller, 1980) or DMSO (Carstens et al., 1980) shock treatments after transfection, which the reader may consult if desired.

Materials

1. 60 mm tissue-culture dishes (see Chapter 11)
2. SF cells at 1×10^6 cells/ml
3. Parental virus and transfer plasmid DNA
4. Transfection buffer: 25 mM HEPES, pH 7.1; 140 mM NaCl; 125 mM $CaCl_2$. Filter-sterilize and store at 4° C. The pH of this solution is critical for the formation of the precipitate.
5. Grace's or IPL-41 medium, supplemented with 10% fetal bovine serum. (Note that we specifically recommend the use of one of these media rather than TC-100 at specific stages of this protocol.)
6. Completed tissue-culture medium (TC-100, IPL-41, or TNM-FH; see Chapter 11 for details of media.)

Method

1. Seed 60 mm tissue-culture dishes with 2×10^6 cells per plate. The final volume of tissue-culture medium should be 4 mls per plate; add tissue-culture medium first, or dilute the cells to their final volume before plating. Incubate at 27° C for 30 minutes to one hour to allow the cells to attach. Alternatively, the cells may be plated at a density of 1×10^6 cells per dish and incubated at 27° C overnight.
2. Aspirate the tissue-culture fluid from the plates and refeed the cells with 750 μl complete Grace's or IPL-41 medium. Do *not* use TC-100 at this step. Incubate at room temperature.
3. In a tube, add 1 μg of viral DNA and 2 μg of transfer plasmid DNA to a final volume of 750 μl transfection buffer. Mix gently.
4. Add the DNA plus transfection buffer dropwise to the cells, and rock gently. The $CaCl_2$ in the transfection buffer and the sodium phosphate in the medium will form a calcium phosphate precipitate, which will coprecipitate with the DNA and adsorb to the cells. Incubate at 27° C for four hours.
5. Aspirate the medium plus transfection buffer. Rinse the cells once with complete tissue-culture medium and refeed with 4 mls complete tissue-culture medium. Incubate at 27° C for four to five days.
6. A few cells with PIBs should be visible by three days, and many cells should contain PIBs by four to five days after transfection. Harvest the tissue-culture medium and centrifuge at $1000 \times g$ for five minutes at 4° C to remove cellular debris. Determine the titer of virus produced and store this stock at 4° C. A good transfection will yield a viral stock of $> 10^7$ pfu/ml.

DEAE-Dextran Mediated Transfection

The facilitation of DNA uptake by treatment of the cells with diethylaminoethyl-dextran (DEAE-dextran) is also a common method for the introduction of DNA and RNA into animal cells (McCutchan and Pagano, 1968). DEAE-dextran is a polycation that binds both DNA and the cell surface, thus promoting adsorption of the DNA to the cell monolayer. It may also stimulate the uptake of DNA in some way.

DEAE-dextran-mediated transfection has not been used extensively in the generation of recombinant baculoviruses, and Potter and Miller (1980) have reported that it is significantly less efficient than calcium phosphate coprecipitation for the transfection of SF-21 cells. However, it may be useful for the transfection of other insect cell lines, as they report for TN-368 cells.

Materials

1. Complete tissue-culture medium. See Chapter 11 for details of various media.
2. 60 mm tissue-culture dishes (see Chapter 11)
3. SF cells at 1×10^6 cells/ml
4. Parental virus and transfer plasmid DNA
5. PBS, pH 6.2
6. DEAE-dextran solution: 500 μg/ml high molecular-weight DEAE-dextran (Sigma # D 9885) in PBS

Method

1. Seed 60 mm tissue-culture dishes with 2×10^6 cells per plate. The final volume of tissue-culture medium should be 4 mls per plate (add tissue-culture medium first). Incubate at 27° C for 30 minutes to 1 hour to allow the cells to attach. Alternatively, the cells may be plated at a density of 1×10^6 cells per dish and incubated at 27° C overnight.
2. Aspirate the tissue-culture medium from the cells and wash the monolayers once with PBS. *Do not let the monolayers dry out.*
3. Incubate the cells in 100 μl DEAE-dextran solution per plate for 30 minutes at room temperature. Rock the cells gently during this period.
4. Aspirate the DEAE-dextran solution and incubate the cells for a further 30 minutes at room temperature with 20 ng to 100 ng of DNA in 100 μl PBS. Use approximately a 25- to 50-fold molar excess of transplacement plasmid to viral DNA (e.g., 30 ng viral DNA and 70 ng plasmid DNA). Again, rock the cells during this incubation.
5. Wash the cell monolayer once with PBS and refeed with complete tissue-culture medium. Incubate at 27° C for four to five days. PIBs should start to become visible by three days after transfection.

Liposome-Mediated Transfection

A recently developed approach to the introduction of nucleic acids into eukaryotic cells is the use of cationic liposomes to facilitate transfer. This method employs the synthetic lipid *N*-[1-(2,3-dioleyloxy)propyl]-*N,N,N*-triethylammonium chloride (DOTMA), which was specifically designed to form positively charged liposomes that spontaneously interact with DNA or RNA (Felgner et al., 1987). In the commercially available Lipofectin reagent, DOTMA is mixed 1:1 with the phospholipid dioleoyl phosphatidylethanolamine. When this reagent is mixed with a DNA solution, essentially all the DNA is bound by the liposomes. Upon addition to the cells, the liposome-DNA complex binds to the negatively charged cell surface. The liposomes then fuse with the cell membrane, efficiently introducing the DNA into the cells.

A number of factors are important for the successful liposome-mediated transfection of DNA. First, the process is known to be inhibited by serum. Presumably, the DNA-lipid complexes bind to serum components and thereby interact less efficiently with the cell surface. Thus, the technique is ideally used with cells that have been propagated in serum-free medium. However, it may also be used with cells grown on serum-containing medium, provided the transfection itself is carried out in incomplete medium and the cells are thoroughly rinsed in incomplete medium prior to cotransfection. An additional consideration that is important for the successful transfection of DNA by this method is the use of polystyrene tubes. Polystyrene tubes are important because the DNA-lipid complexes have a tendency to adsorb to both glass and polypropylene surfaces. Finally, it is important that neither the DNA nor the liposomes are too concentrated in the transfection mix. Concentrations of DNA greater than approximately 100 μg/ml, or of liposomes greater than 500 μg/ml, will lead to the formation of large aggregates that will reduce the efficiency of DNA transfer.

The protocol outlined here works well with SF cells grown in serum-free medium. However, we recommend that researchers optimize the method in their own laboratory. Parameters that may be adjusted are the concentration of liposomes used, the concentration of DNA used, and the length of time the cells are exposed to the transfection mixture.

Materials

1. Serum-free tissue-culture medium (see Chapter 11)
2. 60 mm tissue-culture dishes or six well plates (see Chapter 11)
3. SF cells at 1×10^6 cells/ml in serum-free medium
4. Virus and transfer plasmid DNA
5. Lipofectin reagent (BRL # 82925A)
6. Polystyrene tubes (Falcon # 2027)

Method

1. Seed 60 mm tissue-culture dishes with 8×10^5 cells per plate. The final volume of serum-free medium should be 3 mls per plate (add tissue-culture medium first). Alternatively, seed 6.6×10^5 cells into each well of a six well plate. Incubate at 27° C for 30 minutes to 1 hour to allow the cells to attach.
2. Mix 30 μl of lipofectin reagent with 1.5 mls of serum-free medium in a polystyrene tube. In another tube, mix 1 μg of virus DNA and 2 μg to 4 μg of plasmid DNA in 1.5 mls of serum-free medium.
3. Mix the diluted DNA and lipofectin solutions and swirl gently.
4. Aspirate the serum-free medium from the cells. *Do not let the monolayers dry out.*
5. Add the lipofectin/DNA mixture to the cells.
6. Incubate for four to five hours at 27° C.
7. Aspirate the transfection mix from the cells. Wash once with 2 mls of serum-free medium. Refeed with fresh medium (either serum-free or serum-supplemented).
8. Incubate for five to six days at 27° C. Infected cells should become visible by four to five days after transfection.

IDENTIFICATION AND PURIFICATION OF RECOMBINANT VIRUS

In the first part of this chapter, methods for the cotransfection of parent virus and transfer plasmid DNA were described. You should now have at least two independent virus stocks derived from the cotransfection. These stocks contain the required recombinant virus mixed with a large excess of non-recombinant and single-crossover recombinant viruses. You may purify your recombinant virus from these cotransfection stocks by plaque purification, by end-point dilution, or by a combination of these methods. Recombinants may be identified during this purification procedure in several ways, including visual screening, DNA hybridization, PCR amplification, immunoblotting, and assay of enzyme activity of the expressed gene product. The first three of these options are described here. Included is the identification of recombinant viruses based on the detection of β-galactosidase activity within the protocols for the visual screening of plaques. However, this is an example of the identification of a recombinant based on the enzyme activity of an expressed gene product. Readers interested in identifying their recombinant by immunological means may consult a recent paper describing this approach (e.g., Capone, 1989).

If you have not already done so, you will need to titer your cotransfection-derived virus stocks (see Chapter 12). The yield of virus obtained after cotransfection may vary considerably (especially if you cleaved the parental viral DNA prior to transfection), and it is not possible to efficiently screen for your recombinant without knowing the titer of the cotransfection stock. When you are titering the cotransfection mixes, it is worth screening for recombinant virus using the protocols outlined in this section. While the total number of plaques or infected wells screened will be quite small (and, therefore, the probability of finding a recombinant quite low), it is occasionally possible to find a recombinant at this stage. The probability of finding a recombinant at this stage is higher if linearized viral DNA was used in the cotransfection.

Plaque Purification of Recombinant Virus

The principal methods used with currently available vectors for identifying recombinants based on plaque phenotype employ one or both of two markers. These markers are occlusion-body formation and the formation of blue plaques reflecting β-galactosidase activity due to expression of *lacZ*. The method you will need to use depends on the particular transfer plasmid and parent virus you used. Table 13-1 summarizes the expected phenotypes of recombinant viruses according to the transfer plasmid and parent virus used. A detailed description of how to identify and distinguish occ$^+$ and occ$^-$ plaques, and how to simultaneously evaluate *lacZ* expression is provided in this section.

Vectors that allow you to screen either of occ$^-$ plaques in an occ$^+$ background, or vice versa, are available (see Table 13-1). Identification for occ$^-$ plaques in an occ$^+$ background was the original screen employed in baculovirus expression vector work (Smith et al., 1983b; Pennock et al., 1984), and this method is still employed for many of the currently available transfer plasmids. This screening method is based on the fact that viruses that lack a polyhedrin gene (*polh*) are incapable of forming occlusion bodies and, therefore, form plaques that are visually distinguishable from wt (occ$^+$) virus plaques. This screen has the advantage of being based on the loss of a gene; therefore, it only identifies double recombinants. Its principal disadvantage is that, while the identification of occ$^-$ plaques is easy with some practice, inexperienced workers often find this step difficult and somewhat tedious.

Screening for occ$^+$ plaques is a much easier screen because occ$^+$ plaques are very visually striking on a background of occ$^-$ plaques. This screening method requires the use of a recombinant occ$^-$ virus as parent virus, in combination with a transfer

TABLE 13-1 Recombinant Virus Phenotypes

Transfer plasmid	*Parent Virus*	*Recombinant Phenotype*	*Background*
All AcMNPV occ$^-$ plasmids*	wt AcMNPV	occ$^-$	occ$^+$
All AcMNPV occ$^-$ plasmids*	vDA26Z	occ$^-$, blue	occ$^+$, blue
pAcDZ1	wt AcMNPV	occ$^-$, blue	occ$^+$, white
pBlueBac2	wt AcMNPV	occ$^-$, blue	occ$^+$, white
pAcUW2B	AcRP6-SC	occ$^+$	occ$^-$
pSynXIV VI$^+$	vSynVI$^-$gal	occ$^+$, white	occ$^-$, blue
pSynXIV VI$^+$ X3 series	vSynVI$^-$gal	occ$^+$, white	occ$^-$, blue
pAcAS2	AcAS3	occ$^+$, white	occ$^+$, blue
pAcUW1	AcUW1-*lacZ*	occ$^+$, white	occ$^+$, blue
pEP252	Ac228z	occ$^+$, white	occ$^+$, blue
BmNPV-based plasmids	wt BmNPV	occ$^-$	occ$^+$

*AcMNPV occ$^-$ plasmids include pAcYM1, pEVmXIV, pVL1393, pAcC4, pAc360, pVT-Bac, pc/pS1, pAcMP1, pAcUW3, and p2XIV VI$^-$

plasmid that reintroduces *polh* as well as the gene of interest. Screening for occ$^+$ plaques has the disadvantage of being based on a gain of phenotype and therefore does not distinguish between single- and double-crossover recombinants. This disadvantage can be avoided if a parent virus carrying *lacZ* at the insertion site is used and a screen for the loss of *lacZ* expression is also employed.

Activity of β-galactosidase can be easily detected by inclusion of a chromogenic substrate such as X-gal (5-bromo-4-chloro-3-indolyl-β-D-galactopyranoside) in the agarose overlay. Active enzyme cleaves this substrate, yielding galactose and an indoxyl derivative. The latter is rapidly oxidized to an indigo derivative, which is dark blue and forms an insoluble precipitate. Thus, plaques formed by viruses expressing *lacZ* turn blue in the presence of X-gal.

Because these screens are based on plaque phenotype, the considerations we have discussed previously for the formation of good, easily visualized plaques are critical here (see Chapter 12). An important difference between a plaque assay and screening for recombinant viruses is the use of neutral red stain. We *do not* recommend the use of neutral red as an aid to the visualization of occ$^-$ plaques. This is because it is extremely difficult to determine whether a plaque is occ$^+$ or occ$^-$ if the cells have first been stained with neutral red. If you are not able to identify occ$^-$ plaques without staining, it is preferable to use a *lacZ*-expressing virus, in which *lacZ* is *not* at the replacement site (e.g., vDA26Z), as the parent virus in the cotransfection. In this case, all the plaques will turn blue in the presence of X-gal (see Table 13-1).

The first plaque assay of the screen is the most critical phase of isolating a recombinant virus, and it is important to make the effort to examine each plate diligently. You should pick as many putative recombinants as possible during the first screen. Replaque five to ten of these isolates and keep the rest in media at 4° C in case they are needed. It will be quite clear after the first round of purification which isolates have the correct phenotype, and during subsequent rounds, it is only necessary to process three to five of these isolates.

CONTROLS: We strongly recommend that researchers who are not familiar with the morphology of occ$^+$ and occ$^-$ plaques obtain known occ$^+$ and occ$^-$ viruses; then plaque them alone and in combination to become familiar with plaque appearance. Similarly, if you are screening on the basis of *lacZ* expression, you should plaque out known *lacZ*-positive and *lacZ*-negative viruses. Finally, it is always useful to include an uninfected plate so that you can monitor the condition of the monolayer.

Materials

1. Complete tissue-culture medium. (See Chapter 11 for details concerning different media.)
2. 60 mm tissue-culture dishes (see Chapter 11)
3. SF cells grown to 1×10^6 cells/ml
4. Recently titered virus stocks from cotransfection, and appropriate control viruses
5. Agarose [SeaKem ME (see note 1); FMC # 50014]
6. X-gal stock: 20 mg/ml 5-bromo-4-chloro-3-indolyl-β-D-galactopyranoside (X-gal; Gold Biotechnology # X4281C) in dimethylformamide (Sigma # D 8654). Store at –20° C in the dark in a glass container.
7. Stereo dissecting microscope and light source. We use an Olympus model SZ-III microscope and model TGHM light source.

Method

Primary Screen

1. Seed 60 mm or 100 mm tissue-culture plates with 2×10^6 or 5×10^6 cells per plate respectively (see note 2). We recommend using at least twenty 60 mm plates or eight 100 mm plates for each screen. (You may not need as many plates if you have linearized the parental viral DNA before cotransfection; see Step 3.) The final volume of tissue-culture medium should be 4 mls and 10 mls per plate respectively. To help disperse the cells evenly, either add some tissue-culture medium to the plates before adding the cells, or dilute the cells to their final volume before seeding the plates. Rock the plates gently *immediately* after adding the cells to ensure an even monolayer. Incubate at 27° C for 30 minutes to 1 hour to allow the cells to attach. They will attach faster if plated out in medium lacking serum. Alternatively, the cells may be plated out at half these densities (in complete medium) and incubated at 27° C overnight.
2. Dilute the virus stocks in tissue-culture medium so that there are approximately 200 plaque-forming units (pfu)/ml.
3. Aspirate the tissue-culture medium from the cell monolayers and infect with 0.5 ml of the diluted virus per 60 mm plate, or 1.25 mls per 100 mm plate. *Be careful not to let the plates dry out before addition of the virus.* Be sure to use virus derived from at least two independent cotransfections. Incubate at room temperature for one hour with gentle rocking. If you dilute the virus to 200 pfu/ml and use twenty 60 mm or eight 100 mm plates, you will get approximately 2000 plaques for each cotransfection, without having too many plaques on each plate. Since the required recombinant should represent 0.2% to 1.0% of the total virus population, this is generally adequate. For cotransfections in which the parental viral DNA was first linearized, the proportion of recombinants should be significantly higher, and it may be sufficient to screen a lower number of plaques.
4. While the virus is adsorbing to the cells, prepare the agarose overlay as described here:
   a. Prepare a solution of 5% agarose in ddH_2O and autoclave. Make up 0.5 mls for each 60 mm dish, or 1.25 mls for each 100 mm dish, to be overlaid. (The agarose solution can be prepared in advance and microwaved when required.) This solution of agarose will subsequently be diluted tenfold, so prepare it in a container that is large enough to accommodate 10 times the starting volume.
   b. Heat complete tissue-culture medium (5 mls per 60 mm plate or 10 mls per 100 mm plate) to 60° C in a water bath.
   c. Cool the melted agarose to 60° C.
   d. Dilute the agarose to a concentration of 0.5% with the heated tissue-culture medium. For example, for ten 60 mm dishes, prepare 5 mls of 5% agarose and dilute with 45 mls tissue-culture medium. Mix well by swirling.
   e. Cool the 0.5% agarose to 40° to 42° C.
   f. If you need to screen for *lacZ* expression, add X-gal to a final concentration of 120 μg/ml and mix well.
5. Remove the virus inoculum from the cells and overlay with 4 mls of agarose for each 60 mm plate, or 10 mls for each 100 mm plate. Try to remove as much of the virus inoculum as possible without allowing the cells to dry. It is important that the agarose cool sufficiently before you overlay the cells. Do not let the agarose drop on top of the cells. Instead, let it run gently down the side of the plate to avoid disturbing the cell monolayer.

6. Although occ$^+$ plaques will be visible from three to four days postinfection (pi), we do not recommend beginning the screen until five to seven days pi. This allows more time for plaque development and therefore makes the occ$^-$ plaques easier to visualize. It is also important to allow the development of the blue color of *lacZ*-expressing plaques.
7. Occlusion phenotype is initially determined using a stereo dissecting microscope, with a piece of black velvet on the platform as background and a strong light source at a very acute angle to the platform (i.e., the light beam should be almost parallel to the platform). With a razor blade, etch a series of parallel lines, approximately 7 mm apart, on the underside of each plate. Using these lines as guides, scan the entire plate under the dissecting microscope (the plate should be inverted). Occ$^+$ plaques will appear as clusters of very yellow or white, "shiny" cells against a background of light-grey, uninfected cells. This bright, "shiny" appearance is due to the high refractivity of the occlusion bodies. Occ$^+$ plaques are therefore very easy to detect.

   Due to the absence of occlusion bodies, occ$^-$ plaques are not as refractive as occ$^+$ plaques. However, because cells become rounded during infection, occ$^-$ plaques are more refractive and appear whiter than surrounding uninfected cells. When you screen for occ$^-$ recombinants, we recommend that you search the entire plate for clusters of cells that are more refractive (whiter) than the background. Even though well-developed plaques will often have a center area that is devoid of cells, you should avoid deliberately screening for such features as there will be many similar discontinuities in the monolayer not associated with plaques.

   To screen for *lacZ* expression, scan each plate under the dissecting microscope using a yellow or white background to more easily visualize the blue color. It may help to move the plate slightly as you scan it; the motion makes faint blue plaques more noticeable. If the blue color is very faint and difficult to see, you may use a twofold-higher concentration of X-gal in the overlay (make a stock of X-gal that is two times more concentrated so that the concentration of dimethylformamide in the overlay is not changed). Alternatively, it may help to subject the plates to two to three freeze-thaw cycles. This procedure lyses the cells and ensures that the X-gal is exposed to the β-galactosidase.

   Circle all plaques that display the appropriate phenotype (see Table 13-1) using a sharp instrument (we routinely use dental probes) or a fine pen. When screening for blue plaques using vDA26Z, we have occasionally observed a small number of white plaques. These are viruses that have incurred nonspecific mutations in the *lacZ* gene and should be disregarded.
8. All plaques circled above should be examined carefully using an inverted phase-contrast tissue-culture microscope. For occ$^+$ plaques, you should confirm that many occlusion bodies are visible in the nuclei of the infected cells. The occlusion bodies are quite characteristic and appear as tightly packed clusters of dark, refractive bodies, resembling small crystals, in each nucleus. Occasionally, it is possible to detect Brownian motion of the occlusion bodies around the periphery of the nucleus. It is best to select plaques in which the cells have many occlusion bodies per nucleus. The blue color, due to β-galactosidase activity in *lacZ*-expressing plaques, may obscure the occlusion bodies somewhat, so extra care is needed when examining such plaques.

   If you are screening for occ$^-$ plaques, you should attempt to verify both that the plaque lacks occlusion bodies and that it is indeed a plaque. Again, extra care is needed when examining *lacZ*-expressing plaques. Disregard any plaques in which there is even a *single* cell with *many* occlusion bodies in its nucleus. It has been reported that overexpression of certain proteins can give rise to refractive bodies in the infected nucleus or cytoplasm that may superficially resemble viral occlusion bodies (Martens et al., 1990; Matsuura et al., 1987). Therefore, it may be prudent not to disregard plaques in which the cells appear to have one (or a small number of) refractive bodies. However, we recommend that you treat such plaques with suspicion because they may represent plaques formed by viral mutants and not true recombinants. Be careful not to select areas of the monolayer where the cells are denser or piled up on one another (such areas look rather like occ^- plaques under the dissecting microscope). Infected cells display characteristic cytopathic effects that you should be able to recognize with some practice. Generally, the cells will appear wrinkled and darker than the surrounding cells. Occasionally, the large fibrillar structures due to *p10* expression (see Part I) can be detected as refractive, elongated, and slightly curved or wavy bodies. The virogenic stroma can be distinguished as a granular region in the center of the cell. Some cell lysis may have occurred in the center of well-developed plaques (although, as discussed before, this feature should not be used as a screen). If cell lysis has occurred, there may be numerous small vesicles visible in the central space.

9. It is often helpful to screen all the plates as described in Steps 7 and 8, then incubate them for an additional day and reexamine the selected plaques under the tissue-culture microscope. This is especially useful when screening for a *lacZ*-negative virus. All plaques that appear to be recombinants should be picked. Try to pick some plaques derived from each independent cotransfection stock. This ensures that you will have more than one truly independent plaque isolate and is particularly important when the titer of progeny virus from the cotransfection was low (e.g., when the parental virus DNA was linearized prior to cotransfection). Mark the precise position of each plaque with a dark pen on the bottom of the plate.
10. To pick a plaque, place the tip of a sterilized Pasteur pipet or glass micropipet directly onto the plaque. Carefully apply *gentle* suction until a small plug of agarose is drawn into the pipet. It helps to lift the pipet very slightly while applying gentle suction.
11. Place the agarose plug in 1 ml tissue-culture medium. Vortex well to release the budded virus particles from the plug. It is essential to name each pick carefully at this stage so that you can easily trace the origin of each isolate during subsequent rounds of purification. These isolates may now be rescreened by replaquing and identifying recombinant plaques using the same methodology as used here. This is described in Steps 1 through 8 following. Alternatively, you can now screen these isolates by DNA hybridization analysis or by PCR amplification, as described later in this chapter.

Replaquing Primary Screen Isolates

1. Dilute five to ten of the putative recombinants in tissue-culture medium. Dilutions of 10^{-1}, 10^{-2}, and 10^{-3} will be appropriate.
2. Infect cells in 60 mm plates (2×10^6 cells/plate) with 0.5 ml of the diluted virus stocks and incubate at room temperature for one hour with gentle rocking.
3. Remove the inocula and overlay with agarose as before (Steps 4 and 5 of the preceding protocol). Remember to include X-gal in the overlay if you are also screening for *lacZ* expression. Incubate at 27° C for five to seven days.
4. Identify recombinant plaques using the appropriate screen and methodology outlined in the preceding protocol. While there will almost certainly be non-recombinant virus plaques in these plates, you should have greatly enriched for the recombinant. It should be quite obvious which of the original isolates were, in fact, recombinants (these plates will have a high proportion of plaques with the appropriate phenotype).

5. Select at least three *independent* isolates for further purification and amplification. Choose well-isolated plaques on plates with less than 10 plaques and mark their precise position with a dark pen. At this stage, there is little point in picking more than one plaque for each independent isolate (i.e., each original pick from Step 9 of the primary screen) because such duplicate plaques will almost certainly contain identical viruses.
6. Mark the positions of the selected plaques with a dark pen. Carefully pick these plaques with a sterile Pasteur pipet or glass micropipet and place agarose plug in 1 ml of tissue-culture medium.
7. Replaque 10^{-1}, 10^{-2}, and 10^{-3} dilutions of these isolates as described in Steps 1–4.
8. Plaques derived from this third plaque purification should now all have the recombinant phenotype. If there are still nonrecombinant plaques present, purify the isolate again. When you are confident that the isolate is pure, pick one well-isolated plaque and use this to initiate amplification.

Notes:

1. We recommend SeaKem ME agarose for overlays. We do not recommend low-gelling-temperature agarose, such as SeaPlaque, because the lower gel strength necessitates a higher concentration of agarose, which tends to obscure the plaques.
2. Proper cell density is critically important to the success of a plaque assay. Generally, 2×10^6 cells/60 mm tissue-culture dish is the best cell density to start with if you are unfamiliar with the properties of a particular cell line and its growth in different insect cell culture media. If the plaques are too small after five days, the cell density used was too high. If the plaques are large and diffuse, the density used was too low. Repeat the plaque assay using a series of cell densities to find the optimal cell density for your particular cell line and plaque overlay media.

End-Point Dilution Cloning of Recombinant Virus

If you prefer not to use a plaque purification approach, you may identify and purify your recombinant by screening 96 well plates infected with serial dilutions of the cotransfection stock. End-point dilution is most often used in conjunction with DNA hybridization to identify recombinant infected wells, but it may also be used with a variety of other screens. These include screens for occlusion phenotype and *lacZ* expression, a PCR-based screen, and screens based on the properties of the expressed protein, such as enzyme assays, immunological techniques, and so on.

The number of plates and multiplicity of infection (MOI) used at each stage of screening will depend on whether you choose to look for a gain of phenotype or a loss of phenotype in the primary screen. Both approaches are described here; first, the basic virus infection and cloning procedures are covered, then specific recommendations for the different types of screen that might be used at each stage are provided.

CONTROLS: You should always include some wells infected with wt or the parent virus in each screen. It is also helpful to infect some wells with a virus that displays the phenotype you are screening for, as a positive control. Depending on the screen you use, this may or may not be feasible. Finally, remember to leave some wells uninfected.

Materials

1. Complete tissue-culture medium. (See Chapter 11 for details concerning different media.)
2. 96 well plates (see Chapter 11)
3. SF cells grown to 1×10^6 cells/ml
4. Recently titered virus stocks from cotransfection, and appropriate control viruses
5. 20 mg/ml X-gal stock

Method

Primary Screen for a Gain of Phenotype

The screen begins with the infection of at least one 96 well plate at an MOI of approximately 20 pfu (35 $TCID_{50}$)/well. You will identify a large number of positive clones, most of which will represent infection by single-crossover recombinants. You should rescreen many of these positives (we recommend 20 to 25 if possible), this time infecting with serial dilutions of each one. Because some wells will now be infected at a low MOI, and therefore may not have any nonrecombinant virus, it is possible to screen for the loss of a marker in this second screen. You should pick one well that displays the appropriate phenotype at the highest dilution possible for each original recombinant. (Because the first screen was only for the acquisition of a marker, many of the selected picks will not have given rise to any wells containing virus with the appropriate phenotype at the second screen.) The isolates that do look promising should be purified by further limiting dilution, and then rescreened. Remember that to be at least 95% certain that a well was infected by a single virus particle, you will have to dilute the stock so that only 10% of the wells become infected (see end-point dilution protocol in Chapter 12).

1. Dilute the cells to a concentration of 1×10^5 cells/ml with complete tissue-culture medium.
2. Dilute each cotransfection stock to a concentration of 2×10^3 pfu/ml.
3. Mix 1 ml of the diluted virus with 10 mls of cells, and seed 110 µl of the mix into each well of a 96 well plate. Remember to reserve some wells for the controls. Infection of 90 wells with the cotransfection mix will permit the screening of approximately 1800 pfu from each stock. Because the recombinant virus should represent from 0.2% to 1.0% of the virus population, screening this number of pfu should be adequate.
4. Incubate the plate at 27° C for five to seven days. The individual wells may now be screened in a variety of ways, depending on your particular application. Screening by DNA hybridization and by PCR amplification are described later in this chapter. If you have an antibody against your heterologous gene product, you may use an immunological screen (Capone, 1989). If you used any of the pSynXIV VI$^+$ series of plasmids or pAcUW2B (recombinants are occ$^+$ in an occ$^-$ background), you may examine the wells closely for the presence of occlusion bodies. Alternatively, if you are using the transfer plasmids pBlueBac2 or pAcDZ1 (recombinants are blue in a white background), you may screen for blue coloring in the tissue-culture fluid and the cells (provided X-gal was added at the time of infection).
5. Select 20 to 25 positive isolates for further purification and screening. For each isolate, prepare serial dilutions in tissue-culture medium. Dilutions of 10^{-3}, 10^{-4}, 10^{-5}, and 10^{-6} should be appropriate.
6. Dilute the cells to 1×10^5 cells/ml in complete tissue-culture medium.
7. Mix 250 µl of each virus dilution with 2.5 mls of the cell suspension, and inoculate the wells with 110 µl each of the mix. Infect 22 to 24 wells per dilution. As before, include the appropriate controls.

8. Incubate at 27° C for five to seven days and screen the wells using the appropriate methods. At this stage, you should also be able to screen for a loss of phenotype, especially in wells infected with more dilute samples. For most transfer vectors, this will mean screening for wells that are infected but do not contain any cells with occlusion bodies. The identification of infected wells may be easier if vDA26Z was used as the parent virus. Provided X-gal was included in the tissue-culture fluid, all infected cells will appear blue. Alternatively, if you used any of the pSynXIV VI$^+$ series plasmids, pEP252, pAcAS2, pAcUW1 (with AcUW1-*lacZ*), or pAcUW2B (with AcUW2B-*lacZ*), screen for infected wells that lack any blue color.
9. If you were able to identify positive isolates that had lost the appropriate marker (i.e., double recombinants with no nonrecombinant virus) in Step 8, you will only need to process four or five of these isolates. Otherwise, you should continue with all 20 to 25 of your isolates. Select the virus from *one* well from each plate at the highest dilution possible. There is little point in taking more than one well from each plate at this stage, because there is a high probability that these will represent the same virus.
10. Seed and infect the cells as before using one plate per isolate and 10^{-7} and 10^{-8} dilutions of the stock. Infect 46 to 48 wells with each dilution.
11. Incubate at 27° C for five to seven days and screen as described in Step 8 for both positive and negative markers (if possible). At this stage, it is important to try to obtain a pure stock of each putative recombinant. Therefore, try to ensure that no more than 10% of the wells are infected at the dilution from which you wish to select your isolate.
12. We recommend that you subject the isolates to one further round of purification by end-point dilution before you begin amplification. This is especially important if you only have a gain-of-function screen, because you cannot tell whether a stock is contaminated with nonrecombinant virus. These purified stocks can then be amplified as described later in this chapter.

Primary Screen for a Loss of Phenotype

If you wish to screen for a loss of phenotype in the primary screen, you need to ensure that a large number of wells are infected with a single infectious unit. At an MOI of 1 pfu/well, only 36% of the wells will actually be infected with a single infectious unit (see note). The remainder will either be uninfected or will be infected by more than one infectious unit. Infecting thirty 96 well plates at an MOI of 1 pfu/well should yield enough individually infected wells to ensure the identification of the recombinant. It should only be necessary to process four to five potential isolates in subsequent screens.

1. Dilute the cells to a concentration of 1×10^5 cells/ml with complete tissue-culture medium.
2. Dilute each cotransfection stock to a concentration of 1×10^2 pfu/ml.
3. For each plate, mix 1 ml of the diluted virus with 10 mls of cells and seed 110 µl of the mix into each well. Thirty 96 well plates should allow the screening of approximately 1000 individually infected wells.
4. Incubate the plate at 27° C for five to seven days. Screen the wells for the loss of the marker, as appropriate. For most transfer vectors, this will mean screening for wells that are infected but do not contain any cells with occlusion bodies. The identification of infected wells may be easier if vDA26Z was used as the parent virus. Provided X-gal was included in the tissue-culture fluid, all infected cells will appear blue. Alternatively, if you used any of the pSynXIV VI$^+$ series plasmids, pEP252, pAcAS2, pAcUW1 (with AcUW1-*lacZ*), or pAcUW2B (with AcUW2B-*lacZ*), screen for infected wells that lack any

blue color. At an MOI of 1, approximately 36% of the wells will be uninfected (see note); therefore be careful to distinguish between these wells and wells displaying the required phenotype.

5. Select four to five positive isolates and continue with the further purification and screening from Step 9 of the previous protocol.

Note:

According to the Poisson distribution, the proportion (p) of wells receiving either 0 or 1 infectious unit at a mean concentration of virus (μ) of 1 is given by the equation

$p = \mu e^{-\mu} = e^{-1} = 0.367$

Thus, at an MOI of 1, 36% of the wells will be uninfected, and 36% will have been infected with a single particle.

Screening by DNA Hybridization

DNA hybridization is a simple and popular method of identifying virus isolates that contain the heterologous gene integrated into their genome. DNA hybridization may be employed either in a direct plaque-hybridization procedure or in a simple dot-blot method. In the plaque hybridization method (Summers and Smith, 1987), the agarose overlay from a plaque purification step is removed and saved and the cells blotted onto a membrane. The DNA on the membrane is hybridized with an appropriate probe. Positive signals are lined up with the agarose overlay, and plugs of agarose containing the putative recombinant virus are picked and used for further screening and purification. The major disadvantage of this method is that the cells often smear during blotting, enlarging the apparent diameter of the plaque. Thus, large plugs of agarose have to be picked, making it difficult to achieve any effective purification. Because of this disadvantage, we feel that a dot-blot-type approach, such as the following one, is preferable.

In the protocol described here, putative recombinants (either from plaque purification or end-point dilution) are first inoculated into a 96 well plate. Total viral and cellular DNA is obtained from the infected cells in each well, denatured and immobilized on a membrane support. Any wells that were infected by virus containing the heterologous gene of interest are identified by hybridization of an appropriate radiolabeled probe to the immobilized DNA. Plasmid DNA or linear gel-purified DNA containing the foreign gene may be used as a probe. The probe should not contain sequences homologous to baculovirus DNA. The transfer of DNA to a membrane support, random primer radiolabeling of the probe, and hybridization of the probe and immobilized DNA are described here. These protocols are also discussed in more detail later in this chapter (see Southern hybridization protocols).

CONTROLS: You should always include some wells infected with wt or the parent virus in each screen. It is also helpful to infect some wells with a virus that displays the phenotype you are screening for, as a positive control. Depending on the screen you use, this may or may not be feasible. Finally, remember to leave some wells uninfected.

Transfer of DNA

Materials

1. 96 well tissue-culture plates (see Chapter 11)
2. SF cells at 1×10^6 cells/ml
3. Dot-blot hybridization apparatus (e.g., Schleicher and Schuell # SRC096/0)
4. Nylon membrane cut to fit the dot-blot apparatus (e.g., Zeta-probe, BioRad). Wear gloves at all times when handling the membrane, as grease from your fingers will block transfer of the DNA.
5. Filter paper cut to fit the dot-blot apparatus (S&S # 31480 or Whatman 3MM paper)
6. Multi-(12)-channel pipet (e.g., Costar 12-pette)
7. Disposable channel trays to hold buffers
8. 0.5 N NaOH
9. 20X SSC: 3 M NaCl; 0.3 M sodium citrate, pH 7.0. Store at room temperature. 20X SSC is also available commercially (e.g., Sigma # S 6639).

Method

1. Seed a 96 well plate with 1.5×10^4 SF cells/well.
2. Pick putative recombinant viruses identified by visual screening into separate wells. Try to pick at least 20 plaques from the series of dilutions done from the first round of plaque assays for each transfection mixture. Remember to inoculate some wells with the appropriate controls.
3. Incubate the plate in a humidified incubator at 27° C for at least two but no more than three days. This time period allows sufficient viral replication that will give rise to a strong hybridization signal without allowing the infection to proceed to the point where all the cells lyse and the DNA is lost in the tissue-culture fluid.
4. Transfer the viral supernatants with the multichannel pipet to a sterile 96 well plate. Store the viral supernatants at –20° C if they cannot be processed immediately. While freezing the virus will result in a decrease in titer (see Chapter 12), it is preferable in this case to storing the virus at 4° C as it avoids problems due to condensation inside the lid of the plate leading to cross-contamination of the wells.
5. Transfer the viral and chromosomal DNA to the nylon membrane.
   a. Lyse the infected cells in the plate and denature the DNA by adding 200 µl of 0.5 N NaOH.
   b. Cut nylon and filter paper to fit the dimensions of the dot-blot apparatus and wet them both in ddH_2O.
   c. Assemble the dot-blot apparatus with the filter paper, supporting the membrane from below. Cut one corner of the membrane to help orient the blot with respect to the wells.
   d. Connect the apparatus to a vacuum, and filter 500 µl ddH_2O through each well (apply with the multichannel pipet).
   e. Apply the cell lysates to the wells. Use clean pipet tips for each lysate. If an individual well filters slowly, a bubble may be stuck just above the membrane. Try to remove the bubble by rapid pipeting with a single pipet.
   f. Rinse by filtering 500 µl 0.5N NaOH through each well. Remove any bubbles by the method described in Step e. If the solution still does not filter through, remove any liquid remaining above each well with a pipet.
   g. Disassemble the apparatus and rinse the membrane in 2X SSC for several minutes. Let the filter air-dry, staple between two sheets of Whatman 3MM paper, and bake under vacuum at 80° C for two hours. Store in a sealed plastic bag until ready for hybridization.

Random Primer Labeling

Materials

1. DNA: 25 ng of the DNA to be labeled
2. Random Primers DNA Labeling System (BRL # 8178SA). This kit includes nucleotide stocks, random primers, Klenow fragment (large fragment) of *E. coli* DNA polymerase I, and stop buffer. Alternatively, you may prepare the individual reagents yourself. See the Southern hybridization protocol later in this chapter.
3. [α–^{32}P]dCTP (3000 Ci/mmol, NEN # NEG 013C). Store at –20° C.
4. Whatman DE 81 filters
5. 0.15 M NaCl

Method

1. In a 1.5 ml microcentrifuge tube dispense:
   1 µl DNA (25 ng/µl)
   23 µl ddH_2O
2. Boil the DNA for five minutes, place on ice, and add:
   2 µl each 0.5 mM dATP, dGTP, and dCTP
   15 µl Random Primers Mixture
   5 µl [α–^{32}P]dCTP (10 µCi/µl)
   1 µl Klenow fragment (3 units/µl)
   Incubate at 25° C for at least one hour. The reaction may be allowed to proceed longer (as long as overnight) if convenient.
3. Add 5 µl stop buffer.
4. It is not necessary to remove unincorporated nucleotides. The incorporation of radiolabel may be determined as follows:
   a. Place 1 µl of the labeled reaction mix onto each of two DE81 filters and allow to dry.
   b. Wash one filter extensively in several changes of 0.15 M NaCl. During this step, unincorporated nucleotides are washed off the filter. Allow to dry.
   c. Measure the radioactivity on both filters. The ratio of the radioactivity on the washed to the unwashed filter is the proportion of the radiolabel that was incorporated. From this, you may calculate the specific activity of the labeled probe.
5. Store the probe at –20° C until use. Very high specific-activity probes degrade quite rapidly due to radiochemical decay, so the probe should be used soon after labeling.

Hybridization

Materials

1. Formamide, predistilled (BRL # 5515UA)
2. 50X Denhardt's reagent: 1% (w/v) Ficoll (Type 400, Pharmacia # 170400-01); 1% (w/v) polyvinylpyrrolidone (Sigma # P 5288); 1% (w/v) bovine serum albumin (Sigma # A 7030) in ddH_2O. Filter-sterilize and store at –20° C. Alternatively, Denhardt's reagent is available commercially (e.g., Sigma # D 2532).
3. 1 M NaH_2PO_4, pH 7.2
4. 1 M NaCl
5. 0.5 M EDTA
6. 10% SDS
7. Hybridization solution: 50% formamide; 0.25 M NaH_2PO_4, pH 7.2; 0.25 M NaCl; 7% SDS; 1 mM EDTA; 5X Denhardt's

8. Heat-sealable bags (e.g., seal-a-meal bags)
9. 10 mg/ml denatured salmon-testes DNA (Sigma # D 7656). This is supplied as a sonicated denatured solution, which has been phenol-chloroform-extracted and ethanol-precipitated.
10. 2 M Tris-HCl, pH 8.0
11. 1 N HCl
12. 10 N NaOH
13. 20X SSC
14. 2X SSC, 0.1% SDS
15. 0.1X SSC, 0.1% SDS

Method

1. Prehybridize the filter for at least one hour to reduce spurious background signals. Place the filter in a heat-sealable plastic bag and add ≈150 µl hybridization solution per cm^2 of the membrane. Place the bag in a tray and agitate in a shaking incubator at 42° C.
2. Denature all of the probe by adding *in the following order:*
   50 µl radiolabeled DNA probe
   100 µl salmon-testes DNA (10 mg/ml)
   50 µl 10 N NaOH
   300 µl 2 M Tris-HCl, pH 8.0
   475 µl 1 N HCl
3. Immediately add the denatured probe directly to the bag containing the filter and hybridization solution. Incubate by shaking at 42° C overnight.
4. Wash the filter twice at 55° C with 2X SSC, 0.1% SDS for 15 to 30 minutes.
5. Wash the filter at 55° C with 0.1X SSC, 0.1% SDS for 15 to 30 minutes.
6. Dry the membrane, tape to stiff cardboard, cover with Saran Wrap, and expose to X-ray film.
7. Identify the wells corresponding to the most intense spots, to use as an enriched source of recombinant viruses for subsequent rounds of purification by plaque assay or end-point dilution. As before, the number of wells you should pick depends on the approach you have used to identify your recombinants. If you are confident that your isolates are double-crossover recombinants, continue with virus from three to four wells only. However, if there may be single-crossover recombinants among your isolates, you should continue with 20 to 25 of them.

Screening by PCR Amplification

The amplification of specific segments of DNA by the polymerase chain reaction (PCR) can be extremely useful, both for identifying putative recombinants and for confirming that selected recombinants have the appropriate genome structure (Daugherty et al., 1990; Malitschek and Schartl, 1991; Webb et al., 1991). The technique involves the use of opposing pairs of primers homologous to the template DNA. The template DNA is denatured, and the primers are annealed to it and extended with DNA polymerase. The template is then denatured again and the process repeated many times. This results in the exponential amplification of the sequence flanked by the two primers. The procedure is facilitated by the use of a thermostable DNA polymerase, which is not inactivated by the multiple denaturation steps. See Ehrlich (1989) or Innis et al. (1990) for further information and protocols.

Depending on the primers used, this technique can therefore confirm that specific sequences are present in the template DNA and indicate the distance between the primer binding sites on the template DNA. For baculovirus expression vector work, the judicious choice of primers can enable the researcher to differentiate the appropriate recombinant virus from single-crossover recombinants and/or contaminating parent virus. In the example shown in Figure 13-1, primers BV1 and BV2 are homologous to sequences flanking the cloning site in the transfer plasmid. The size band amplified from a recombinant virus will depend on both the size of the gene inserted, and on the transfer plasmid used, because different plasmids delete *polh* to varying degrees. (In the example in Fig. 13-1, a band of 1300 bp is expected.) Both wt and recombinant bands will be observed if the recombinant isolate is contaminated with wt virus, or if a single-crossover recombinant has been isolated. The size of the amplified fragment can provide some assurance that the appropriate segment of DNA has been cloned into the virus. (A caveat to this simple approach is that, if the inserted DNA is very large, a recombinant-specific fragment may not be amplified efficiently.)

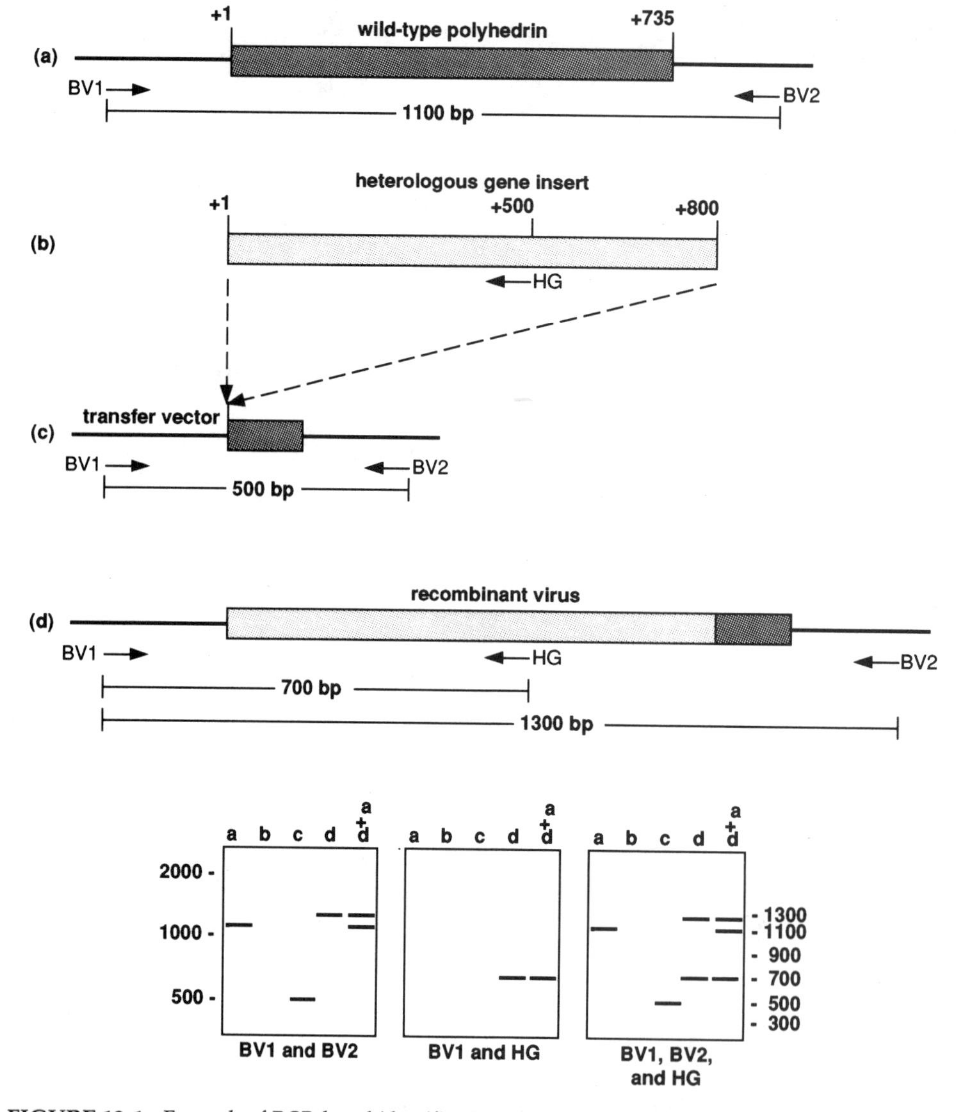

FIGURE 13-1 *Example of PCR-based identification of a recombinant virus.* The figure illustrates the structures of the transfer plasmid, wt, and recombinant virus and shows the positions of three primers that could be used in a PCR-based screening procotol. The fragments that would be amplified from each template are also shown, and the expected gel patterns obtained with different combinations of templates and primers are illustrated at the bottom.

Obviously, one can confirm that the appropriate gene has been cloned into the virus by carrying out a second amplification experiment using a pair of primers specific for the heterologous gene. However, a more convenient approach is a combination of the above two; that is, using three primers, two of which flank the cloning site (like primers BV1 and BV2), and the third of which is homologous to sequences in the heterologous gene (like primer HG in Fig. 13-1). [Alternatively, primer BV2 can be within *polh* sequences that are removed during allelic replacement (Malitschek and Schartl, 1991).] Primer HG, which is specific for the heterologous gene, is chosen so that the size fragment amplified by primers BV1 and HG is distinct from those amplified by primers BV1 and BV2 from wt or recombinant virus (700 bp vs. 1100 and 1300 bp in the example in Fig. 13-1). Amplification of a fragment by primers BV1 and HG indicates that the appropriate gene has been cloned at the correct location and in the expected orientation. Contaminating wt or single-crossover recombinants will be revealed by amplification of the wt-virus-specific fragment by primers BV1 and BV2. This approach also avoids the problem associated with the insertion of very large genes, when the distance between primers BV1 and BV2 might be too great to permit the efficient amplification of a recombinant-specific fragment.

The PCR amplification of viral DNA extracted directly from a plaque is not efficient enough to permit the routine use of this approach. Instead, it is necessary to scale up the plaque pick by infection of cultured cells (e.g., in 24 or 96 well plates). This is not necessary when recombinant virus is being purified by end-point dilution. Essentially, any of the techniques described at the beginning of this chapter for the preparation of viral DNA are compatible with the PCR procedure. The miniprep procedure is particularly useful for the preparation of DNA from small viral stocks derived from plaque picks. The following protocol describes the preparation of DNA from infected wells on a 96 well tissue-culture plate.

The problem most likely to be encountered in any PCR amplification procedure is contamination of test samples with small amounts of amplifiable template, resulting in artifactual amplified products. This is due to a combination of the extreme sensitivity of the technique, together with the large amounts of amplified product present in each reaction. If due care is not taken, it is quite easy to inadvertently contaminate a PCR reaction with a minute amount of the product of a previous reaction. The most important precautions one can take to avoid this problem are to ensure that the reagents used for the PCR reaction (e.g., primers, enzyme, buffer, etc.) are always kept physically separate from PCR reaction products, and that the same utensils (e.g., pipets, etc.) are never used for both reagents and products. It is also important to include the appropriate negative controls (see box). Readers should consult Erlich (1989) or Innis et al. (1990) for further discussion of possible PCR artifacts and how to avoid them.

CONTROLS: An aliquot of wt and/or parental viral DNA (if different) should be amplified in parallel as a control for the wt-specific amplification product. Similarly the transfer plasmid containing the cloned gene should be amplified to provide a control for the recombinant-specific amplification product(s). *Be extremely careful to avoid contaminating any other samples with the plasmid DNA.* Finally, include a sample to which no template DNA has been added. This serves as a control for the contamination of any reagents with amplifiable template. This reaction should be set up last.

Materials

1. 96 well tissue-culture plates (see Chapter 11)

2. SF cells at 1×10^6 cells/ml
3. Multi-(12)-channel pipet (e.g., Costar 12-pette)
4. Disposable channel trays to hold buffers
5. 0.5 N NaOH
6. 5 M ammonium acetate
7. Sterile ddH_2O
8. 0.1X TE
9. 400 µl microcentrifuge tubes certified for use in a thermal cycler
10. PCR reagent kit such as Genamp (Perkin Elmer Cetus # N801-0055). This includes AmpliTaq DNA polymerase, 10 mM stock solutions of all 4 dNTPs, and a 10X PCR buffer (100 mM Tris-HCl, pH 8.3; 500 mM KCl; 15 mM MgCl2; 0.01% gelatin).
11. Primers: The primers used will obviously depend on the particular application. We recommend the use of three primers (see Fig. 13-1), two flanking the cloning site in the transfer plasmid, and the third within the heterologous gene oriented in the antisense direction (to give an amplified product spanning the 5′ end of the gene). Dilute the primers to a concentration of 20 µM in sterile ddH_2O and store at –20° C.
12. Mineral oil (Sigma # M 5904)
13. Thermal cycler

Method

Preparation of DNA

1. Seed a 96 well plate with 1.5×10^4 SF cells/well.
2. Pick putative recombinant viruses identified by visual screening into separate wells. Try to pick at least 20 isolates in the first round of screening for each transfection mixture. Remember to inoculate some wells with the appropriate controls.
3. Incubate the plate in a humidified incubator at 27° C for at least two but no more than three days. This time period allows sufficient viral replication that will give sufficient template, without allowing the infection to proceed to the point where all the cells lyse and the DNA is lost in the tissue-culture fluid.
4. Transfer the viral supernatants with the multichannel pipet to a sterile 96 well plate. Store the viral supernatants at 4° C.
5. Lyse the cells in the plate by adding 200 µl of 0.5 N NaOH to each well.
6. Neutralize the lysate by adding 40 µl of 5 M ammonium acetate to each well.
7. Take 10 µl from each well and dilute with sterile ddH_2O or 0.1X TE. A tenfold dilution will probably work well, but this may be adjusted as necessary. The reaction requires ≈2 ng to 10 ng of template DNA. Do not use 1X TE as the diluent as the EDTA will chelate Mg^{++} ions necessary for optimal polymerase activity.

PCR Amplification

1. Prepare the PCR reaction mix as follows:
   56.5 µl ddH_2O
   10.0 µl 10X buffer
   2.0 µl each dNTP
   0.5 µl (2.5 u) AmpliTaq DNA polymerase
   5.0 µl each primer
   10.0 µl template DNA
   Overlay with 50 µl mineral oil per tube. Remember to set up all appropriate control reactions.
2. Amplify by subjecting to 25 to 35 cycles of the following series of conditions:

Denature at 95° C for one minute
Anneal at 55° C for one minute (see note)
Extend at 72° C for one minute

3. After the last cycle, incubate at 72° C for 10 minutes, then place at 4° C.
4. Withdraw 10 µl of the reaction mix and analyze on a 1.5% agarose gel.

> **Note:**
>
> The annealing temperature used will depend on the sequences of the primers used. Generally, a temperature 5° to 15° C below the predicted melting temperature (T_m) of primer/template duplex is suitable. The melting temperature may be estimated using the equation $T_m = 4(G + C) + 2(A + T)$ (Wahl et al., 1987). The specificity of the reaction can be altered when necessary by changing the annealing temperature.

RECOMBINANT AMPLIFICATION AND CONFIRMATION OF IDENTITY

Once you are satisfied that you have several pure isolates of a potential recombinant virus, you will need to generate large stocks of these isolates so that you can analyze them to confirm that you have obtained the required recombinant. Many of the considerations discussed in Chapter 12 for virus amplification are relevant here (e.g., avoiding defective particles, etc.). We recommend that you amplify and characterize at least three putative recombinants. This will help ensure that you find the required recombinant, as occasionally you will obtain viruses that do not express the cloned gene, although they display the correct phenotype. In addition, it is useful to check the levels of gene expression from several independent recombinants to find the recombinant that expresses best. Remember that if you only screened for a gain of phenotype, the majority of your isolates will be single-crossover recombinants. You should process a much greater number of these until you can ascertain which are double recombinants.

Recombinant Virus Amplification

Materials

1. Complete tissue-culture medium. (See Chapter 11 for details concerning media.)
2. SF cells grown to approximately 1×10^6 cells/ml
3. 35 mm and 100 mm tissue-culture plates (see Chapter 11)
4. Recombinant viruses purified by serial replaquing or by end-point dilution

Method

1. Seed a 35 mm tissue-culture dish with 1×10^6 cells in a final volume of 2 mls tissue-culture medium. Let attach for 30 minutes to 1 hour at 27° C. Alternatively, the cells may be seeded at 5×10^5 cells per 35 mm dish and used the following day.
2. Aspirate the tissue-culture medium from the plate. Infect with 0.5 ml purified inoculum and incubate at room temperature for one hour with gentle rocking. Dilute isolates purified by end-point dilution to 1 ml before beginning amplification.
3. Add 1.5 ml complete tissue-culture medium (there is no need to remove the virus inoculum) and incubate at 27° C for four days. Because this is a very low MOI infection, it will be difficult to see any cytopathic effects early after

infection. However, with an occ^+ virus, it should be possible to see some cells with occlusion bodies by three days pi.

4. Collect the tissue-culture medium and centrifuge at 1000 × g for five minutes at 4° C to remove cells and debris. Be careful to keep the tissue-culture fluid sterile. Store the supernatant at 4° C. This is designated the passage one stock.
5. Seed four 100 mm tissue-culture dishes with 5×10^6 cells each. The final volume in each plate should be 10 mls.
6. Dilute 1 ml of the passage one stock (from Step 4) with 3 mls of complete tissue-culture medium. Save the other 1 ml of passage one stock at 4° C as an emergency back-up.
7. Aspirate the tissue-culture medium from the plates and infect with 1 ml each of the diluted passage one stock. Incubate at room temperature for one hour with gentle rocking.
8. Feed with 9 mls complete tissue-culture medium each. Incubate at 27° C for four days.
9. Harvest the tissue-culture medium and centrifuge it to remove debris. Determine the titer of this passage two stock (Chapter 12) and take several 1 ml aliquots for storage at –80° C as long-term back-ups (Chapter 12). You may use 10 mls of this stock as a source of DNA to verify the structure of the recombinant (see the following protocols). The remainder of the stock should be kept at 4° C and only used as inoculum for working stocks. The cells remaining after production of the passage two stock may also be used as a source of DNA to verify the structure of the recombinant (Chapter 12).
10. A large-scale working stock of virus may be prepared by repeating Steps 5 through 9, using as many tissue-culture dishes as required and infecting with an MOI of 0.1 pfu/cell. Alternatively, cells may be infected in suspension as follows:
    a. Prepare a 200 ml spinner culture (Chapter 11) and grow until the cells reach a density of 1×10^6 cells/ml.
    b. Collect the cells by centrifugation at 1000 × g for five minutes at room temperature and resuspend very gently in 20 mls of complete tissue-culture medium.
    c. Add passage two virus (from Step 9) to give an MOI of 0.1 (therefore, add 1×10^7 pfu of virus). Incubate at room temperature for one hour with gentle rocking. Swirl periodically to mix.
    d. Return the cells and virus to the spinner bottle or flask and feed with 180 mls complete tissue-culture medium. Incubate at 27° C with stirring for three to four days. Harvest by centrifugation to remove cells and debris, and titer the supernatant as described in Chapter 12. Store at 4° C and use as a working stock.

Restriction Analysis and Southern Blotting

This section describes how to characterize the structure of your putative recombinant viruses at the DNA level. Your primary goal here is to confirm that the required segment of DNA has recombined into the appropriate site in the viral genome via an allelic replacement reaction. To do this, you will need to prepare viral DNA for each putative recombinant and analyze its structure by restriction endonuclease analysis and Southern blotting. Techniques for the preparation of viral DNA were described earlier in this chapter and are not reiterated here.

When characterizing a recombinant virus, you should try to confirm the following three features of the DNA structure:

1. The DNA should have the appropriate insert at the correct site in the genome.
2. The insert should have arisen via an allelic replacement reaction, not simply by the insertion of the entire transfer plasmid into the viral genome.

3. There should be no other alterations in the DNA restriction pattern as compared to wt virus.

These requirements dictate the particular restriction enzymes that you should choose to characterize your recombinants. Generally, you will want to cleave with enzyme(s) that flank the insert as well as with one or two enzymes that cut within the insert. It is useful if at least one of these enzymes cuts the genome frequently, such as HindIII or EcoRI. By examining the pattern obtained after cleavage with such an enzyme, you can gain some assurance that the viral genome has not been altered outside the recombination site. It will be easiest to characterize the recombinant virus if the physical map is known for all the enzymes you use.

Even if you have partially characterized your recombinant virus isolates by PCR analysis during screening, we recommend that you characterize the viral DNA by restriction endonuclease analysis. This helps ensure that you are not working with a recombinant that has become altered in some unexpected way, or that you are not working with a stock contaminated with defective interfering particles.

A linearized map of the viral genome, showing the cleavage sites of several common restriction enzymes, is shown in Figure 13-2. Figure 13-3 shows typical banding patterns obtained after agarose-gel electrophoresis of wt AcMNPV DNA, cut with some of these enzymes. The HindIII patterns of both L-1 and E2 variants of the virus are shown. This is the only restriction enzyme that is known to yield different patterns for the two variants (see Chapter 5). Table 13-2 lists the estimated sizes of the fragments observed in Figure 13-3. Figures 13-4 and 13-5 show expanded restriction maps of the regions flanking the AcMNPV *polh* and *p10* loci. The sequences of these regions are provided in Appendices 1 and 2.

TABLE 13-2 Sizes of AcMNPV Restriction Fragments (kbp)

Fragment	*HindIII*	*EcoRI*	*BamHI*	*XhoI*	*PstI*
A	20.00	14.20	86.50	29.20	25.70
B	22.00*	13.30	23.30	23.20	21.60
C	11.10	12.20	8.50	14.40	17.10
D	9.95	10.40	3.57	14.40	10.50
E	9.95	9.00	3.33	10.40	9.40
F	8.40	8.78	1.94	7.40	8.44
G	8.15	8.71	1.01	7.30	7.10
H	5.60	8.71		6.00	5.50
I	5.02	7.33		5.99	5.10
J	4.74	6.66		3.43	3.70
K	2.90	5.38		2.40	3.22
L	2.62	3.78		2.20	2.95
M	2.30	3.65		1.30	2.70
N	2.23	2.50		0.35	2.62
O	2.23	2.25			1.66
P	2.22	2.03			
Q	2.19	1.93			
R	1.79	1.48			
S	1.65	1.46			
T	1.04	1.33			
U	1.02	0.94			
V	0.93	0.94			
W	0.78	0.52			
X	0.77	0.51			

*In AcMNPV L-1, HindIII-B is divided into fragments B1 and B2 of ≈14.5 and 7.5 kbp.

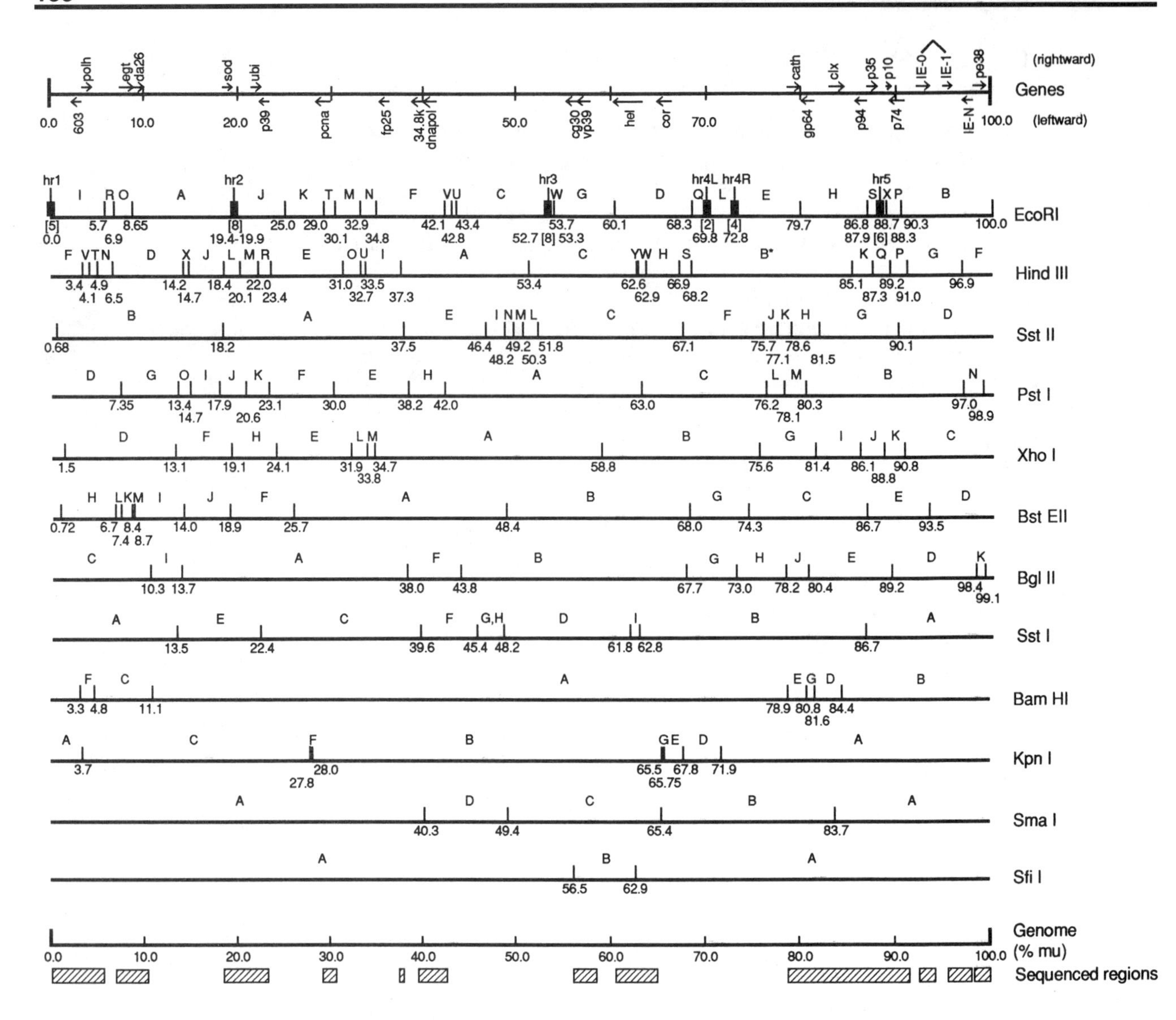

FIGURE 13-2 *Restriction map of AcMNPV genome.* This map shows the positions of the cleavage sites of a variety of different restriction endonucleases on a linear map of the AcMNPV (E2) genome. The positions of the cleavage sites are given in % map units (mu). A scale in mu is presented at the bottom of the figure. The entire genome is 128 kbp so that 1 mu = 1.28 kbp. The approximate locations of regions of the genome that have been sequenced are indicated at the bottom, while the locations and directions of transcription of characterized genes are presented at the top. The HindIII fragments are labeled as they were in Vlak and Smith (1982). However, it seems likely that HindIII-B (marked with an asterisk in the figure) is in fact larger than HindIII-A. In AcMNPV L-1 and C-6, an extra HindIII site occurs within HindIII-B, dividing it into fragments B1 and B2. Cochran et al. (1982) report the existence of a small XhoI fragment of ≈350 bp (XhoI-N). However, its location has not been clearly established and it is not marked on this map.

Illustrating the type of strategy used to confirm the identity of a recombinant virus, the following approach is the one we used to characterize the structure of vEV55p53, a recombinant virus expressing mouse p53 (O'Reilly and Miller, 1988). To construct this virus, a cDNA clone of p53 was inserted into the transfer plasmid pEV55, yielding pEV55p53. (The plasmid pEV55 is very similar to pEVmXIV; see Chapter 7.) The structure of pEV55p53 is shown at the top of Figure 13-6. Below

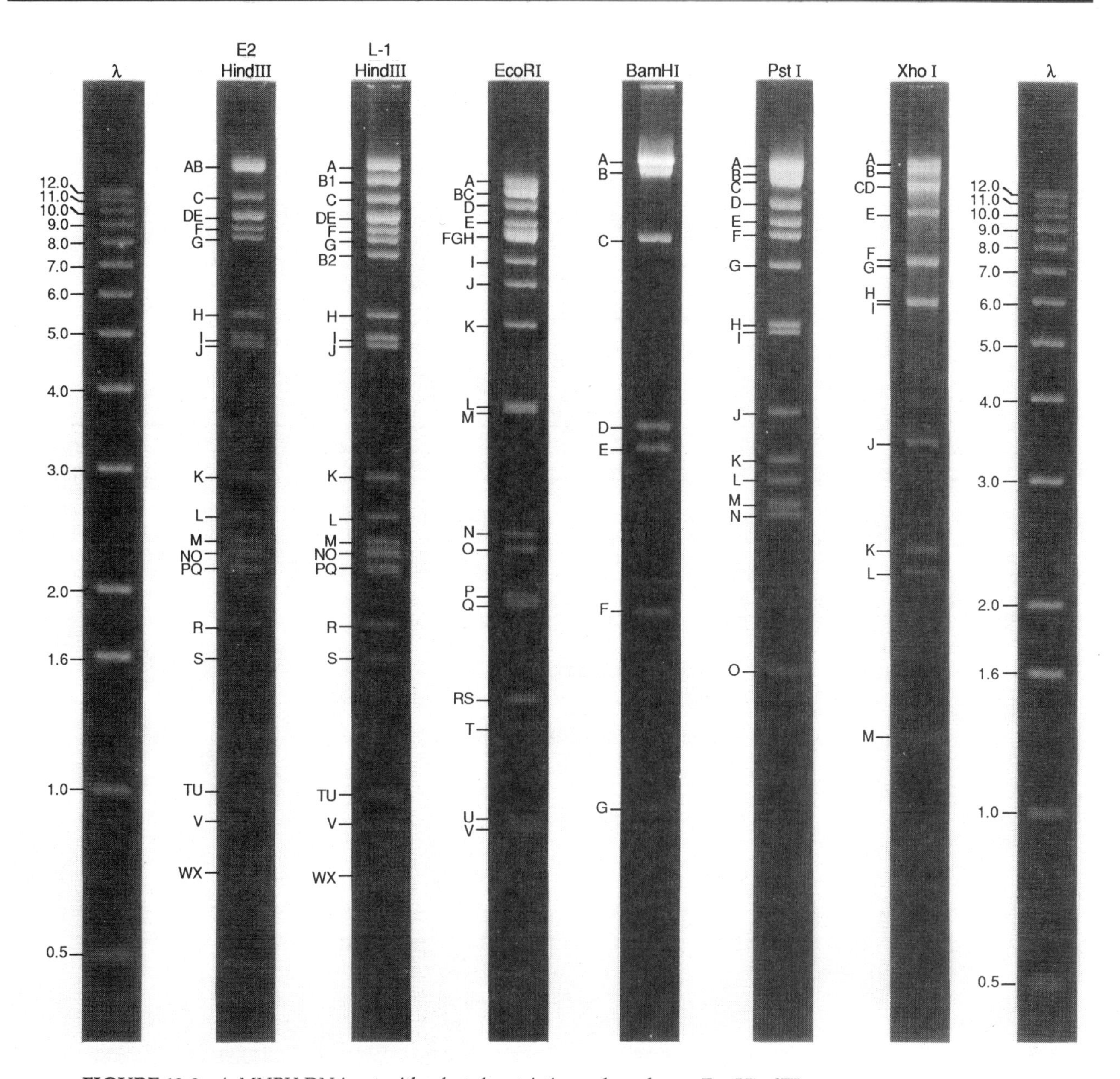

FIGURE 13-3 *AcMNPV DNA cut with selected restriction endonucleases.* For HindIII digestion, the patterns observed after cleavage of both AcMNPV L-1 and E2 are shown. All other cleavage patterns are of AcMNPV L-1. The fragment names are indicated beside each lane. Fragment sizes are given in Table 13-2.

this, a map of the wt viral DNA around *polh* is presented. The expected structures of the required double-crossover recombinant (vEV55p53), as well as both possible single-crossover recombinants, are shown underneath. In this case, we cleaved the DNAs with EcoRI and PstI. The sizes of the fragments expected are shown at the right side of the figure. Both EcoRI and PstI digestion show that the insertion has taken place in the correct region of the genome (EcoRI-I and PstI-D are disrupted). In addition, PstI cleavage results in a new band of 130 bp, which derives from within the cDNA. The presence of this band, which is also present in the transfer plasmid digest, confirms that the required fragment has been inserted. Both PstI and EcoRI distinguish between the required recombinant and either single-crossover recombinants or wt virus.

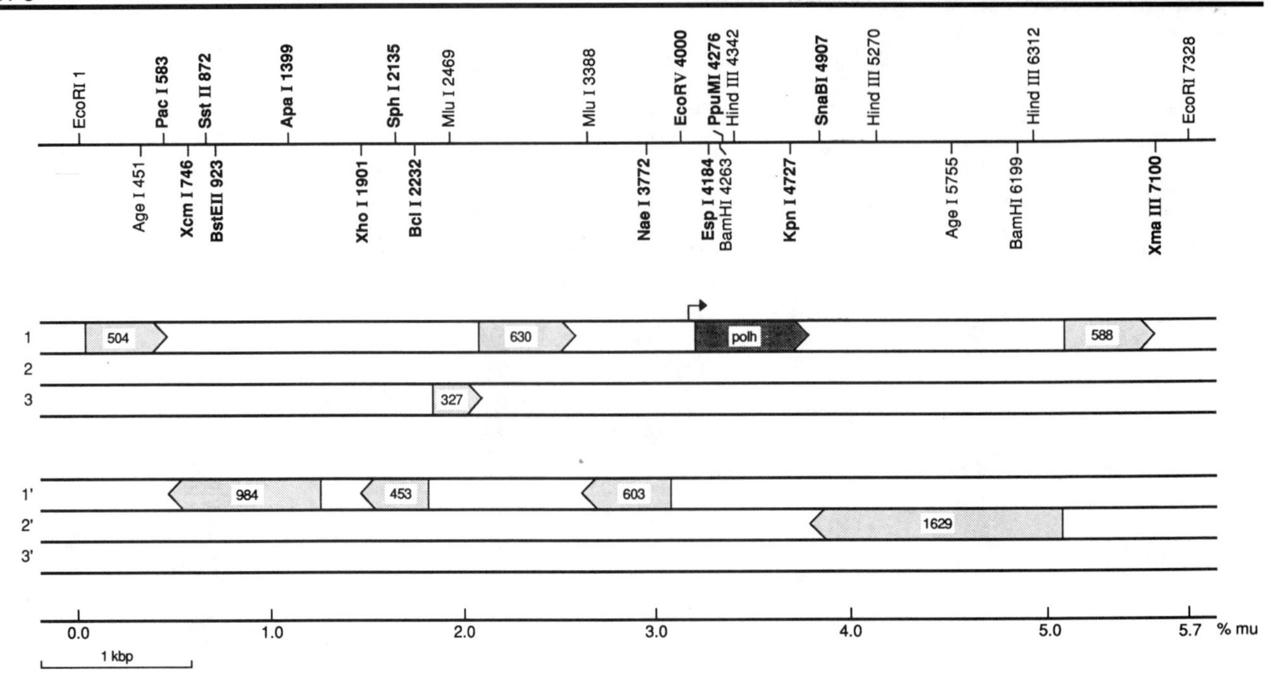

FIGURE 13-4 *Restriction map of the AcMNPV polyhedrin region.* This figure presents a map of the EcoRI-I fragment of AcMNPV. It is based on the sequence data of Possee et al. (1991). Unique restriction sites are in bold; nt 1 is the G of the EcoRI site at 0.0 mu. A scale in mu is presented at the bottom of the figure. Major ORFs in the region are illustrated underneath the restriction map. The arrow indicates the *polh* transcriptional start site; the *polh* coding region is represented by the darkly shaded box. Other ORFs are named according to Possee et al. (1991). The sequence of this region is presented in Appendix 1.

It is always worthwhile to analyze the restriction digests by Southern blotting and DNA hybridization. This analysis provides further confirmation that the appropriate fragment has been inserted, and it allows the visualization of bands that might otherwise not have been seen (e.g., because they were too faint or because they were obscured by wt bands of similar mobility). Generally, the transfer plasmid itself can be labeled and used as a probe. Because it contains viral sequences flanking the inserted gene, it will hybridize to the insertion site in the parent virus, as well as to the novel bands in the recombinant virus. In our example, all the bands discussed will hybridize with pEV55p53. Alternatively, the heterologous gene sequences themselves can be used as a probe. In this case, only the bands not enclosed in brackets in the figure will hybridize to p53 probe.

In the sections that follow, we provide guidelines for the restriction cleavage of viral DNA, agarose gel electrophoresis, Southern blotting, and hybridization to a labeled probe. All of the techniques used are basic molecular biology methods, and further details can be found in any standard molecular biology manual (e.g., Sambrook et al., 1989).

CONTROLS: You should always compare the restriction patterns of your recombinant virus with those of wt virus (and/or the parent virus if different), as well as with those of the transfer plasmid.

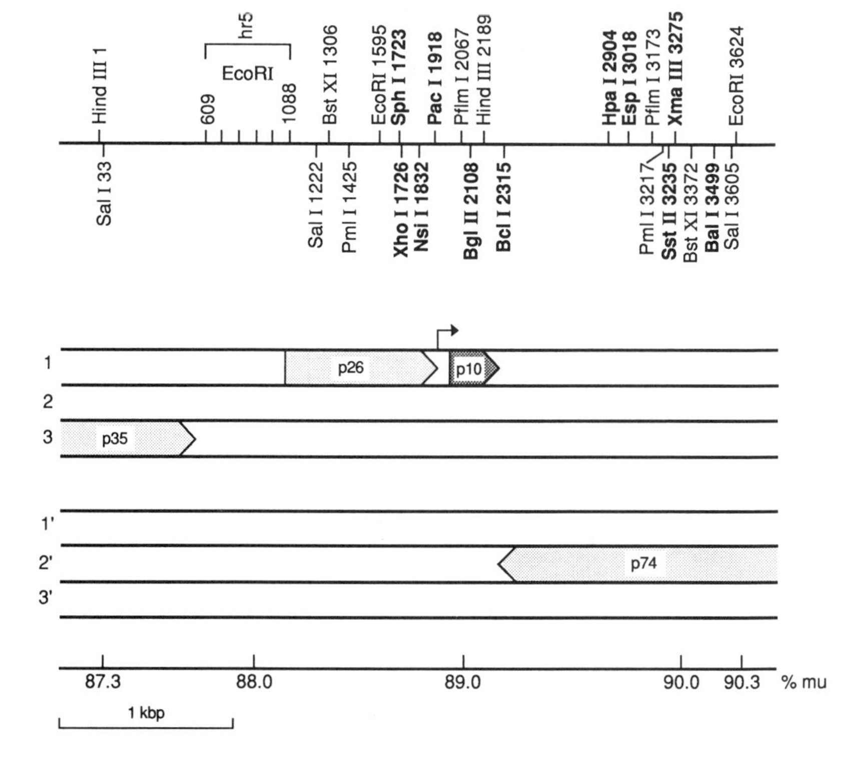

FIGURE 13-5 *Restriction map of the AcMNPV p10 region.* A map of the AcMNPV genome from the HindIII site at 87.3 mu to the EcoRI site at 90.3 mu is presented. A scale in mu is provided at the bottom of the figure. Unique restriction sites are in bold; nt 1 is the first A of the HindIII site at 87.3 mu. The map was prepared from the sequence of this region as compiled from data published by Lübbert and Doerfler (1984), Kuzio et al. (1984), Liu et al. (1986), Friesen and Miller (1987), and Kuzio et al. (1989). Guarino et al. (1986) have also published sequence data for the *hr*5 sequences within this region. Their sequence differs in a number of positions from that of Liu et al. (1986). It is not clear whether this reflects the different AcMNPV strains used. The differences do not alter the restriction map. ORFs in the region are indicated below the restriction map; *p10* coding sequences are represented by the darkly shaded box. The arrow indicates the *p10* transcriptional start site.

Restriction Endonuclease Digestion and Agarose Gel Electrophoresis

Materials

1. DNA: DNA from your putative recombinants and appropriate controls
2. Appropriate restriction endonucleases: Follow the manufacturer's instructions for correct storage conditions. Most, but not all, enzymes should be stored at –20° C.
3. 10X buffers for restriction enzymes: These are generally supplied with the enzyme.
4. Phenol-chloroform-IAA
   Caution: Phenol is highly corrosive and causes severe burns. Wear protective clothing and safety glasses. In case of contact with skin, apply 80% glycerol to the affected area and wash extensively with water.
5. 95% and 70% ethanol in ddH_2O
6. 3 M sodium acetate, pH 5.2
7. Agarose (SeaKem ME FMC # 50014)
8. 5X TBE: 450 mM Tris-borate, pH 8.0; 10 mM EDTA. Store at room temperature.
9. 10X TAE: 400 mM Tris-acetate, pH 8.0; 10 mM EDTA. Store at room temperature.

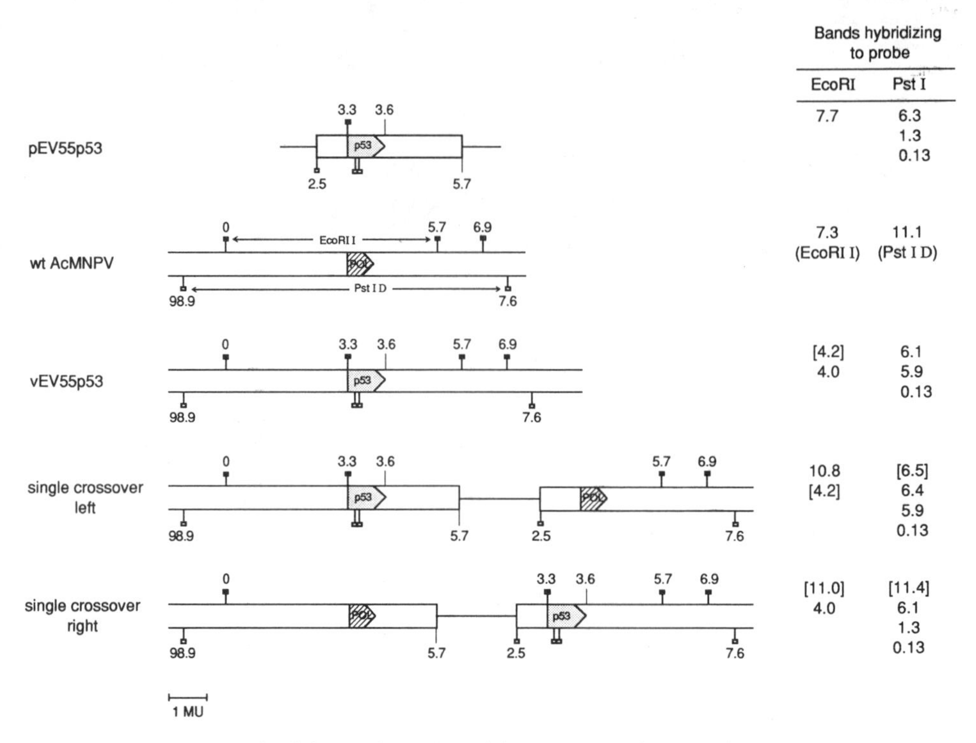

FIGURE 13-6 *Example of the confirmation of the structure of a recombinant virus.* The figure presents maps of the transfer vector pEV55p53, wt AcMNPV, the required recombinant vEV55p53, and the two possible single-crossover recombinants. The open rectangles represent AcMNPV sequences flanking the polyhedrin region; the thin line represents plasmid vector sequences; *polh* is denoted by the hatched arrow; the p53 gene is indicated by the dotted arrow; EcoRI sites are represented by the solid boxes above each map; PstI sites are represented by open boxes below each map; and the units given are map units. The numbers on the right indicate the sizes, in kbp, of the bands that would hybridize to the plasmid pEV55p53 following cleavage with either restriction enzyme. The brackets indicate bands that would not be detected if only p53 sequences were used as a probe.

10. 10 mg/ml ethidium bromide stock
    Caution: Ethidium bromide is moderately toxic and a powerful mutagen. Wear gloves when handling reagents containing this dye.
11. 5X loading buffer: 0.25% Bromophenol Blue (Bio-Rad # 161-0404); 0.25% Xylene Cyanole FF (Bio-Rad # 161-0423); 15% Ficoll (Pharmacia # 17-0400-01). Store at room temperature.
12. DNA slab-gel electrophoresis unit and generator
13. UV transilluminator with a Polaroid camera
    Caution: UV light is harmful, particularly to the eyes. Wear protective goggles or a safety mask that efficiently blocks UV light.

Methods

1. First, calculate the quantities of the reaction mixture components required for the cleavage reaction. Use 2 μg to 5 μg of viral DNA or 100 ng to 200 ng of plasmid DNA per digest. For viral DNA, we recommend cleavage with 3 to 4 units of enzyme per μg of DNA, three times for 2 to 3 hours each (i.e., a total of 10 to 12 units per μg DNA). One addition of 3 to 4 units of enzyme per μg of plasmid DNA should be adequate. The final volume of the reaction mix should be at least 10 times greater than the total volume of restriction enzyme to be added. (Restriction enzymes are typically supplied in 50% glycerol, which must be diluted to 5% or less during the cleavage reaction.) The volume of

10X reaction buffer used will be one-tenth the final reaction volume. Bring the reaction mix to the final volume with ddH_2O.

If you need to digest the DNA with two different enzymes, you should determine whether the enzymes cut efficiently in the same buffer (consult the manufacturer's specifications to see the compositions of the different buffers). If so, remember to allow for the amount of the second enzyme when calculating the final volume. Sometimes, it is possible to convert from one restriction buffer to another simply by adding salt. When this is feasible, make sure you start by cutting with the enzyme using the lower ionic strength buffer first. Again, allow for the addition of the second enzyme when calculating the total volume. If neither of these options are possible, you will need to precipitate the DNA after the first digestion. In this case, make up the reaction mix for one enzyme only.

2. Mix the DNA, 10X buffer and the appropriate volume of ddH_2O on ice. Add the enzyme last. If you need to dilute the enzyme, do so in ice-cold 1X reaction buffer immediately prior to use.
3. Incubate at the appropriate temperature for two to three hours. (For most enzymes, this will be 37° C, but it is important to check the manufacturer's instructions.)
4. For viral DNA only, add the same amount of enzyme twice more, and incubate for two to three hours each time.
5. If you are only cutting the DNA with a single-restriction enzyme, the DNA is now ready for analysis by agarose gel electrophoresis. For digestion with a second enzyme, adjust the salt concentration if necessary, add the first aliquot of the second enzyme, and proceed as in Step 3. If the second enzyme requires different conditions and the buffer cannot be easily converted, extract the DNA once with phenol-chloroform-IAA and precipitate by addition of two volumes of 95% ethanol plus 0.1 volumes of sodium acetate to the aqueous phase. Centrifuge at 12,000 × g for 10 minutes in a microfuge at 4° C. Discard the supernatant and carefully rinse the pellet with 70% ethanol. Air-dry briefly, resuspend in a small volume of TE, and prepare a second reaction mix as described in Steps 1 and 2. (Digested viral DNA will resuspend rapidly.)
6. While the DNA is being digested, you may prepare the agarose gel. Seal the ends of the gel tray of the electrophoresis apparatus in the appropriate manner. (This differs with different gel apparatus designs, but usually it is only necessary to tape the ends with 3 M scotch tape.) Prepare a solution of 0.6% agarose in 1X TBE, containing 200 μg/l ethidium bromide. (TBE is usually satisfactory for all applications, but some workers find that TAE provides better resolution of higher molecular weight bands. The buffering capacity of TAE is exhausted more rapidly, so it should be recirculated if electrophoresis is carried out for long periods of time.)

   Melt the agarose by microwaving for 2 to 5 minutes. Ensure that the solution is completely clear. Cool to 65° to 70° C and pour into the sealed gel tray with the comb in place. Let harden for 30 to 45 minutes at room temperature. Carefully remove the comb. Remove the seals (tape) from the gel and place it in the electrophoresis tank. Fill the tank with 1X TBE plus 200 μg/l ethidium bromide so that the gel is submerged. We recommend submerging the gel to a volume of at least 1 cm. This helps prevent viscous DNA samples from coming out of the well after loading.
7. Often, the volume of the DNA sample after digestion is too large to load into a single slot in the gel. In this case, extract the sample once with phenol-chloroform-IAA and ethanol precipitate. Resuspend in a small volume of TE.

8. Add 0.2 volumes loading buffer to each sample and mix briefly. Carefully pipet the samples into the wells in the gel. The Ficoll in the loading buffer will cause the sample to sink to the bottom of the well. Try to avoid breaking the surface tension of the buffer immediately above the wells when you withdraw the pipet tip. This can cause viscous samples to come back out of the well. It is preferable to move the pipet tip away from the wells before withdrawing it from the buffer.
9. Connect the electrophoresis unit to a power supply. The DNA will migrate towards the anode, so the *red lead* should be connected to the end of the gel furthest from the wells.
10. Run the gel at a voltage of 1 to 2 V/cm. Generally, optimal resolution is obtained by electrophoresis at low voltages for long periods of time. (Remember that it may be advisable to recirculate the buffer if you are using TAE during prolonged electrophoresis.) The optimal length of time for electrophoresis will depend on the size of the fragments you wish to resolve and will need to be determined empirically. Bromophenol blue migrates at a position corresponding to a DNA fragment of approximately 300 bp, whereas xylene cyanole corresponds to a fragment of approximately 4 kbp. However, these dyes are generally difficult to see after prolonged electrophoresis.
11. Photograph the gel using a Polaroid camera and UV transilluminator. Closely study the restriction pattern obtained, identifying the bands by comparison to the patterns shown in Figure 13-3. Attempt to verify that you can see all the novel bands you predicted for your recombinant, and that all viral bands that should have been disrupted are no longer visible.

    Occasionally, you may see weaker bands of unexpected sizes in the restriction patterns. These either represent partial digestion products or they may indicate the presence of a second viral genome type in the original DNA preparation. These possibilities may be quite difficult to distinguish. If you only observe such bands with one of the restriction enzymes, they most likely represent partial digestion products and can be discounted. Otherwise, we recommend doing the digestions again to see if they are consistently present. Southern blot analysis may also help to clarify the nature of such bands (see the following protocol). If you determine that there is more than one virus type in your recombinant stock, it will be necessary to further purify the virus (Chapter 12) and prepare fresh DNA.

Southern Transfer

The analysis of DNA fragments by hybridization is greatly facilitated by transferring the DNA in a denatured form to a solid support. This technique was pioneered by Southern (1975) and is therefore known as Southern transfer or Southern blotting. Originally, nitrocellulose filters were used as the solid support matrix. More recently, modified nylon membranes have become the material of choice. The transfer protocol is very simple and involves the denaturation of the DNA in the gel, followed by its transfer to the membrane by one of a number of methods. The most commonly used method, described here, is transfer by capillary action. However, vacuum transfer is becoming increasingly popular as a means for the rapid quantitative transfer of DNA to the membrane. If you choose to use vacuum transfer, we recommend you follow the manufacturer's instructions supplied with the transfer apparatus.

A wide range of modified nylon membranes are available from different suppliers. The protocols for using these membranes may differ slightly, so the user is therefore advised to consult the information provided by the manufacturer. The following protocol is one that we use successfully with Zeta-probe (BioRad) membranes:

Materials

1. DNA samples electrophoresed on an agarose gel
2. 0.25 N HCl
3. Denaturation solution: 0.5 N NaOH, 1.5 M NaCl. Store at room temperature.
4. 20X SSC
5. Nylon membrane (Zeta-probe or equivalent) cut to the size of the gel. Wear gloves at all times when handling the membrane as grease from your fingers will block transfer of the DNA.
6. Several sheets of Whatman 3MM paper cut to the size of the gel

Method

1. Soak the gel in several volumes of 0.25 N HCl for 15 minutes with gentle rocking. This treatment will partially depurinate the DNA and helps in the transfer of larger fragments.
2. Rinse the gel in ddH_2O and soak the gel in several volumes of denaturation solution for 30 minutes with gentle rocking. During this Step, the DNA is denatured, and the depurinated residues are hydrolyzed.
3. Lay a sheet of Saran Wrap or other plastic film on a flat surface. Place the gel *upside down* on the plastic (see note).
4. Wet the membrane in ddH_2O. (If the membrane does not saturate completely, do not use it; or, at least, place the unwetted area over a part of the gel that does not contain bands of interest.)
5. Flood the surface of the gel with denaturation buffer and lay the membrane on top of the gel, taking care to exclude air bubbles.
6. Wet one sheet of Whatman paper and lay it on top of the membrane.
7. Roll a pipet across the Whatman paper, applying constant gentle pressure. This pressure helps force out any air bubbles that may have become trapped between the layers.
8. Place the remaining sheets of dry, Whatman filter paper over the wet sheet, and fold the Saran Wrap over the edges so that it covers approximately 5 mm of the top of the pile on all four sides. This coverage ensures that the buffer can transfer from the gel to the paper towels only by passing through the membrane.
9. Place a stack (5 cm–10 cm high) of paper towels over the Whatman filters. Put a glass plate on the stack and place a weight of approximately 500 g on top. The weight serves to ensure close contact between all layers of the assembly. Although the transfer will probably be largely complete in three to four hours, we recommend letting it proceed overnight to ensure efficient transfer of larger DNA fragments.
10. With a pencil, mark the position of the wells on the membrane. These marks will help later in orienting the blot. Peel the membrane from the gel (which is now extremely thin) and soak the membrane in 6X SSC for five minutes. The efficiency of transfer can be evaluated by staining the gel in a 200 μg/l solution of ethidium bromide for 20 to 30 minutes and then examining it under a UV light. The bands should be barely visible. Alternatively, it is possible to see the transferred bands by UV illumination of the membrane.
    Caution: Ethidium bromide is moderately toxic and a powerful mutagen. Wear gloves when handling reagents containing this dye.
    Caution: UV light is harmful, particularly to the eyes. Wear protective goggles or a safety mask that efficiently blocks UV light.
11. Bake the membrane at 80° C for two hours under vacuum, and store dry at room temperature until use.

Note:

You can achieve more efficient transfer of the DNA from the gel by placing the gel on a wick composed of Whatman 3MM paper immersed at the ends in 6X SSC. However, for Southern blotting of viral DNA, the capillary action, generated from the buffer in the gel, adequately transfers enough DNA to be very easily detected by hybridization. Thus, the more rapid method of placing the gel directly on a plastic film works well in this case.

Probe Labeling and Hybridization

Specific DNA fragments immobilized on nylon membranes may be detected by hybridization to a DNA or RNA probe that has been labeled in some manner. When confirming the structure of a recombinant baculovirus, the most convenient molecules to use as probes are either the entire transfer plasmid, or the heterologous DNA fragment originally cloned into that plasmid (see Fig. 13-6). Thus, we will not describe the use of RNA or oligonucleotide probes further in this section.

The basic principle of the protocol described here is to carry out the hybridization under conditions that allow the probe to bind to complementary sequences on the membrane, while minimizing binding to noncomplementary sequences. These requirements are generally satisfied by hybridizing approximately 20° to 30° C below the melting temperature (T_m) for the hybrid in question. The T_m is affected by the incubation conditions, the size and composition of the probe, and the degree of sequence identity between the probe and the target molecule (see note).

The applications discussed here deal with probes that are always 100% identical to the target sequence and moderately large. The small differences between the lengths and the G + C contents of the probes that are likely to be used will not affect the T_m very dramatically. Thus, it is possible to describe a set of hybridization conditions that should work well in all cases. Readers interested in a more extensive discussion of the parameters of hybridization of nucleic acids on a solid support are directed to the review by Wahl et al. (1987).

A wide range of techniques is available for labeling DNA, with either radioactive or nonradioactive probes. Following is a description of a single protocol for probe labeling, using the random, oligonucleotide priming method. However, many of the other techniques are also suitable for labeling probes for blot-hybridization experiments, and researchers may use whatever protocol they prefer. Other labeling methods are described extensively in molecular cloning manuals (Sambrook et al., 1989).

Random Primer Labeling

Random primer labeling of DNA, developed by Feinberg and Vogelstein (1983), involves the hybridization of a mixture of short heterogeneous oligodeoxyribonucleotides to the DNA to be labeled. These oligodeoxyribonucleotides bind at many positions on the template DNA, and function as primers for the initiation of DNA synthesis when DNA polymerase and deoxynucleotide triphosphates (dNTPs) are included. Use of a labeled dNTP as one of the precursors allows the synthesis of very high specific-activity, uniformly labeled probes. Typically, specific activities of 5×10^8 to 2×10^9 cpm/μg are obtained.

Materials

Many kits, such as the "Random Primers DNA Labeling System" from BRL, are commercially available and have all the reagents required for the labeling reaction, with the exception of the [α–^{32}P]dNTP. Alternatively, the individual reagents required are described here.

1. DNA: 10 ng to 100 ng of the DNA to be labeled
2. dNTPs: dATP, dGTP, and dTTP may be obtained as 100 mM aqueous solutions from Sigma (# D 4788, D 5038, and T 9656). Prepare 10 mM solutions in 3 mM Tris-HCl, pH 7.0, 200 μM EDTA and store at –20° C.
3. Random primers (e.g., BRL # 8190SA). These should be stored at a concentration of approximately 90 OD units/ml (3 mg/ml) in 3 mM Tris-HCl, pH 7.0, 200 μM EDTA at –20° C.
4. 10 mg/ml BSA
5. Reaction buffer: 300 mM Tris-HCl, pH 8.0; 30 mM $MgCl_2$; 62.5 mM β-mercaptoethanol; 120 μM each dGTP, dATP, and dTTP; 1.25 M HEPES, pH 6.6; and 33 OD units/ml random primers. Store at –20° C.
6. [α–^{32}P]dCTP (3000 Ci/mmol, NEN # NEG013C). Store at –20° C.
7. Klenow fragment (large fragment) of *E. coli* DNA polymerase I (BRL # 8012SA). Store at –20° C.
8. 10% SDS
9. TE
10. Whatman DE81 filters
11. 0.15 M NaCl

Method

1. Dilute 10 ng to 100 ng of the template DNA to 30 μl with ddH_2O. Denature by incubation at 100° C for three to five minutes.
2. Cool to 42° C and add the following reagents:
   10 μl reaction buffer
   2 μl 10 mg/ml BSA
   5 μl (50 μCi) [α–^{32}P]dCTP
   3 units Klenow DNA polymerase
   ddH_2O to a final volume of 50 μl
3. Incubate at room temperature for at least three to four hours. Because the reaction is carried out under conditions that minimize the 3′ to 5′ exonuclease activity of the enzyme (i.e., at pH 6.6), the reaction can be allowed to proceed for longer periods of time. Generally, an overnight incubation is convenient.
4. Stop the reaction by adding the following:
   2 μl 10% SDS
   1 μl 0.5 M EDTA
   47 μl TE
5. There is no need to purify the unincorporated label away from the probe. Determine the incorporation of radiolabel by adsorption to Whatman DE81 filters, as follows:
   a. Place 1 μl of the labeled reaction mix on each of two DE81 filters and allow to dry.
   b. Wash one filter extensively in several changes of 0.15 M NaCl. During this step, unincorporated nucleotides are washed off the filter. Allow to dry.
   c. Measure the radioactivity on both filters. The ratio of the radioactivity on the washed to the unwashed filter is the proportion of the radiolabel that was incorporated. From this, you may calculate the specific activity of the labeled probe.
6. Store the probe at –20° C until use. Very high specific-activity probes degrade quite rapidly due to radiochemical decay, and so the probe should be used soon after labeling.

Hybridization

Materials

1. DNA immobilized to membrane
2. 20X SSC
3. 10% SDS
4. 50X Denhardt's reagent: 1% (w/v) Ficoll (Type 400, Pharmacia # 170400-01); 1% (w/v) polyvinylpyrrolidone (Sigma # P5288); 1% (w/v) bovine serum albumin (Sigma # A7030) in ddH_2O. Filter-sterilize and store at –20° C. Alternatively, prepared Denhardt's reagent is available commercially (e.g., Sigma # D 2532)

   OR

   1X BLOTTO: 5% (w/v) nonfat dried milk; 0.02% sodium azide. Store at 4° C. **Caution:** Sodium azide is poisonous. Label all solutions clearly and handle with care.
5. 10 mg/ml denatured salmon-testes DNA (Sigma # D7656). This is supplied as a sonicated, denatured solution that has been phenol-chloroform-extracted and ethanol-precipitated.
6. Hybridization solution: 6X SSC; 5X Denhardt's or 0.05X BLOTTO; 0.5% SDS
7. Heat-sealable bags (e.g., seal-a-meal bags)
8. Wash solution: 2X SSC; 0.1% SDS

Method

1. Wet the blotted membrane by floating on a solution of 2X SSC, 0.1% SDS.
2. Seal the membrane in a heat-sealable plastic bag. Cut one corner and add 150 µl of hybridization solution/cm^2 of membrane. Seal the bag, trying to eliminate all the air bubbles. Submerge the bag in a 68° C water bath and incubate for at least 10 minutes (it can be much longer—several hours to overnight).
3. Prepare the probe as follows:
   a. The probe will be used at a concentration of 1 ng to 2 ng/ml of hybridization solution. Put the required amount of probe in a microfuge tube.
   b. Add denatured salmon-testes DNA to the probe so that the concentration in the hybridization solution will be 100 µg/ml.
   c. Denature the probe and salmon-testes DNA by incubation at 100° C for five minutes. Place immediately on ice.
4. Remove the hybridization solution from the membrane and add an equal volume of fresh hybridization solution (preheated to 68° C) to the bag. Add the denatured probe/salmon-testes DNA mix and quickly seal the bag, removing all air bubbles. To minimize the possibility of a leak of radioactive solution, seal the bag in a second heat-sealable bag. Submerge in a 68° C water bath and incubate overnight with gentle rocking.
5. Carefully remove the hybridization solution from the bag. Then, remove the membrane and place it immediately in a dish containing a large volume of 2X SSC, 0.1% SDS at 68° C. Incubate at 68° C with agitation for 30 minutes.
6. Repeat this wash step twice more, then rinse briefly in 2X SSC at room temperature.
7. Label the membrane asymmetrically with radioactive ink (waterproof ink with old ^{32}P added to give approximately 1000 cpm/µl). Wrap the membrane in Saran Wrap or other plastic film. *Do not* allow the membrane to dry after washing.
8. Expose the membrane to X-ray film (Kodak X-OMAT AR or equivalent) at room temperature. The appropriate exposure time will have to be determined empirically. Develop the film after one to two days, and then reexpose as necessary. You can reduce the exposure time by using an intensifying screen.

Place the film between the membrane and the screen and expose at –70° C. If you use an intensifying screen, exposure from several hours to overnight should be adequate. The signal from the radioactive ink will allow you to line up the autoradiograph with the membrane to facilitate the identification of the bands.

Note:

The T_m of a DNA-DNA hybrid (the temperature at which 50% of the hybrid is dissociated into single strands) can be estimated from the equation:

$$T_m = 81.5^{o} + 16.6(\log[Na^{+}]) + 41(\text{mole fraction } G + C) - 500/L - 0.62(\%\ \text{formamide})$$

where L is the length in base pairs of the hybrid (Wahl et al., 1987). This relationship is valid for hybrids greater than approximately 50 bp. The T_m decreases by about 1° C for every 1% of base pairs that are mismatched between the two hybrids.

14

CHARACTERIZING RECOMBINANT GENE EXPRESSION

Once you have successfully constructed your recombinant baculovirus expression vector and have confirmed that it has the appropriate genome structure, you may want to characterize the protein being synthesized and evaluate the levels and kinetics of its production. This chapter provides a series of protocols for the analysis of heterologous gene expression and protein production in infected insect cells. These include the characterization of gene transcription by Northern blot analysis and the analysis of the protein product by metabolic labeling, immunoprecipitation, SDS-polyacrylamide gel electrophoresis, immunoblotting, and immunofluorescence. The next chapter describes some methods for studying the extent of post-translational modification and the processing of the over-produced protein. Appendix 3 provides a list of all recombinant proteins that have been produced in the baculovirus system up to late 1991. This list is worth consulting to see if any proteins similar to the one you are interested in have been produced in the system. If so, the published report might include useful protocols and other information relevant to the particular protein you wish to characterize.

IDENTIFYING PROTEINS AND KINETICS OF EXPRESSION

Several approaches may be used to evaluate whether your recombinant virus is producing the required protein and, if so, to determine the levels and kinetics of expression. The simplest of these approaches is the analysis of total infected cell proteins by staining SDS-polyacrylamide gels. Information about the kinetics of expression can be obtained by radiolabeling the cells for a brief period prior to lysis, and then analyzing the samples by SDS-PAGE. These protocols can be used in combination with various immunological techniques (immunoprecipitation and immunoblotting) to confirm that the required protein is being expressed. Alternatively, rapid confirmation of the identity of the expressed protein can be obtained by immunofluorescence staining of infected cells. This technique also provides information on the subcellular localization of the expressed protein. If you can assay specifically for an activity of the expressed protein, you can confirm that active protein is being produced as well as obtain information on the levels and kinetics of expression.

Even if you confirm the identity of the expressed protein by immunofluorescence analysis or a specific assay, we recommend also evaluating protein expression by one or more SDS-PAGE-based approaches. Analysis by SDS-PAGE provides important information on the size of the expressed protein (which may reflect post-translational modification, processing, and so on; see discussion in Chapter 15). In addition, characterization of whole-cell lysates (either by Coomassie blue staining or by pulse-labeling and fluorography) is a useful method to monitor the course of the viral infection and to verify that your recombinant virus does not display any aberrant or unusual features.

Analysis of transcription of the cloned gene may also be informative. This approach provides an additional view of heterologous gene expression and may also be useful for the diagnosis of problems when protein synthesis is low. For example, Carbonell et al. (1988) found extremely high levels of scorpion-toxin gene transcripts in infected cells even though the protein product was barely detectable. The isolation and Northern blot analysis of mRNA from baculovirus-infected cells are described at the end of this chapter.

Time Course of Total Cell Protein Production

This section describes the analysis of total proteins present at various times after infection. For most proteins that are not secreted, expression levels are generally such that the protein can be easily visualized by Coomassie blue staining of total protein on SDS-polyacrylamide gels. On the other hand, if the protein is secreted, the extra dilution in the tissue-culture fluid generally makes it quite difficult to detect by Coomassie blue staining. More sensitive staining procedures, such as nickel staining or silver staining, can be used. However, these protocols are often technically difficult, and we find it easier to circumvent the problem by radiolabeling the cells prior to analysis by immunoprecipitation and/or SDS-PAGE. If information is needed on the steady-state levels of a secreted protein (such information cannot be derived from the analysis of pulse-labeled proteins), we recommend immunoblot (Western blot) analysis, provided a suitable antibody is available.

The following protocol describes the infection of cells and the collection of whole-cell lysates at various times postinfection (pi). These lysates are suitable for further characterization by immunoprecipitation and/or SDS-PAGE, followed by Coomassie blue staining or immunoblot analysis (described in subsequent sections). A critical factor in carrying out these time course experiments is the multiplicity of infection (MOI) used. To obtain reproducible time courses, it is essential that all cells in the population are infected synchronously. For 99% of the cells in a population to receive at least one infectious unit, an MOI of at least five plaque-forming units (pfu)/cell must be used (see note). We recommend using an MOI of 10 to 20, which not only ensures a synchronous infection but also drives viral replication at its fastest rate. It is important to remember that the titer of a virus must be determined on the cell line used in the time-course experiment. The same virus stock will not necessarily have the same titer on different cell lines. We recommend that you retiter your virus stocks at fairly frequent intervals (e.g., every two to three months).

The condition and density of the cells are also important parameters to monitor for reproducible kinetics and optimal levels of foreign gene expression. The cells should be healthy and actively growing at the time of infection. If the cells are too densely spaced on the plate, the late phase of the viral infection may be compromised. As the infection proceeds, the cells will not adhere as well to the substrate. Thus, extra care is needed to avoid excessive loss of cells late in the infection cycle.

The times at which you collect your samples will depend on the promoter driving expression of your foreign gene. The time courses in Figure 14-1 are typical of what you can expect with polyhedrin (*polh*) or *p10* promoter-driven expression. In this experiment, recombinant viruses expressing SV40 T antigen and mouse p53 under the control of the *polh* promoter were compared to wt AcMNPV during infection of SF21 cells (O'Reilly and Miller, 1988). In panel A, whole-cell lysates were made at 12, 24, 36, and 48 h pi and analyzed by Coomassie blue staining of SDS-polyacrylamide gels. Polyhedrin migrates as a 32 kDa protein on SDS gels and

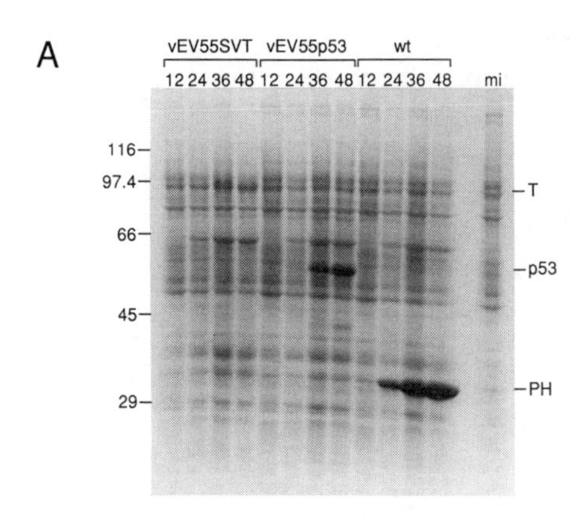

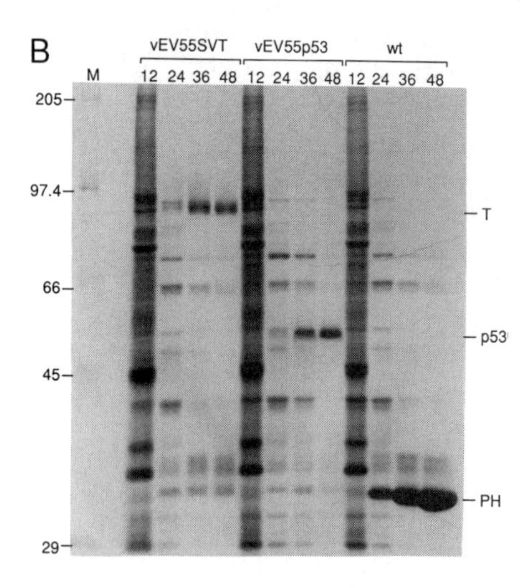

FIGURE 14-1 *Time course of protein synthesis in baculovirus-infected cells.* SF-21 cells were infected with AcMNPV L-1 (wt) or with recombinant viruses expressing SV40 T antigen (vEV55SVT) or mouse p53 (vEV55p53). The cells were lysed at 12, 24, 36, and 48 h pi. Panel A shows total cell protein as revealed on a Coomassie blue-stained gel whereas panel B is a fluorograph of [^{35}S]methionine-labeled samples. The positions of T antigen, p53, and polyhedrin (PH) are shown. An uninfected cell lysate is shown in lane mi and molecular weight markers are in lane M. Reprinted with permission from O'Reilly and Miller (1988), © ASM.

is easily detectable by 24 h pi with wt AcMNPV. It accumulates to extremely high levels at 36 h pi and 48 h pi. In some experiments, discrete bands of higher mobility than polyhedrin (10-25 kDa) are also observed in occ$^+$ virus-infected samples. These bands probably represent polyhedrin breakdown products. Although not expressed to the same high levels, both recombinant proteins show a similar pattern of expression. In panel B, the cells were pulse-labeled with [^{35}S]methionine prior to lysis, and the samples were analyzed by fluorography of SDS-polyacrylamide gels. Synthesis of all three proteins has begun by 24 h pi and continues through 48 h pi. Note that the background of cellular proteins that are labeled decreases dramatically from 12 h pi to 24 h pi with all three viruses. This is due to the virus-mediated shut-off of host-cell protein synthesis (see Chapter 3).

If expression of the foreign gene is driven by a late promoter or by a late-very-late fusion promoter (in cases where the transfer plasmids pAcMP1 or pc/pS1 were used), the time of onset of expression of the foreign gene should be advanced by 6 to 12 hours. In these cases, it might be desirable to include an 18 hour time point in the time course.

CONTROLS: The controls performed in a typical time course experiment should include an identical time course performed with wt virus (and/or the parent virus, if different), as well as one dish of mock-infected cells.

Materials

1. Cells: SF cells grown to approximately 1×10^6 cells/ml
2. Virus: recently titered stocks of your recombinant virus and control viruses
3. Complete tissue-culture medium. (See Chapter 11 for details concerning the various media available.)
4. 35 mm tissue-culture dishes (see Chapter 11)
5. PBS, pH 6.2
6. Protease inhibitors
   a. PMSF (Phenylmethylsulfonyl fluoride, Fluka # 78830); 10 mg/ml in isopropanol. Store at room temperature. Use at a final concentration of 100 μg/ml. PMSF is not stable in aqueous solution and so should only be added to the lysis buffer *after* lysis of the cells.
   Caution: PMSF is very harmful and may be fatal if inhaled, swallowed, or by skin contact. Wear protective clothing and a mask when handling PMSF.
   b. Pepstatin A (Fluka # 77170); 100 μg/ml in methanol. Store at –20° C. Use at a final concentration of 1 μg/ml.
   c. Leupeptin (Fluka # 62070); 50 μg/ml in triple-distilled deionized water (ddH_2O). Store at –20° C. Use at a final concentration of 0.5 μg/ml.
7. NP40 Lysis buffer: 1% NP40; 150 mM NaCl; 50 mM Tris-HCl, pH 8.0. Store at 4° C.
8. 2X SDS gel loading buffer: 4% SDS; 125 mM Tris-HCl, pH 6.7; 30% (v/v) glycerol; 0.002% (w/v) bromophenol blue. Store at 4° C in small aliquots. Before use, add β-mercaptoethanol to a concentration of 2% (v/v). Use each aliquot within two to three weeks of adding the β-mercaptoethanol.

Method

1. Seed 35 mm tissue-culture dishes with 1×10^6 cells per dish. You will need one dish per time point for each virus you analyze, as well as one dish for a mock-infected control. The final volume of tissue-culture medium should be 2 mls per dish. To help disperse the cells evenly, it is best to add some tissue-culture medium to the dishes before the cells, or dilute the cells to their final volume before plating. Rock each dish gently *immediately* after addition of the cells to ensure an even monolayer. Incubate at 27° C for 30 minutes to one hour to allow the cells to attach. Alternatively, the cells may be seeded at a density of 0.5×10^6 cells per dish and incubated at 27° C overnight.
2. Dilute the viruses in tissue-culture medium to give 2×10^7 pfu in 250 μl to 500 μl. This will give an MOI of 20 pfu/cell. Use the same final volume for all viruses.
3. After the cells have attached, aspirate the tissue-culture medium from the cell monolayers and infect with the appropriate diluted virus. *Be careful not to let the cells dry out before addition of the virus.* For the mock-infected control, add tissue-culture medium instead of virus. Incubate at room temperature for one hour with gentle rocking.
4. Aspirate the inoculum from the cells and refeed with 2 mls complete tissue-culture medium. Incubate at 27° C.
5. Examine the cells at intervals under the inverted tissue-culture microscope to monitor the course of the infection. Between 6 h pi and 12 h pi, you should see many blebs appearing on the surface of infected cells, giving them a ruffled, bumpy appearance. You may be able to see enlargement of the nucleus, although this is more difficult to visualize. The blebbing disappears from 12 h pi onwards and, by 24 h pi, the cells will be smooth again. The nucleus will appear more granular by 12 h pi due to the formation of the virogenic stroma.

Occlusion bodies will begin to appear in cells infected by occlusion-positive (occ^+) viruses from 18 h pi to 24 h pi. Initially, they will appear as tiny dark bodies in the nuclei of only some of the cells. However, by 36 h pi, all cells should have many crystal-like refractive bodies clearly visible in the nucleus. Cells infected by occ^- viruses will not show much morphological change from 24 h pi onwards. In some cases, the fibrous sheets (due to *p10* expression) in infected cells may be seen as refractive, elongate, and slightly curved or wavy bodies. From 36 h pi, some infected cells will begin to detach from the substrate. Lysis begins to occur from 60 h pi to 72 h pi, with the cells breaking up into numerous small vesicles.

It is important to monitor the progress of the infection carefully during the time-course experiment. Any problems (e.g., incorrectly titered virus, unhealthy cells, etc.) will often be apparent at this stage, and you can save yourself a considerable amount of work by identifying the problem early.

6. At each time point, aspirate the extracellular fluid from one dish infected by each virus. Process the mock-infected cells at the same time as the first time point. Place the dishes on ice, and rinse twice with 2 mls of ice-cold PBS. After the second rinse, stand the dishes on edge for a few minutes, then remove any PBS that has drained to the bottom.
7. The lysis procedure you use will depend on the protein you are expressing. The simplest approach, suitable for most purposes, is to lyse the cells in SDS gel-loading buffer. Add 50 µl of 1X SDS gel-loading buffer to each dish. Scrape the lysate off the dish with a rubber policeman or similar implement, and stand the dish on edge for a few minutes. Collect the lysate and store at –20 or –80° C. This procedure solubilizes most of the proteins in the cell.

   If your recombinant virus is an occ^+ virus, you may wish to choose lysis conditions in which polyhedrin is not solubilized, especially if your expressed protein is similar in size to polyhedrin. The NP40-based lysis buffer described in the Materials list is an example of a buffer that is suitable for this purpose, but you may wish to use some other lysis buffer, as appropriate, for your protein. Add 50 µl of lysis buffer to each dish and incubate on ice for 30 minutes. Scrape the lysates off the dishes, and stand the dishes on edge for a few minutes. Collect the lysates, vortex extensively, and spin for two minutes in a microfuge to remove the bulk of the polyhedrin. Store at –20 or –80° C. (Obviously, you can also omit the centrifugation step and add an aliquot of the entire lysate to the SDS gel-loading buffer before electrophoresis.) Note that, in addition to polyhedrin, certain other proteins may be insoluble or not efficiently extracted by the NP40 lysis buffer. This is an important concern with any lysis buffer, and we recommend that you give careful thought to your choice of buffer. If you are having trouble visualizing your protein or appear to have very low levels of expression, it may be worthwhile trying a different lysis buffer.

   While postlysis proteolysis will probably not be a problem if the cells are lysed in SDS gel-loading buffer, you may need to include protease inhibitors in other lysis buffers. Whether a protein is susceptible to postlysis degradation is a property of each individual protein. Thus, it is wise to include protease inhibitors in the lysis buffer during initial experiments. If you are doing repeated time course experiments, it may be worthwhile to carry out a trial experiment omitting the inhibitors from the lysis buffer.
8. The samples are now ready for further processing by immunoprecipitation and/or analysis by SDS-PAGE as described later in this chapter. For direct analysis by SDS-PAGE, we recommend loading 5 µl to 10 µl of each sample per well.

Note:

If 99% of the cells in the population receive at least 1 pfu, then 1% will receive no infectious virus particles. According to the Poisson distribution (see Chapter 12), the proportion of cells (p) receiving no infectious virus is given by the equation

$$p = e^{-\mu}$$

where μ is the mean concentration of infectious units in the sample. Hence,

$$0.01 = e^{-\mu}$$

which implies that $\mu = 4.6$. Thus, to ensure that 99% of the cells in a population receive at least one infectious particle, an MOI of at least 4.6 pfu/cell should be used.

Metabolic Labeling of Infected Cells

The Coomassie blue staining or immunoblot analysis of whole-cell lysates analyzed on SDS gels (described later in this chapter) provides information on the steady-state levels of proteins during the infection process. If the cells are incubated with a radiolabeled amino acid for a short period of time prior to lysis, information can also be obtained on the rates of synthesis of the different proteins. This is because the strength of a signal on an autoradiograph or fluorograph depends on the net amount of radioactivity incorporated by a protein during the labeling period, which, in turn, is partly dependent on the rate of synthesis of the protein.

An additional advantage of pulse-labeling the cells before lysis is the extra sensitivity provided by this approach. This is not only due to the incorporation of radiolabeled amino acid into the protein during synthesis, but also because the background of host protein synthesis is very low late in infection. As we discussed already, pulse-labeling is often the best way of detecting secreted proteins in the extracellular fluid.

The most important parameter to consider when planning a pulse-labeling experiment is which radiolabeled amino acid to use. The amino acid [^{35}S]methionine is most commonly used, but [^{35}S]cysteine is also used quite frequently. Recently, ICN introduced a mixture of the two, known as Tran^{35}S-label, which is considerably cheaper than either amino acid individually. Other amino acids, labeled with a variety of different isotopes, can also be obtained. Therefore, your first step is to examine the amino acid sequence of the protein you are expressing and decide whether methionine or cysteine (or some other amino acid) is most appropriate to use as a metabolic label. Remember that you will need tissue-culture medium deficient in the particular amino acid you choose. Methionine-deficient medium is commercially available, but for other amino acids you may have to prepare the medium in the laboratory. (Tran^{35}S-label is largely [^{35}S]methionine and works well with methionine-deficient medium.) The following protocol describes labeling with Tran^{35}S-label. Included are the labeling and collection of both intracellular and secreted proteins. The steps that may be omitted are indicated if you are only interested in intracellular proteins.

Yamada et al. (1990) have reported the presence of a cysteine protease in the extracellular fluid of baculovirus-infected cells. These workers achieved a dramatic increase in the yield of their heterologous gene product by including the cysteine protease inhibitor p-chloromercuribenzene sulphonic acid (pCMBS) in the extracellular fluid. Thus, it may be worth including pCMBS in the protease inhibitors you use, especially if you are working with a secreted protein. However, SF cells will not tolerate prolonged exposure to pCMBS, so it should only be added to the extracellular fluid one to two hours before collecting the sample.

The following metabolic labeling techniques are generally used in conjunction with a time course experiment and the reader is directed to the time course protocol in the preceding section for other relevant protocols and cautionary statements.

CONTROLS: The controls used should include an identical time course performed with wt virus (and/or the parent virus if different), as well as one dish of mock-infected cells.

Materials

1. Cells: SF cells grown to approximately 1×10^6 cells/ml
2. Virus: recently titered stocks of your recombinant virus and control viruses
3. Complete tissue-culture medium. (See Chapter 11 for details concerning media available.)
4. Tissue-culture medium lacking methionine (or other amino acid, as appropriate)
5. 35 mm tissue-culture dishes (see Chapter 11)
6. Tran^{35}S-label (> 1000 Ci/mmol in aqueous solution, ICN # 51006). Store in small aliquots under nitrogen at –80° C.
7. PBS, pH 6.2
8. Protease inhibitors
   a. 10 mg/ml PMSF stock
   Caution: PMSF is very harmful and may be fatal if inhaled, swallowed, or by skin contact. Wear protective clothing and a mask when handling PMSF.
   b. 100 µg/ml Pepstatin A stock
   c. 50 µg/ml Leupeptin stock
9. NP40 lysis buffer or other lysis buffer of your choice
10. 2X SDS gel-loading buffer

Method

1. Seed the cells onto tissue-culture dishes and infect as described for the standard time course experiment in the preceding section.
2. Two hours before each time point, remove the medium from the plates and replace with the same volume of methionine-deficient medium. Incubate at 27° C for one hour. This depletes the endogenous pools of methionine and ensures more efficient incorporation of the labeled amino acid.
3. One hour before each time point, remove the methionine-deficient medium from the cells. Replace with 0.5 mls methionine-deficient medium containing 25 µCi Tran^{35}S-label. Incubate at 27° C or room temperature for one hour (see note) with gentle rocking.
4. At the time point, remove the tissue-culture fluid containing the radioactive amino acid and dispose of appropriately. Rinse the cells *gently* three times with 2 mls each of PBS.

   If you do not want to analyze secreted proteins, the cells may be lysed at this point, as described in the standard time course protocol. Then proceed to the immunoprecipitation protocol immediately following, or directly to the SDS-PAGE protocol following that. If you are loading labeled whole-cell lysates directly onto SDS-polyacrylamide gels, we recommend loading 1 µl to 3 µl of each lysate per well.

   To examine both intracellular and extracellular fractions, continue as follows.

5. Incubate the cells in 0.5 mls PBS for one hour at room temperature with gentle rocking.
6. Collect the PBS, spin for five minutes in a microfuge at 4° C, and store the supernatant at –20 or –80° C. This is termed the extracellular fraction.
7. Lyse the cells as described in the standard time course protocol. You may now subject these samples to immunoprecipitation analysis or analyze them directly by SDS-PAGE. For SDS-PAGE analysis, we recommend first trying 1 μl to 3 μl of each lysate per well and the maximum possible volume of each extracellular fraction. In subsequent experiments, it may be possible to reduce the volume of extracellular fraction loaded.

Note:

You may wish to label the cells for longer periods of time. This will increase the amount of label that is incorporated by stable proteins. However, it will become more difficult to detect proteins that are rapidly turned over. Longer labeling periods will also decrease the resolution with which you can characterize the kinetics of protein synthesis during the infection process. You may incubate the cells in the presence of the radiolabel for up to 12 hours, according to the protocol described here, without any obvious ill effects on the cells. You should not need to supplement the medium with serum.

Immunoprecipitation

Immunoprecipitation involves the incubation of the protein-containing sample with a specific antibody and collection of the immune complexes formed between the antibody and its cognate antigen. These complexes are then analyzed by SDS-PAGE. The protocol can thus provide immunological confirmation of the identity of the protein you have expressed. In addition, it may be used in conjunction with metabolic labeling techniques to provide information on the kinetics of synthesis or on the post-translational modification of a protein (discussed in Chapter 15). Immunoprecipitation can also provide information on the steady-state levels of the specific protein recognized by the antibody (e.g., by staining the SDS gel with Coomassie blue), but immunoblot analysis is better suited to this purpose.

The immunoprecipitation protocol presented here, derived from the original protocol of Kessler (1975), comprises the following stages:

- Sample preparation
- Preclearing the sample
- Formation of antigen-antibody complex
- Precipitation of antigen-antibody complex
- Elution of bound antigen and analysis by SDS-PAGE

Samples for immunoprecipitation analysis are prepared essentially as described in the preceding time course and metabolic labeling sections. The choice of buffer used to lyse the cells is critical. The NP40-based lysis buffer works well, *provided* it is compatible with your protein of interest. However, do *not* use the SDS gel-loading buffer. [See Harlow and Lane (1988) for further discussion of lysis buffers compatible with immunoprecipitation.]

The aim of preclearing the samples prior to formation of the antigen-antibody complex is to remove proteins that will nonspecifically precipitate during the immunoprecipitation step. This is achieved by incubating the sample with nonimmune serum and removing antibodies and antibody-antigen complexes with protein A. All proteins in the sample that bind nonspecifically to the protein A reagent are also removed in this procedure. Preclearing the sample may not be necessary when a major protein in the sample is being immunoprecipitated, as is often the case in baculovirus expression vector work.

Following preclearing, the specific antibody-antigen complex is formed simply by adding the antibody to the protein sample. Either polyclonal antisera or monoclonal antibodies may be used effectively. Generally, polyclonal antisera will display a very high affinity for the specific antigen. In addition, they have the advantage of binding well to protein A (discussed later). However, they often give rise to higher backgrounds. Monoclonal antibodies are highly specific and usually do not give rise to generalized high-background problems. However, individual monoclonal antibodies may cross-react with unrelated proteins in a specific manner. This is because the epitope recognized by a monoclonal antibody may be as short as four to five amino acids so that there is a reasonable probability of a given epitope occurring on a protein by chance alone. Different monoclonal antibodies also display widely varying affinities for their cognate antigen and for protein A. Thus, while some monoclonal antibodies work extremely well in immunoprecipitation protocols, others are very unsatisfactory, giving undesirable cross-reactions and poor sensitivity. Ideally, several monoclonal antibodies against the same antigen should be tested to identify which works best in the immunoprecipitation procedure.

Once formed, the antibody-antigen complex is collected by binding to a protein from the bacterium *Staphylococcus aureus* cowan I (SAC) known as protein A. Protein A is a component of the bacterial cell wall that binds to the Fc portion of the heavy chains of certain antibody classes [see Harlow and Lane (1988) or another basic immunology manual for a description of antibody structure and nomenclature]. The biological significance of this interaction is not clear. To facilitate collection of the immune complexes, the protein A must be bound to some insoluble matrix. In practice, either fixed SAC bacteria or protein A covalently bound to beads is used. The former approach is considerably cheaper but may give rise to higher backgrounds.

Clearly, the success of this method for the collection of immune complexes is dependent on the strength of the binding between the antibody and protein A. As alluded to earlier, this parameter varies greatly depending on the class and subclass of the antibody used. For mouse antibodies, IgG_{2a} antibodies bind with high affinity to protein A, but IgG_3, IgG_{2b}, and IgG_1 antibodies all bind with progressively lower affinities. [See Harlow and Lane (1988) for information on antibodies of other species.] For IgG_3 and IgG_{2b} antibodies, the interaction is often strong enough for at least some of the antigen-antibody complex to be precipitated, but this process is not quantitative. With IgG_1 antibodies, it may not be possible to collect the antibody-antigen complex using protein A alone. Because a polyclonal antiserum will contain many antibodies of different classes, this differential affinity for protein A is not problematic. However, with a monoclonal antibody, it is important to know the particular class of the antibody being used. If it is an IgG_1, IgG_{2b}, or IgG_3, the problem may be circumvented by use of an appropriate anti-immunoglobulin secondary antibody. This secondary antibody (generally a polyclonal antiserum) will bind the monoclonal antibody-antigen complex, and then the entire complex can be efficiently bound by protein A.

CONTROLS: The controls used should include both the immunoprecipitation of control samples with the test antibodies and immunoprecipitation of the test samples with control antibodies. Appropriate control samples include samples derived from a time course using wt virus (or the parent virus, if different), as well as a sample derived from mock-infected cells. The best control antibody for a polyclonal antiserum is serum bled from the same animal prior to immunization. If this is not available, use nonimmune serum from the same species. For a monoclonal antibody, the best control is a monoclonal antibody of the same type raised against an unrelated protein.

Materials

1. Cell lysates or extracellular fluid obtained as described in the preceding time-course and/or metabolic labeling protocols. Cell lysates should have been prepared with the NP40-based lysis buffer or a similar buffer. *Do not use SDS gel-loading buffer.*
2. Antibodies
   a. Appropriate test and control antibodies
   b. Normal serum for preclearing the samples, if necessary
   c. Anti-immunoglobulin antiserum; this may be required if the primary antibody does not bind well to SAC. Be sure to get an antiserum specific for the species from which your primary antibody was derived. Thus, if you are using a mouse monoclonal antibody as the primary antibody, goat antiserum raised against mouse immunoglobulins (Sigma # M 8642) would be appropriate as secondary antibody.

   Store antibodies at –20° C in small aliquots. They are stable for many years in this form. However, do not freeze and thaw repeatedly. Working stocks may be stored for at least six months at 4° C. Sodium azide should be added to a concentration of 0.02% (w/v) to stocks stored at 4° C.
   Caution: Sodium azide is poisonous. Label all solutions clearly and handle with care.
3. NET buffer: 0.14 M NaCl; 50 mM Tris-HCl, pH 8.0; 5 mM EDTA; 0.05% (v/v) NP40. Store at 4° C.
4. Protein A (SAC) wash buffer: 50 mM Tris-HCl, pH 7.5; 5% (v/v) β-mercaptoethanol; 2% (w/v) SDS. Prepare fresh.
5. RIPA: 10 mM Tris-HCl, pH 8.0; 1 mM EDTA; 150 mM NaCl; 1% NP40; 1% sodium deoxycholate; 0.1% SDS. Store at 4° C.
6. Protein A reagent: You may use either fixed *S. aureus* cowan I (SAC) bacteria or protein A beads (e.g., Sigma # P 3391). The beads are generally more expensive but often give lower backgrounds. Prepare and use protein A beads as described by the manufacturer. SAC may be prepared in the laboratory but this is rather tedious, and commercially available preparations are quite inexpensive (e.g., Insoluble Protein A; Sigma # P 9151). Resuspend the lyophilized bacteria at a concentration of 10% (w/v) in NET buffer plus 1 mg/ml BSA, and store in small aliquots at –80° C. SAC may be stored for several months in this form. If you are going to stain the immunoprecipitate gel with Coomassie blue, you should pretreat the SAC to eliminate bacterial proteins that will elute into the SDS gel-loading buffer, as follows:
   a. Resuspend lyophilized SAC in SAC wash buffer at a concentration of 2% (w/v).
   b. Incubate at 100° C for 20 minutes.
   c. Centrifuge for two minutes in a microfuge and discard the supernatant.
   d. Resuspend the pellet in SAC wash buffer and repeat steps b and c.
   e. Resuspend the pellet in RIPA.
   f. Centrifuge for two minutes in a microfuge and discard the supernatant.
   g. Repeat steps e and f two more times.

h. Resuspend at a final concentration of 10% (w/v) in NET plus 1 mg/ml BSA. Store at –80° C in small aliquots. Use within two weeks.
7. 10 mg/ml BSA. Store at –20° C.
8. 2X SDS gel-loading buffer

Method

The volume of each sample you use in the following immunoprecipitation experiment will need to be determined empirically. We recommend first trying 5 µl of each cell lysate and 50 µl of extracellular fluid samples.

1. Preclearing the sample
   a. Add 2 µl normal rabbit serum to each sample and incubate at 4° C for one hour. You may add NET to adjust the mix to a convenient volume if desired. It is important to use NET or a similar buffer to ensure the salt concentration is maintained at appropriate levels.
   b. Add 10 µl 10% SAC and incubate at 4° C for 30 minutes. This SAC does not need to be washed prior to use.
   c. Spin in a microfuge at 4° C for two minutes. Carefully remove the supernatant and discard the pellet. Add a further 10 µl 10% SAC to the supernatant and incubate at 4° C for 30 minutes.
   d. Spin at 4° C for two minutes. Again, discard the pellet and spin the supernatant once more at 4° C for two minutes.
   e. The final supernatant is the precleared sample and is ready for addition of the specific antibody.
2. Formation of antigen-antibody complex
   a. Mix the sample (either cell lysate, extracellular fraction, or precleared sample prepared as in Step 1) with the antibody directed against your protein of interest. The quantity of antibody you use will also have to be determined empirically. To start, we recommend trying 2 µl to 5 µl of a polyclonal antiserum and 50 µl to 100 µl of a monoclonal antibody. For quantitative immunoprecipitations, you will need to test a series of antibody concentrations to ensure that your antibody is in excess.
   b. Incubate for at least three hours at 4° C. As before, you may dilute the sample with NET to a more convenient volume if required. For immunoprecipitations with a monoclonal antibody, we recommend you add BSA to a final concentration of 1 mg/ml.
3. Precipitation of antigen-antibody complexes
   a. Add 1 µl to 5 µl of goat antimouse immunoglobulin antiserum to the sample containing your protein of interest and specific antibody. Incubate at 4° C for at least one hour. Omit this step if your primary antibody was a polyclonal antiserum or an IgG_{2a} subclass mouse monoclonal antibody.
   b. Add 50 µl of 10% SAC to the sample and incubate at 4° C for one hour. If you plan to stain the immunoprecipitate gel with Coomassie blue, you should use extensively washed SAC (see the Material list).
   c. Collect the SAC-antibody-antigen complex by centrifugation in a microfuge for two minutes at 4° C.
   d. Wash the SAC-antibody-antigen complexes by resuspending vigorously in 0.5 ml NET buffer. The SAC pellet can be quite difficult to resuspend. It may help to let the pellet plus buffer sit on ice for a few minutes before vortexing, or to first resuspend in 50 µl NET, and then bring the volume to 500 µl and vortex extensively.
   e. Repeat steps c and d twice more.
   f. Collect the washed SAC-antibody-antigen complexes by centrifugation for two minutes. The complexes are dissociated by resuspending in 50 µl

1X SDS gel-loading buffer and heating at 100° C for three minutes. The samples may be heated either immediately after resuspending the final SAC pellet or just before SDS-PAGE.

g. Analyze the immunoprecipitated samples by SDS-PAGE as described in the next section. Initially, load as much of each sample as possible and adjust as needed in subsequent experiments. If you stain the gel with Coomassie blue, the immunoglobulins will be apparent as two bands corresponding to the antibody heavy chains (55 kDa) and light chains (25 kDa) respectively. These bands will be very intensely stained if you used a polyclonal antiserum as primary or secondary antibody. If you used a monoclonal antibody, they may be quite faint, but are usually visible. However, if your sample was metabolically labeled and you analyze the gel by autoradiography or fluorography, these bands will not be detected. When interpreting your data, identify specifically immunoprecipitated proteins by careful comparison with the control immunoprecipitations.

4. Eliminating background

If there appears to be a high background (i.e., a large number of proteins that are nonspecifically precipitated by the antibody), some or all of the following procedures may help:

a. If you did not include a preclearing immunoprecipitation, repeat the experiment including this step.
b. Carefully titrate all the antibodies used in the experiment. Use only enough to ensure the quantitative precipitation of your antigen.
c. If you were using a polyclonal antiserum, preadsorb the antiserum against a protein preparation that does not contain the specific antigen, or affinity-purify the antiserum using immobilized antigen [see Sambrook et al. (1989) for protocols].
d. If you were using a monoclonal antibody, use a different antibody to the same protein if available.
e. If you were using SAC, use protein A beads.
f. Increase the number and/or stringency of washes of the SAC-antibody-antigen pellet. A suitable wash buffer for high-stringency washes is RIPA buffer.

SDS-Polyacrylamide Gel Electrophoresis

Denaturing polyacrylamide-gel electrophoresis is a simple and powerful method for the separation of proteins according to size. In the most commonly used procedure, the protein samples are first denatured by heating in the presence of a reducing agent and the strong, anionic detergent, SDS. This treatment dissociates virtually all protein complexes. The denatured proteins bind SDS and acquire a negative charge. The amount of SDS bound is proportional to the molecular weight of the protein but independent of its amino acid sequence, so the migration of the SDS-polypeptide complex during electrophoresis is dependent on the size of the polypeptide. Note, however, that post-translational modification of a protein, such as glycosylation or phosphorylation, can also affect its migration, so the apparent molecular weight is not necessarily a true reflection of the mass of a polypeptide.

SDS-PAGE is generally carried out using a discontinuous buffer system originally devised by Ornstein (1964), Davis (1964), and Laemmli (1970). In this system, a bipartite gel is used, comprising a resolving gel of low porosity and a stacking gel of high porosity. The resolving gel is buffered with Tris-HCl at pH 8.8 while the stacking gel and the samples are buffered with Tris-HCl at pH 6.8. The running buffer in the reservoirs is different again, comprising Tris-glycine at pH 8.3. All components contain 0.1% SDS. During electrophoresis through the highly porous stacking gel, the proteins are trapped between the chloride ions and glycine molecules and are deposited in a thin, concentrated layer on the resolving gel. The higher pH in the resolving gel causes the ionization of the glycine. The glycine ions can now migrate along with the chloride ions at the running front so the proteins are no longer trapped and migrate through the resolving gel according to their size.

Materials

1. Protein-gel electrophoresis apparatus (e.g., Hoefer Mighty Small electrophoresis apparatus)
2. Acrylamide stock: 29% (w/v) acrylamide; 1% (w/v) bis-acrylamide. Use electrophoresis grade reagents (Bio-Rad # 161-0100 and # 161-0200) and store at 4° C in a dark bottle. These reagents break down slowly, so prepare fresh solutions every few months.
   Caution: Acrylamide and bisacrylamide are potent neurotoxins. Their effects are cumulative. Wear gloves and a mask when handling these reagents.
3. 1 M Tris-HCl, pH 8.7. (Do not use Tris·Cl or Trizma because the salt concentration will be too high.)
4. 1 M Tris-HCl, pH 6.8. (Do not use Tris·Cl or Trizma.)
5. 10% (w/v) SDS. Use electrophoresis grade reagent (Bio-Rad # 161-0300).
6. 10% (w/v) Ammonium persulfate (Fisher # BP179-25). Store at 4° C for no more than a week.
7. TEMED (N,N,N′,N′-tetramethylethylenediamine, Sigma # T 8133)
8. 10X running buffer: 250 mM Tris base; 1.92 M glycine; 1% SDS; pH 8.3
9. Protein molecular weight markers: any commercially available markers should be suitable. Prestained markers (e.g., BRL # 6041LA) are particularly useful, especially if you do not otherwise need to stain the gel.
10. Power supply (capacity: 200V, 500 mA)
11. Coomassie blue stain: 0.5% Commassie blue R-250 (Sigma # B 0630); 20% acetic acid; 50% methanol
12. Destain: 1% formic acid; 31.25% methanol
13. Whatman 3MM paper
14. Gel drying film (e.g., Promega # v713A)

Method

1. Pouring the gel
   a. Assemble the gel sandwich according to the manufacturer's instructions. Pay careful attention to sealing the edges and corners of the plates. Leaks of the acrylamide solution from the plate sandwich are the most common problem when preparing protein gels.
   b. Prepare an appropriate volume of the resolving gel mix based on the following table. The volume you need will depend on the particular gel unit being used. The volumes (in mls) given are for a final volume of 50 mls. The polymerization reaction will be initiated upon addition of the TEMED, so do not add it until the gel is ready to be poured.

	% Acrylamide			
Reagents	*5.0*	*8.0*	*10.0*	*15.0*
Acrylamide stock	8.3	13.3	16.6	25.0
H_2O	22.2	17.2	13.9	5.5
1M Tris-HCl, pH 8.7	18.7	18.7	18.7	18.7
10% SDS	0.5	0.5	0.5	0.5
10% Ammonium persulfate	0.25	0.25	0.25	0.25
TEMED	0.04	0.04	0.04	0.04

The concentration of acrylamide you use in the resolving gel depends on the size range you want to separate. As a rough guide, the following table (from Sambrook et al., 1989) indicates the *linear* range of separation with gels of different concentrations of acrylamide. (Proteins outside these size ranges will also be resolved. However, their migration will not be directly proportional to the logarithm of their apparent molecular weight.)

% Acrylamide	*Range (kD)*
5.0	57–212
8.0	30–84
10.0	16–68
15.0	12–43

c. Mark the depth of the comb on one of the plates and then remove the comb. Pour the acrylamide solution between the glass plates so that it will be 0.5 cm to 1 cm (depending on the size of the gel) below the bottom of the wells.
d. Overlay the acrylamide carefully with 0.1% SDS (for gels with acrylamide concentrations of 8% or less) or isobutanol (for gels with greater than 8% acrylamide). Stand the gel vertically at room temperature.
e. The gel will have polymerized within 20 to 30 minutes. However, it may be left at room temperature for several hours to overnight without any adverse effects. Alternatively, it may be stored for several weeks at 4° C if rinsed (Step f) and wrapped in Saran Wrap.
f. Pour off the overlay and unpolymerized acrylamide. Rinse several times with H_2O and drain well. Remove residual water with the corner of a paper towel.
g. Prepare an appropriate volume of stacking-gel mix based on the quantities in the following table. The volumes given are for a final volume of 10 mls. Again, add the TEMED just before pouring.

Reagents	*Volume (mls)*
Acrylamide stock	1.7
H_2O	6.8
1M Tris-HCl, pH 6.8	1.3
10% SDS	0.1
10% Ammonium persulfate	0.05
TEMED	0.01

h. Pour the stacking gel on top of the polymerized resolving gel until the acrylamide fills the plates. Immediately insert the comb, being careful to avoid trapping air bubbles under the teeth. Stand vertically at room temperature for 20 to 30 minutes. It is not advisable to store the gel for any period of time at this stage. Storage will lead to diffusion of the pH discontinuity between the running and stacking gels with a consequent decrease in resolving power of the gel.
i. When the stacking gel has polymerized, carefully remove the comb. Rinse the wells several times with H_2O. Mount the gel in the electrophoresis apparatus and fill the upper and lower chambers with 1X running buffer. Flush the wells with running buffer and straighten the teeth with the side of a narrow gauge needle if necessary. Remove any air bubbles trapped underneath the running gel.

2. Running the gel
   a. The volume of sample to be loaded will vary depending on your particular application and generally must be determined empirically. See the individual sample preparation sections for specific recommendations. If the samples were not lysed in SDS gel-loading buffer, add an equal volume of 2X SDS gel-loading buffer to each sample. Incubate all samples (including the markers if necessary) at 100° C for three minutes.
   b. Centrifuge the samples briefly in a microfuge and carefully load them into the wells. Generally, a pipet with narrow tips or a Hamilton microliter syringe will work well.
   c. Because the proteins in the sample are complexed with SDS, they all have a net negative charge and will therefore migrate toward the anode (positive electrode). Thus, connect the anode (red lead) to the bottom chamber, and the cathode to the top. The choice of electrophoresis conditions depends to a large extent on the type of gel unit being used. A constant voltage of approximately 15 V/cm will probably be suitable. The gel should be run at half this voltage until the samples enter the resolving gel.
   d. Continue electrophoresis until the blue dye is at the bottom of the gel. Disconnect the electrophoresis apparatus from the power supply and remove the gel.
   e. Carefully pry the plates apart and cut the gel at one corner to mark its orientation. The gel is now ready for staining, fluorography, or immunoblotting as required.
3. Staining the gel
   a. Remove the gel carefully from the plate and place it in several volumes of stain. Incubate at room temperature with gentle rocking for one to four hours (or longer if convenient). The minimum time necessary depends on the size and thickness of the gel.
   b. Remove the stain (it may be reused many times) and rinse the gel in H_2O. Add a large volume of destain and multiple, small pieces of foam rubber (see note 1). Incubate at room temperature (see note 2) with gentle rocking until the gel is completely destained (usually several hours).
   c. Remove the destain [it may also be reused a number of times (see note 1); replace when it is visibly blue] and discard the foam pieces. The sensitivity of this staining procedure is approximately 0.1 µg to 0.5 µg. The gel may be preserved by storage in 1% formic acid (e.g., in a sealed plastic bag) or it may be dried. Dry the gel onto Whatman 3MM paper or equivalent by incubation at 80° C under vacuum on any conventional gel drier. (If you require a photograph of the gel, take it before drying the gel onto 3MM paper.) Alternatively, the gel may be dried onto cellulose film such as Promega's gel-drying paper. In this case, follow the manufacturer's instructions. We find it helpful to lay the wet gel on some aluminum foil when transferring it to 3MM paper or into a plastic bag. The foil does not stick to the gel and aids in manipulating the gel without tearing it.

Notes:

1. The foam-rubber pieces accelerate the destaining process by binding the stain as it leaches from the gel. They are not essential. However, if they are not used, it will probably be necessary to replace the destain at least once while destaining. In addition, the destain will rapidly become saturated with stain and then cannot be reused.

2. The destaining process can be accelerated by incubation at 45° C instead of at room temperature.

Fluorography

The isotopes most frequently used to label amino acids (^{35}S, ^{14}C, and ^{3}H) are all relatively weak β emitters. The signal cannot be enhanced using an intensifying screen, as the particles emitted are of insufficient energy to pass through the X-ray film to activate the screen. Instead, it is necessary to impregnate the gel with a fluor so that the fluor is in immediate proximity to the isotope. The β particles from the isotope are adsorbed by the fluor, which then fluoresces, emitting many photons for each adsorbed quantum of radiation. Fluorography will decrease the exposure time of ^{35}S- or ^{14}C- containing gels five- to tenfold, and is essential for gels containing ^{3}H.

There are many commercially available kits for the fluorography of protein gels. Alternatively, two simple protocols, based on the techniques of Bonner and Laskey (1974) and Chamberlain (1979) respectively, are provided here. It is not necessary to stain the gel with Coomassie blue prior to fluorography (although you can do so if required; for example, to visualize markers, etc.). If the gel has not been stained with Coomassie blue, the proteins must be fixed within the gel matrix prior to fluorography.

PPO Fluorography

In this protocol, devised by Bonner and Laskey (1974), the fixed gel is thoroughly soaked in DMSO to remove all the water. It is then impregnated with the fluor 2,5-diphenyloxazole (PPO) in DMSO. Finally, the fluor is precipitated within the gel by soaking the gel in water (PPO is insoluble in water).

Materials

1. Destain: 1% formic acid; 31.25% methanol
2. DMSO, reagent grade (Fisher # D128-500)
3. 22.5% (w/v) PPO (Sigma # D 4630) in DMSO. Store at room temperature in a dark bottle with a tightly fitting lid.
4. Whatman 3MM paper

Method

1. If the gel has not already been fixed, soak it in destain for at least 15 to 20 minutes (the bromophenol blue at the running front will turn yellow as the acid penetrates the gel).
2. Remove the destain and drain the gel well. Soak in 10 volumes of DMSO for 30 minutes at room temperature.
3. Remove the DMSO, drain well, and repeat Step 2. The DMSO may be reused once at each step. Thus, you may use fresh DMSO twice for the second wash, and then twice for the first wash before discarding.
4. Remove the DMSO, drain well, and soak the gel in 22.5% PPO in DMSO. Rock gently at room temperature for 30 minutes.
5. Remove and save the PPO solution. This solution can be reused many times provided it is kept moisture-free. Replace it when you feel that the intensity of the fluorographs has begun to decrease. Rinse the gel several times in a large volume of H_2O. As soon as you add the water, the PPO will precipitate within the gel, which will become opaque and white.

6. Soak the gel for at least one hour in a large volume of water. It is essential that the gel is thoroughly rehydrated before drying. Otherwise, it will continuously absorb moisture from the atmosphere and will be exceedingly difficult to dry completely.
7. Dry the gel onto Whatman 3MM paper at 80° C under vacuum, using a standard gel dryer.
8. Expose the dried gel to X-ray film at –80° C. An overnight exposure will generally be a useful first exposure to try for gels containing ^{35}S-labeled cell lysates. When exposing the gel, it is important that the gel and film are kept firmly pressed against one another. Do not wrap the gel in Saran Wrap.

Salicylate Fluorography

In this procedure, the fixed gel is soaked in a solution of sodium salicylate prior to drying. This method is considerably faster than PPO impregnation, but the bands obtained may be slightly more diffuse.

Materials

1. Destain: 1% formic acid; 31.25% methanol
2. 1 M sodium salicylate (Sigma # S 3007), pH 6.0
   Caution: Salicylate can be allergenic and is readily absorbed through the skin. Wear gloves when handling this reagent.
3. Whatman 3MM paper

Method

1. If the gel has not already been fixed, soak it in destain for at least 15 to 20 minutes (the bromophenol blue at the running front will turn yellow as the acid penetrates the gel).
2. Remove the destain and drain the gel well. Rinse several times with water to remove all traces of the destain.
3. Soak the gel in 10 volumes 1 M sodium salicylate for 30 minutes at room temperature with gentle rocking.
4. Remove and discard the salicylate solution. Dry the gel onto Whatman 3MM paper at 80° C under vacuum, using a standard gel dryer.
5. Expose the dried gel to X-ray film at –80° C. An overnight exposure will generally be a useful first exposure to try for gels containing ^{35}S-labeled cell lysates. When exposing the gel, it is important that the gel and film are kept firmly pressed against one another. Do not wrap the gel in Saran Wrap.

Immunoblot Analysis

Immunoblot analysis of proteins, like immunoprecipitation, allows you to obtain immunological confirmation of the identity of the protein produced by your recombinant baculovirus. The intensity of a band stained by immunoblotting is dependent on the total amount of protein in the band. Thus, immunoblotting is well suited to the evaluation of steady-state levels of a particular protein in a sample. The following protocol is a modification of the methods of Towbin et al. (1979) and Burnette (1981) and comprises four stages:

- Separation of the proteins in the sample by SDS-PAGE
- Transfer of the separated proteins from the gel to a membrane
- Reaction of the primary antibody with the proteins bound to the membrane
- Detection of the specifically bound primary antibody

Gels obtained by standard SDS-PAGE using the preceding protocols are suitable for use in immunoblot experiments. There are no procedures specific to immunoblotting that must be carried out. Following electrophoresis, do not attempt to fix or stain the gel in any way.

Transfer of the proteins from the gel to a membrane support is most often accomplished by electrophoretic elution. A variety of apparati are commercially available for this purpose. The membrane used may be either a nitrocellulose or an activated nylon membrane. Nitrocellulose membranes generally work well; however, they are brittle and difficult to manipulate when dry. Although activated nylon membranes do not have this problem and do not change size during the experiment, they tend to give higher backgrounds. For immunoblot detection of baculovirus-produced proteins, where the protein of interest is a major component of the sample, this higher background is generally not problematic, and activated nylon membranes are the support of choice. However, researchers who experience problems with high background levels may wish to consider switching to nitrocellulose membranes. (Activated paper has also been used as a support for transferred proteins. However, it is not compatible with SDS gels run in a Tris-Glycine buffer system and does not offer any substantial advantages for our purposes.)

It is important to remember that the proteins in the sample were denatured during SDS-PAGE and are transferred to the membrane in a denatured form. Thus, the primary antibody used should be able to recognize the proteins in a denatured form. (This is a significant difference between immunoblotting and immunoprecipitation.) If you are using a polyclonal antiserum, this will not be a concern, because the antiserum will contain a wide range of antibodies against the protein, many of which will bind to denaturation-resistant epitopes. However, when using a monoclonal antibody as primary antibody in immunoblot analysis, it is important to verify before starting that the antibody recognizes a denaturation-resistant epitope.

There are several options for the detection of the primary antibody bound to the membrane. These include enzyme-linked or radiolabeled secondary antibodies or protein A, as well as directly labeled primary antibody. The following method (Blake et al., 1984) involves the use of an alkaline-phosphatase-conjugated secondary antibody directed against the primary antibody. This secondary antibody is allowed to bind to the primary antibody on the membrane, and then is visualized by virtue of the alkaline phosphatase activity. The substrate used is 5-bromo-4-chloroindoxyl phosphate (BCIP)/nitro blue tetrazolium. Cleavage of the phosphate from BCIP results in the release of the indoxyl group and hydrogen ions. The hydrogen ions cause the reduction of the nitro blue tetrazolium, yielding an intensely blue, insoluble diformazan.

CONTROLS: Appropriate control samples are samples derived from a time course using wt virus (or the parent virus if different), as well as a sample derived from mock-infected cells. All samples should be run on duplicate gels, one of which should be probed with the test antibody and one with the control antibody. The best control antibody for a polyclonal antiserum is serum bled from the same animal prior to immunization. If this is not available, use nonimmune serum from the same species. For a monoclonal antibody, the best control is a monoclonal antibody of the same type raised against an unrelated protein.

Materials

1. Samples: cell lysates or extracellular fluid electrophoresed on an SDS-polyacrylamide gel as described earlier in this chapter. Be sure to include protein

molecular-weight markers on the gel. Prestained markers (e.g., BRL # 6041LA) are ideal, as they obviate the need to stain the blot after transfer. If you do not use prestained markers, you may need to cut the marker lane from the blot after transfer to stain the markers. Thus, load the markers at the edge of the gel. Do not stain or fix the gel after electrophoresis.

2. Electroblot buffer: 25 mM Tris base; 192 mM glycine; 20% (v/v) methanol; pH 8.3. Store at room temperature.
3. Electroblotting apparatus: e.g., Bio-Rad Trans-Blot Cell
4. Membrane: activated nylon membrane suitable for protein blotting, such as Immobilon (Millipore, # IPVH 000 10). The membrane should be cut to the size of the gel just before blotting. Always wear gloves when handling membranes.
5. Whatman 3MM paper. Cut four pieces the same size as the gel, before blotting.
6. Amido black: 0.1% amido black (Sigma # N 3005) in 45% (v/v) methanol; 10% (v/v) acetic acid. Store at room temperature.
7. Destain: 1% formic acid; 31.25% methanol
8. TBS: 25 mM Tris base; 150 mM NaCl; pH 7.5. Store at room temperature.
9. TBST: TBS plus 0.05% (v/v) Tween 20 (Sigma # P 1379)
10. Blocking buffer: 3% BSA (w/v) in TBS plus 2 mM sodium azide (Sigma # S 2002)
    Caution: Sodium azide is poisonous. Label all solutions clearly and handle with care.
11. Antibody dilution buffer: 2% BSA (w/v) in TBS plus 2 mM sodium azide
12. Antibodies
    a. Appropriate test and control antibodies
    b. Secondary antibody: alkaline-phosphatase conjugated polyclonal antiserum directed against immunoglobulins of the species in which the primary antibody was raised [e.g., alkaline-phosphatase conjugated goat antiserum against rabbit IgG (Sigma # A 8025)]

    Store antibodies at –20° C in small aliquots. They are stable for many years in this form. However, do not freeze and thaw repeatedly. Working stocks may be stored for at least six months at 4° C. Sodium azide should be added to a concentration of 0.02% to stocks stored at 4° C.
13. 0.15 M Tris, pH 9.5
14. NBT stock: Dissolve nitro-blue tetrazolium (Sigma # N 6876) in a minimal volume of methanol (e.g., 100–200 µl for 10 mg) and dilute to 1 mg/ml in 0.15 M Tris-HCl, pH 9.6. Store at 4° C for up to two to three weeks.
15. BCIP stock: 5 mg/ml 5-bromo-4-chloroindoxyl phosphate (Sigma # B 8503) in dimethylformamide (Sigma # D 4254). Store at 4° C.
16. Substrate solution: Just before you develop the blot, dilute the NBT stock tenfold and the indoxyl phosphate 100-fold in 0.15 M Tris-HCl, pH 9.5; 4 mM $MgCl_2$. Use this reagent within one hour.

Method

1. Transfer of proteins to membrane
    a. Equilibrate the gel in electroblot buffer for 30 minutes.
    b. Wet the activated nylon membrane (cut to the size of the gel) by floating in 100% methanol (or follow the manufacturer's instructions if different). When the membrane is wet, rinse in ddH_2O and soak in electroblot buffer.
    c. Place the gel and membrane in the electroblot apparatus according to the manufacturer's instructions. Generally, the gel and membrane will be sandwiched between Whatman 3MM paper that has been soaked in electroblot buffer. Try to ensure that there are no bubbles between any layers of the sandwich. Gently rolling a glass pipet across the sandwich will help expel any bubbles. Pay careful attention to the orientation of the gel

and membrane relative to the electrodes when you are placing the sandwich in the apparatus. The proteins will be transferred towards the anode (red lead), *so the membrane should be closest to the anode.*

d. Fill the transfer chamber with electroblot buffer. If required, run cold water through the cooling coil, place a magnetic stir bar in the tank, place the apparatus on a stir plate, and gently stir the buffer.
e. Transfer according to the manufacturer's instructions. With a Bio-Rad Trans-Blot electroblotting cell, we routinely transfer at 100 mA overnight, and then at 200 mA for an additional hour before removing the blot.

2. Visualizing the markers

If you used prestained molecular-weight markers on the SDS gel, there is no need to stain the blot, and you can proceed to Step 3.

a. Cut off a strip of the blot containing the markers.
b. Incubate in the amido black solution for two hours at room temperature.
c. Destain until the markers are clearly visible.
d. Air-dry and store at room temperature.

3. Blocking nonspecific binding sites on the membrane
   a. Rinse the blot briefly with ddH_2O.
   b. Immerse the blot in blocking buffer and rock gently for one hour at room temperature.
4. Incubation with primary antibody
   a. Remove the membrane from the blocking solution and rinse twice for five minutes each in TBS.
   b. Dilute the primary antibody in antibody dilution buffer. The appropriate dilution of primary antibody will have to be determined empirically. You will judge it based on the signal-noise ratio obtained on the blot. To start, we suggest diluting polyclonal antisera 1:1000 and monoclonal antibodies 1:5.
   c. Immerse the blot in a small volume of the primary antibody solution. The volume used should be the minimum possible to cover the blot. The incubation may be carried out in a shallow tray or in a heat-sealed plastic bag. In the latter case, take care to avoid air bubbles.
   d. Incubate for one to two hours at room temperature with gentle rocking. If desired, the incubation may be carried out for longer periods (up to 18 hours at 4° C). Longer incubations may increase sensitivity.
   e. Remove the membrane from the primary antibody solution, rinse briefly in ddH_2O, and wash twice in a large volume of TBST for 10 minutes each at room temperature, with gentle rocking.
5. Probing the blot with secondary antibody
   a. Dilute the secondary antibody in antibody dilution buffer. We suggest diluting the antibody 1:2000 in initial trials.
   b. Immerse the blot in a small volume of the secondary antibody solution. Again, use the minimum volume required to cover the blot.
   c. Incubate at room temperature for one hour with gentle rocking.
   d. Remove the blot from the secondary antibody solution, rinse briefly in ddH_2O, and wash at room temperature with two changes of TBST for 10 minutes each.
6. Staining the blot
   a. Equilibrate the blot for five minutes in 0.15 M Tris, pH 9.5.
   b. Immerse the blot in substrate solution and incubate at 37° C with constant gentle rocking.
   c. Monitor the intensity of the developing bands. When they appear suitably dark, generally after 15 to 60 minutes, remove the blot from the substrate

solution. Rinse briefly with ddH_2O and air-dry. The blot may be stored indefinitely at room temperature in this form. The bands will not fade.

7. Eliminating background

   If the background (nonspecifically recognized protein bands or general staining of the membrane) is problematic, any or all of the following procedures may help. (Note that polyhedrin is often detected nonspecifically when wt virus-infected samples are analyzed. This is due to the extremely high levels of polyhedrin present in these samples, and little can be done to avoid this problem.)

   a. Titrate the primary and secondary antibodies. Use the minimum amount of antibody required to give quantitative detection of the antigen.
   b. Try using a different blocking buffer. Using 5% (w/v) nonfat dried milk in TBST is a suitable alternative.
   c. Increase the number and/or duration of washes at each step.
   d. If you are using a monoclonal antibody, change the antibody if possible.
   e. If you are using a polyclonal antiserum, try preadsorbing the antiserum with a protein preparation that does not contain the antigen of interest [see Sambrook et al. (1989) for methods].
   f. Carry out a control experiment probing the blot with the secondary antibody alone. If there is any signal on the blot, try reducing the amount of secondary antibody used. Alternatively, try using a secondary antibody with a different label.
   g. Repeat the experiment using nitrocellulose membranes.

Immunofluorescence

Immunofluorescence involves the visualization of an antigen within the cell employing specific fluorochrome-conjugated antibodies. Because it does not require protein separation by SDS-PAGE or transfer of proteins to a solid support, the technique is more rapid than either immunoprecipitation or immunoblotting. It has the added advantage of providing information on the subcellular localization of the specific antigen. A significant disadvantage is that it is not of use with secreted proteins. Immunofluorescence procedures may be classified as either direct, where the primary antibody is conjugated to the fluorochrome, or indirect, where the primary antibody is detected with a secondary fluorochrome-conjugated antibody. Indirect immunofluorescence is the more common procedure because it is simpler and more sensitive. The basic steps are as follows:

- Fixation and permeabilization of infected cells
- Binding primary antibody to antigen
- Binding fluorochrome-conjugated secondary antibody to primary antibody
- Visualization of fluorescent secondary antibody

There are many possible variations of this basic procedure, including the use of different fixation conditions, different methods of visualization of the secondary antibody, simultaneous detection of more than one antigen by double labeling, and so on. This multitude of possibilities has made immunofluorescence an extremely powerful technique with many sophisticated applications. We confine ourselves here to a basic procedure that should be appropriate for the detection of most antigens expressed in the baculovirus expression vector system. Readers interested in other applications of the technique are directed to the manual of Harlow and Lane (1988) and references therein.

In the protocol presented here, the infected cells are fixed and permeabilized by immersion in a mixture of acetone and methanol. These solvents dissolve cellular lipids and dehydrate the cells, causing the proteins to be precipitated onto the cytoskeleton. This procedure allows the antibodies access to the interior of the cell while preventing intracellular antigens from leaking from the cell. It provides reasonably good preservation of the structural features of the cell (although no fixation procedure is perfect in this regard). Fixation and permeabilization are not necessary for visualization of antigens exposed on the exterior of a cell.

Either polyclonal antisera or monoclonal antibodies may be used as primary antibodies for immunofluorescence procedures, but each has its associated constraints. The principal problem likely to be encountered is nonspecific binding of the antibody to the fixed cells. A polyclonal antiserum contains the entire spectrum of antibodies that were present in the immunized animal, in addition to those directed against the specific antigen. Cross-reaction between any of these antibodies and cellular antigens can give rise to spurious signals. Fortunately, such cross-reactive antibodies generally constitute a minor part of the antiserum, and appropriate dilution can often eliminate the problem. The high levels of antigen obtained with the baculovirus expression vector system are helpful in this regard because they permit extensive dilution of the polyclonal antiserum.

Spurious cross-reactions may also be problematic with monoclonal antibodies. Because the epitope recognized by an individual antibody is quite small (as little as four to five amino acids), there is a reasonable probability that any given epitope will occur on a protein by chance alone. Little can be done to avoid this problem other than testing several independent antibodies against the same antigen and using one that does not cross-react. The use of the appropriate controls (see box) is essential to distinguish between real and spurious signals.

An additional problem that may be encountered using a monoclonal antibody is that certain antigenic sites may be destroyed or masked because the fixation and permeabilization procedures will partially denature the cellular proteins. Generally, it is necessary to determine empirically whether a particular monoclonal antibody recognizes its cognate antigen after fixation.

Detection of the primary antibody bound to the antigen is achieved using a second antibody conjugated to a fluorochrome. Fluorochrome-conjugated antibodies against immunoglobulins from a variety of different species are commercially available. Be sure to use an antiserum directed against immunoglobulins of the appropriate species (and class if the primary antibody is a monoclonal antibody). The most common fluorochromes used are fluoroscein and rhodamine. Fluoroscein absorbs light of wavelength 495 nm and emits light of wavelength 525 nm, while rhodamine absorbs at 552 nm and emits at 570 nm. Thus, fluoroscein gives a green stain while rhodamine gives a red stain. Fluoroscein has the disadvantage of being subject to quenching by the exciting radiation more rapidly than rhodamine (so that the signal fades rapidly while viewing the sample). However, compounds are available to retard this photobleaching effect. In addition, the strong signals obtained due to the high levels of antigen produced in baculovirus-infected insect cells generally allow ample time for examination and photography before the signal is reduced to an unusable level. While rhodamine is not prone to such rapid photobleaching, it tends to give higher nonspecific binding of the secondary antibody to the sample, and, consequently, higher backgrounds. Again, the high levels of antigen normally present in this system should permit the dilution of the secondary antibody to the point where this is not a problem.

The visualization of fluorochrome-conjugated antibodies requires a microscope equipped for epifluorescence, that is, one in which the exciting light is transmitted through the objective lens onto the specimen. Special filter sets are required to ensure that the specimen is only irradiated by light of the correct wavelength and that only emitted light is viewed by the observer. Suitable filter sets are readily available for both fluoroscein and rhodamine.

CONTROLS: Controls should include the staining of recombinant virus-infected cells with control antibodies, and the staining of uninfected cells and cells infected with wt virus (and/or parent virus if different) with the test antibodies. An additional useful control is to stain infected cells with only the secondary antibody. This provides information on the background signal due to the secondary antibody. It may also be useful to examine an unstained preparation of infected cells, as they may autofluoresce at low levels. The best control antibody for a polyclonal antiserum is serum bled from the same animal prior to immunization. If this serum is not available, use nonimmune serum from the same species. For a monoclonal antibody, the best control is a monoclonal antibody of the same type raised against an unrelated protein.

Materials

1. 35 mm tissue-culture dishes (see Chapter 11)
2. No. 1 glass coverslips (VWR # 48366 067)
3. Complete tissue-culture medium. (See Chapter 11 for details concerning media.)
4. SF cells grown to 1×10^6 cells/ml
5. Viruses: recently titered stocks of the recombinant virus expressing the protein of interest, wt virus, and/or the parent virus used in the construction (if different)
6. PBS, *pH* 8.0: 8 mM Na_2HPO_4; 137 mM NaCl; 0.5 mM $MgCl_2$; 1.6 mM KH_2PO_4; 2.7 mM KCl; pH 8.0. Note that, for optimal antibody-antigen interaction, we use PBS at pH 8.0 in this experiment rather than the pH 6.2 PBS used in other protocols.
7. Fixative: 70% acetone; 30% methanol. Prepare only small volumes each time and store at –20° C in an explosion-proof freezer.
8. Antibodies
   a. Polyclonal antiserum or monoclonal antibody specific to the protein you are expressing
   b. Control antisera
   c. Secondary antibody; fluorochrome-conjugated (either fluoroscein or rhodamine) polyclonal antiserum directed against immunoglobulins of the species in which the primary antibody was raised [e.g., fluoroscein isothiocyanate (FITC)-conjugated goat antiserum against rabbit IgG (Sigma # F 0382)]

   Store antibodies at –20° C in small aliquots. They are stable for many years in this form. However, do not freeze and thaw repeatedly. Working stocks may be stored for at least six months at 4° C. Sodium azide should be added to a concentration of 0.02% (w/v) to stocks stored at 4° C. **Caution:** Sodium azide is poisonous. Label all solutions clearly and handle with care.
9. Mounting solution: 60% (v/v) glycerol; 2.5% DABCO [1,4-diazanicyclo-[2.2.2] octane—bleaching retardant, Sigma # D 2522] Store at 4° C.
10. Clear nail polish

Method

1. Infection and sample preparation
   a. Sterilize the glass coverslips by dipping in ethanol and flaming. Alternatively, place in a glass dish and sterilize by baking.

b. Place the coverslips in 35 mm tissue-culture dishes (one coverslip per dish). Prepare one dish per virus per time point, as well as extra dishes for the controls.
c. Seed each dish with 1×10^5 cells. The final volume of tissue-culture medium should be 2 mls per dish. To help disperse the cells evenly, it is best to add some tissue-culture medium to the dishes before adding the cells, or dilute the cells to their final volume before plating. Rock each dish gently *immediately* after adding the cells to ensure an even monolayer. Incubate at 27° C overnight to ensure good attachment of the cells to the glass coverslip.
d. Dilute the viruses in tissue-culture medium to give 4×10^6 pfu in 250 μl to 500 μl. This will give an MOI of 20 pfu/cell. You should use the same final volume for all viruses.
e. Aspirate the tissue-culture medium from the cell monolayers and infect with the appropriate diluted virus. *Be careful not to let the cells dry out before addition of the virus.* Instead of virus, add an equal volume of tissue-culture medium to the mock-infected control. Incubate at room temperature for one hour with gentle rocking.
f. Aspirate the inoculum from the cells and refeed with 2 mls complete tissue-culture medium. Incubate at 27° C.
g. Monitor the course of the infection as described in the time course protocol earlier in this chapter.
h. At each time point, remove the tissue-culture medium and rinse the cells three times in cold PBS.
i. Remove the coverslip from the tissue-culture dish and immerse in fixative *in a glass container* for 10 minutes at –20° C. (The fixative may cause the coverslip to irreversibly stick to a plastic dish.) *Ensure that you always know which side of the coverslip the cells are on.* (If you do drop the coverslip and are unsure of the orientation, you can identify the cell-coated side by carefully scraping a corner of both sides of the coverslip while viewing the cells under the tissue-culture microscope. When you scrape the cell-coated side, you will be able to see the cells being removed.)
j. Remove the coverslip from the fixative, place cell side up on a paper tissue, and air-dry.
k. Once dry, the coverslip may be returned to the tissue-culture dish, *cell side up,* and stored at –20° C.

2. Incubation with primary antibody
   a. Incubate each sample in 2 mls PBS for 15 minutes at room temperature. Do not let the coverslips dry at any point from now to the end of the procedure.
   b. Aspirate the PBS and place 50 μl of the (diluted) primary antibody onto each coverslip. The antibody should cover as much of the coverslip as possible. The appropriate dilution of antibody to use should be determined empirically. We suggest trying a 1:100 dilution for a polyclonal antiserum and using monoclonal antibodies undiluted. Incubate for one hour at 37° C in a humidified environment. To ensure that the coverslip does not dry out, we recommend placing a small piece of tissue paper soaked in PBS in each dish.
   c. Rinse each sample once in PBS, then wash twice for 15 minutes each in 2 mls of PBS at room temperature. Rock the samples gently during these washes.
3. Incubation with the secondary antibody
   a. Aspirate the PBS and place 50 μl of diluted secondary antibody onto each coverslip, trying to cover as much of the coverslip as possible. The correct dilution will have to be determined empirically. We suggest initially

trying a 1:100 dilution. Incubate at 37° C in a humid environment. Again, a piece of tissue soaked in PBS can be placed in each dish. From this point onward, protect the samples from light as much as possible.

b. Rinse the samples once with PBS and wash twice for 15 minutes each with 2 mls PBS at room temperature. Rock gently during these washes.

4. Mounting and viewing the sample
   a. Place 20 µl of mounting solution in the center of a clean microscope slide. Remove a coverslip from the PBS, drain excess PBS, and slowly place the coverslip, *cell side down,* onto the mounting solution. Be careful to avoid trapping air bubbles between the coverslip and the microscope slide. Remove any excess mounting solution from the edges of the coverslip with a paper tissue.
   b. Seal the edges of the coverslip with clear nail polish. This will prevent the coverslip from moving on the slide. The sample may be viewed immediately or stored at 4° C in the dark. We do not recommend storing the samples for more than two to three days.
   c. View the samples using a microscope equipped for epifluorescence with a UV source and a camera. Ensure that you are using the appropriate filter set for the particular fluorochrome. Remember to use immersion oil when necessary.
   d. We recommend first examining the samples by phase contrast without exposing the samples to the UV light. When you have focused the microscope and chosen a suitable field of view, switch over to the UV source. Cells expressing the cognate antigen should display strong fluorescence from the particular subcellular region and/or organelle containing the antigen. Nonexpressing cells will display a low level of diffuse fluorescence (they will not be completely invisible). Careful comparison of the sample with the controls will allow you to determine whether cells infected by your recombinant virus are expressing antigen recognized by the specific antibody. Photograph the samples to obtain a permanent record of the results as required.
5. Elimination of background

   High background may derive from a number of sources including autofluorescence, the primary antibody, or the secondary antibody. Careful examination of the controls may identify the source of the problem. Any or all of the following steps may help reduce these problems. See Harlow and Lane (1988) for further suggestions.

Autofluorescence

Try using a fluorochrome with a different excitation and emission spectrum.

Primary Antibody

a. Titer the antibody and use the minimum amount required to get a suitable signal.
b. If using a monoclonal antibody, try using a different antibody if available.
c. Increase the number and/or duration of washes between each step. Try including 0.2% Tween-20 in the PBS for washing.
d. Reduce the incubation time.
e. If using a polyclonal antiserum, preadsorb the antiserum against extracts of uninfected cells.

Secondary Antibody

a. Titer the antibody and use the minimum amount required to get a suitable signal.

b. Use an antibody conjugated with a different fluorochrome.
c. Increase the number and/or duration of washes between each step. Try including 0.2% Tween-20 in the PBS for washing.
d. Reduce the incubation time.

ISOLATION AND NORTHERN BLOT ANALYSIS OF RNA

In cases where recombinant protein synthesis is very low, it may be useful to examine the expression of the heterologous gene at the transcriptional level. This section deals with the isolation and Northern blot analysis of poly(A)$^+$ RNA from AcMNPV-infected SF cells. [Readers who are interested in the analysis of the RNA by other techniques are directed to standard molecular-biology manuals (Berger and Kimmel, 1987; Davis et al., 1986; Sambrook et al., 1989) for specific recommendations.]

By far the most critical parameter for the successful manipulation of RNA is the maintenance of a ribonuclease (RNase)-free working environment. RNases are extremely stable enzymes. They are not destroyed by boiling, and they renature readily after treatment with many denaturants. They have no cofactor requirements and are active in a broad range of conditions. Sources of RNases in the laboratory include the sample being characterized, solutions, glassware, and human beings; RNases from fingers are a major problem when working with RNA.

In the protocol described here (Chirgwin et al., 1979), endogenous RNases present in the infected cells are inactivated by lysis in guanidinium isothiocyanate (GITC). This is an extremely powerful chaotropic protein denaturant, which, when used in combination with a reductant, irreversibly denatures all RNases. To avoid contamination by RNases from exogenous sources, the following recommendations should be followed:

- Maintain separate stocks of reagents exclusively for use in the isolation of RNA. Ideally, certain pieces of equipment, such as pipets and a gel box, should also be reserved exclusively for RNA work.
- Wear gloves at all times during the manipulation of RNA or when handling any reagents or equipment used for RNA work.
- Whenever possible, use sterile disposable plasticware for the preparation and storage of reagents used in RNA manipulation.
- Glassware and spatulas should be sterilized by baking at 180° C overnight.
- All solutions made up in water should be prepared with diethyl pyrocarbonate (DEPC)-treated water (see the following Materials list). DEPC is a nonspecific inhibitor of RNases. Note that DEPC will carboxymethylate purine residues in RNA, and it must be thoroughly removed from all solutions by autoclaving before use, especially if the RNA is to be used for *in vitro* translation experiments.
- Prior to adjusting the pH of an RNase-free reagent, the pH electrode should be immersed in GITC (see Materials) for 30 minutes, then rinsed extensively with DEPC-treated water.
- Prior to determining the absorbance of RNA-containing solutions, soak the quartz cuvettes for one hour in concentrated HCl:methanol (1:1) and rinse copiously with DEPC-treated water.

Scrupulous adherence to these recommendations is essential for the successful isolation and characterization of RNA.

The following protocol is subdivided into sections describing the infection and lysis of SF cells, the purification of poly(A)$^+$ mRNA, and the characterization of specific transcripts by Northern blot analysis.

Infection and Lysis of SF Cells for RNA Isolation

This part of the protocol is simply a time course experiment, and the parameters and guidelines discussed earlier in this chapter are relevant. The choice of appropriate time points depends on the promoter used to drive expression of your heterologous gene. For late and very late promoters, time points at 12, 24, 36, and 48 hours will probably be suitable.

> CONTROLS: The controls used should include a time course carried out with wt virus (and/or the parent virus if different). In addition, RNA derived from mock-infected cells should also be examined.

Materials

1. 100 mm tissue-culture dishes (see Chapter 11)
2. Complete tissue-culture medium. (See Chapter 11 for information concerning different media.)
3. SF cells grown to approximately 1×10^6 cells/ml
4. Recently titered stocks of recombinant and control viruses
5. DEPC-treated water: add DEPC (Sigma # D 5758) to ddH_2O at a final concentration of 0.1% (v/v). Mix well and let stand at room temperature overnight. Autoclave for at least two hours to remove the DEPC.
6. 250 mM sodium citrate, pH 7.0
7. GITC solution: 4.0 M guanidinium isothiocyanate (Fluka # 50990); 25 mM sodium citrate, pH 7.0; 0.5% sodium lauryl sarcosinate (Sigma # L 5125). Filter through a Whatman No. 1 filter and store at room temperature. Add β-mercaptoethanol to a final concentration of 1% (v/v) just before use.
8. 15 ml polypropylene tubes (e.g., Falcon # 2067)

Method

1. Seed 100 mm tissue-culture dishes with 5×10^6 cells per dish. We recommend using two dishes per time point for each virus you analyze, as well as two dishes for a mock-infected control. The final volume of tissue-culture medium should be 10 mls per dish. To help disperse the cells evenly, it is best to add some tissue-culture medium to the dishes before the cells or to dilute the cells to their final volume before plating. Rock each dish gently *immediately* after addition of the cells to ensure an even monolayer. Incubate at 27° C for 30 minutes to one hour to allow the cells to attach. Alternatively, the cells may be plated out at a density of 2.5×10^6 cells per dish and incubated at 27° C overnight.
2. Infect the cells with the appropriate viruses at an MOI of 20 pfu/cell and monitor the course of the infection as described in the time course protocol at the beginning of this chapter.
3. At each time point, aspirate the bulk of the extracellular fluid from two dishes infected by each virus (leave about 2 mls). Scrape the cells off both dishes and combine in a sterile 15 ml polypropylene tube. Process the mock-infected plates at the same time as the first time point.
4. Centrifuge at $1000 \times g$ for five minutes at 4° C.
5. Discard the supernatant and add 2.5 mls GITC solution containing β-mercaptoethanol to each pellet. Vortex *immediately* until the pellet has dissolved.
6. Store at –80° C.

Isolation and Purification of Poly(A)$^+$ RNA

RNA present in the lysates is now separated from the denatured proteins and DNA by centrifugation through a cesium-chloride cushion. The RNA pellet obtained after centrifugation consists of total cellular RNA. [It is often not essential to specifically analyze poly(A)$^+$ RNA; the total RNA sample obtained at this stage can be used instead.] Poly(A)$^+$ RNA is purified from total RNA by affinity chromatography on oligo-dT cellulose. The following protocol is based on that of Davis et al. (1986).

Materials

1. DEPC-treated water
2. 5.7 M CsCl; 0.1 M EDTA, pH 7.5. Autoclave and store at room temperature.
3. Clear ultracentrifuge tubes (e.g., Beckman # 344057)
4. TES: 10 mM Tris-HCl, pH 7.5; 1 mM EDTA. Autoclave and add SDS to 0.1% (w/v). Store at room temperature.
5. 3 M sodium acetate, pH 5.2. Autoclave and store at room temperature.
6. Absolute ethanol
7. 70% (v/v) ethanol
8. Oligo(dT) cellulose (BRL # 5940SA)
9. Disposable column (e.g., Bio-Rad polyprep column # 531-1550)
10. Glass wool: bake at 180° C overnight before use.
11. Disposable 1 ml syringe (Becton Dickinson # 9602)
12. 0.1 M NaOH; 5 mM EDTA
13. Loading buffer A: 40 mM Tris-HCl, pH 7.4; 1 M NaCl; 1 mM EDTA. Autoclave and add SDS to 0.1% (w/v). Store at room temperature.
14. Loading buffer B: 20 mM Tris-HCl, pH 7.4; 0.1 M NaCl; 1 mM EDTA. Autoclave and add SDS to 0.1% (w/v). Store at room temperature.
15. Elution buffer: 10 mM Tris-HCl, pH 7.4; 1 mM EDTA. Autoclave and add SDS to 0.05% (w/v). Store at room temperature.

Method

Isolation of Total RNA

1. Thaw the cell lysates and load each onto a 2 ml cushion of 5.7 M CsCl, 0.1 M EDTA in ultraclear tubes for an SW 55 rotor (Beckman) or equivalent.
2. Centrifuge at 150,000 × g (35,000 RPM in an SW 55) for 15 hours at 20° C.
3. Very carefully remove the supernatant with a pipet. Use an automatic pipet for the last ml.
4. Invert the tube and drain well. The pellet will be transparent and may be difficult to see.
5. Very carefully add 0.5 ml 70% ethanol to the tube, then invert and drain well.
6. Allow the pellet to dry at room temperature. Dissolve in 200 µl TES by drawing up and down many times with an automatic pipet. *Ensure that the RNA pellet has completely dissolved.* It may be quite difficult to resuspend: this is a frequent cause of low recovery of RNA.
7. Transfer the RNA solution to a microfuge tube. Rinse the ultracentrifuge tube with 50 µl TES and transfer to the microfuge tube. Add 25 µl 3 M sodium acetate, pH 5.2, and 625 µl ethanol.
8. Store the RNA as an ethanol precipitate at –80° C until needed.
9. Centrifuge at 12,000 × g in a microfuge for 10 minutes at 4° C to collect the RNA. Rinse the pellet carefully with 70% ethanol and air-dry.
10. Dissolve the RNA pellet in 50 µl sterile DEPC-treated water. Read the OD_{260} and OD_{280} of an aliquot. The OD_{260} should be 1.8 to 2 times the OD_{280}. An OD_{260} of 1.0 corresponds to an RNA concentration of approximately 40 µg/ml.

This RNA sample may be used in the following Northern blot protocol, or in other protocols if you wish to use total RNA. If you are not going to use the RNA immediately, add 0.1 volumes of 3 M sodium acetate, pH 5.2, and 2.5 volumes of ethanol, and store at –80° C.

Purification of Poly(A)$^+$ RNA

1. Preparation of oligo(dT)-cellulose column
   a. Suspend 0.5 g to 1.0 g of oligo(dT)-cellulose in loading buffer B.
   b. Pour into a disposable column or a 1 ml syringe plugged with glass wool to give a packed column volume of 0.5 ml to 1 ml.
   c. Wash with 3 mls of 0.1 M NaOH, 5 mM EDTA.
   d. Wash column with DEPC-treated water until the pH of the effluent is less than 8.0.
   e. Equilibrate the column with 5 mls of loading buffer A.
2. Loading the column

   Poly(A)$^+$ RNA is bound to the column by loading the sample in a high-salt buffer.
   a. Dilute 5 mg to 10 mg of total RNA to a final volume of 500 µl with DEPC-treated water.
   b. Heat at 65° C for five minutes to disrupt any regions of secondary structure that might impede binding.
   c. Add 500 µl of loading buffer A, which has been prewarmed to 65° C, and cool rapidly to room temperature.
   d. Apply the sample to the column. Collect the eluate, heat to 65° C for five minutes, cool rapidly to room temperature, and reapply to the column.
3. Washing the column

   Nonpoly(A)$^+$ RNA is now removed from the column by extensive washing with a medium-salt loading buffer.
   a. Wash the column with 5 mls to 10 mls of loading buffer B.
   b. Collect 1 ml fractions and read the OD_{260} of each. Initial fractions should have a very high absorbance, but this will decrease as all the non-polyadenylated RNA is washed from the column. Continue washing until the eluate has little or no absorbance at 260 nm.
4. Elution of poly(A)$^+$ RNA

   Poly(A)$^+$ RNA is eluted from the column with a buffer lacking salt.
   a. Elute the column with 2 mls to 3 mls elution buffer.
   b. Collect 0.25 ml fractions. Identify the fractions containing the poly(A)$^+$ RNA by measuring the absorbance at 260 nm.
   c. Pool the RNA-containing fractions, heat to 65° C for five minutes, dilute with an equal volume of loading buffer at 65° C, and cool rapidly to room temperature. Carry out a second round of chromatography on the same column as described in Steps 2 through 4.
5. Concentration and storage of poly(A)$^+$ RNA
   a. Pool the final RNA-containing fractions. Add 0.1 volumes of 3 M sodium acetate, pH 5.2, and 2.5 volumes of ethanol. Store the RNA as an ethanol precipitate at –80° C until needed.
   b. To collect the RNA, centrifuge at 12,000 × g for 10 minutes at 4° C in a microfuge. Decant the supernatant and wash the pellet carefully in 70% ethanol. Drain and air-dry.
   c. Resuspend the pellet in a small volume of DEPC-treated water and determine the quantity of RNA present by measuring the absorbance at 260 nm as described earlier. About 1% to 2% of the total RNA loaded should be recovered as poly(A)$^+$ RNA.

Northern Blot Transfer of RNA

Northern blot analysis is the RNA equivalent of Southern blotting of DNA. One of the principal differences between the techniques is that, because RNA is single stranded, it has a propensity to form secondary structures that cause it to migrate aberrantly during agarose gel electrophoresis. Thus, it is necessary to denature the RNA prior to or during electrophoresis. A range of techniques have been developed that accomplish this, notably denaturation by glyoxal/DMSO, formaldehyde, or methyl mercury. We favor the first of these approaches, originally devised by McMaster and Carmichael (1977), because it is significantly less hazardous than the others. Both formaldehyde and methyl mercury are highly toxic.

Because the RNA is already denatured, there is no denaturation step after electrophoresis. There is also no need to fragment the RNA in any way. Instead, the RNA is simply transferred directly to a solid support, either nitrocellulose or activated nylon. As for DNA blotting, nitrocellulose has the advantage of giving lower backgrounds. However, the nylon is considerably stronger and easier to manipulate. For the analysis of highly abundant transcripts, such as those from genes under the control of a very strong viral promoter (e.g., *polh*), nylon membranes are probably the matrix of choice.

Materials

1. Total or poly$(A)^+$ RNA derived from control and test samples
2. Electrophoresis apparatus (reserved for RNA work)
3. DEPC-treated water. Depending on the electrophoresis apparatus used, you may need a large volume of DEPC-treated water (4 l-5 l).
4. Peristaltic pump with tubing (e.g., 1.14 mm internal diameter, Gilson # F1 17 93 9). Reserve tubing for RNA work.
5. Mixed-bed ion-exchange resin [e.g., Bio-Rad #AG 501-X8(D)]
6. Deionized glyoxal
   a. Add 50 mls 40% glyoxal (Sigma # G 3140; this is a 6 M solution) to 5 g ion-exchange resin. Mix well for 10 to 15 minutes, then filter through Whatman No. 1 filter paper or equivalent.
   b. Measure the pH of the filtrate.
   c. Repeat as necessary with fresh resin until the pH is greater than 5.0.
   d. Store in small aliquots at –80° C.
   e. Do not reuse an aliquot once it has been opened.
7. DMSO (Fisher # D128-500): Prepare several small aliquots from a newly opened bottle and store at –80° C. Do not reuse an aliquot once it has been opened.
8. 0.1 M NaH_2PO_4
9. 0.1 M Na_2HPO_4
10. 0.1 M sodium phosphate: Add 0.1 M NaH_2PO_4 to 0.1 M Na_2HPO_4 until the pH is 7.0.
11. 10X running buffer: 0.1 M sodium phosphate; 10 mM EDTA, pH 7.0
12. Markers: A variety of RNA-size markers are commercially available (e.g., BRL RNA Ladder; # 5620SA). You may also use DNA markers. However, it is essential that the DNA markers are RNase-free and resuspended in TE, pH 7.5.
13. Agarose (SeaKem ME, FMC # 50014)
14. RNA loading buffer: 50% (v/v) glycerol; 10 mM sodium phosphate, pH 7.0; 0.25% (w/v) bromophenol blue (Bio-Rad # 161-0404); 0.25% xylene cyanole FF (Bio-Rad # 161-0423)
15. Two magnetic stir plates
16. Teflon coated stir bars: Bake overnight at 180° C before use.
17. Ethidium bromide (Sigma # E 7637): 0.5 µg/ml in 0.1 M ammonium acetate

Caution: Ethidium bromide is moderately toxic and a powerful mutagen. Wear gloves when handling reagents containing this dye.

18. UV transilluminator

    Caution: UV light is harmful, particularly to the eyes. Wear protective goggles or a safety mask that efficiently blocks UV light.
19. Polaroid camera
20. 20X SSC
21. Activated nylon membrane (e.g., Bio-Rad Zetaprobe # 162-0159 or equivalent)
22. Whatman 3MM paper: four to five pieces cut to the size of the gel
23. Plastic film (e.g., Saran Wrap)
24. Paper towels

Method

RNA Electrophoresis

1. Preparation of electrophoresis apparatus
   a. Soak the gel box, tray, and comb in 0.1% (v/v) DEPC overnight. It is a good idea to pump the DEPC solution through the peristaltic pump tubing while doing this.
   b. Rinse the apparatus extensively with DEPC-treated water. Again, pump the DEPC-treated water through the peristaltic pump tubing to rinse the tubing.
2. Denaturation of samples
   a. Mix the following on ice (for a 20 μl well):
      2.7 μl deionized glyoxal
      8.5 μl DMSO
      1.6 μl 0.1 M sodium phosphate (pH 7.0)
      RNA sample [up to 10 μg of total RNA or 0.5 μg to 1 μg of poly(A)$^+$ RNA]
      DEPC-treated water to a final volume of 16 μl
      Remember to denature the markers at the same time. DNA markers must be in TE (pH 7.8) to ensure proper denaturation.
   b. Incubate at 50° C for one hour.
3. Pouring the gel
   a. While the samples are being denatured, prepare the agarose gel. Seal the ends of the gel tray in the appropriate manner for the apparatus you are using. (Most often, this simply involves taping the ends of the tray.) Place the tray on plastic film to avoid RNase contamination of the tray. Make a solution of 1% to 1.6% agarose in 1X running buffer and melt by microwaving for two to five minutes. Do not add ethidium bromide.
   b. Cool the agarose to 65° to 70° C and pour into the sealed gel tray with the comb in place. Let harden for 30 to 45 minutes at room temperature. Cover with plastic film during this time to keep dust off.
   c. When the agarose has set, carefully remove the comb and seals from the gel. Place the gel in the electrophoresis unit, and submerge it in 1X running buffer.
   d. Place a stir bar in the buffer chamber at either end of the gel and place the entire unit on two magnetic stir plates.
4. Running the gel

   Due to its low buffering capacity, the running buffer must be recirculated during electrophoresis. Otherwise, the pH will rise dramatically at the cathode during electrophoresis. At a pH greater than 8.0, the glyoxal dissociates from RNA.

   a. After denaturation, cool the samples rapidly on ice.

b. Add 4 µl of loading buffer to each sample and immediately load onto the gel. If possible, leave an empty lane between the markers and the samples.
c. Connect the electrophoresis unit to the power supply. The RNA will migrate toward the anode, so connect the red lead to the electrode furthest from the wells.
d. Run at 100 V for 10 to 15 minutes until the samples have entered the gel.
e. Circulate the buffer in the electrophoresis unit with the peristaltic pump. Mix the buffer in both chambers using the stir bars.
f. Run the gel at 3 V/cm to 4 V/cm until the bromophenol blue dye has migrated approximately two-thirds of the length of the gel. Depending on the gel apparatus used, this will probably take from 12 to 20 hours.

5. Visualization of markers
   a. After electrophoresis, cut the lane containing the markers off the gel.
   b. Stain in the ethidium bromide solution for 30 to 45 minutes.
   c. View on a UV transilluminator, place a ruler beside the markers, and photograph with the Polaroid camera.

Transfer of RNA to Membrane

1. Fill a large dish with 10X SSC.
2. Place a glass or perspex support on top of the dish. Wet two sheets of Whatman 3MM paper in 10X SSC and place them across the support such that each end is immersed in the 10X SSC. These sheets will act as a wick during transfer.
3. Carefully invert the gel onto the wet sheets of 3MM paper. Be careful to avoid trapping any air bubbles. (It helps to flood each layer of the transfer "sandwich" with 10X SSC prior to placing the next layer on top.)
4. Cut the nylon membrane to the size of the gel, wet it in 10X SSC, and lay it on top of the gel. If the membrane does not wet completely, do not use it, or at least ensure that the unwetted area is not placed over a critical part of the gel. Again, try to avoid trapping any air bubbles.
5. Wet a sheet of 3MM paper (cut to the size of the gel) in 10X SSC, and place it on top of the membrane. At this stage, it is a good idea to roll a sterile pipet across the sandwich, applying gentle pressure, to exclude any bubbles that may have been trapped.
6. Place three to four sheets of dry 3MM paper (cut to the size of the gel) on top of the wet sheet of 3MM paper.
7. Place a sheet of Saran Wrap on top of the 3MM paper. With a razor blade, cut a window out of the Saran Wrap so that the film covers only about 5 mm of the 3MM paper on all sides.
8. Place a stack of paper towels (≈10 cm) on top of the 3MM paper, and, finally, place a suitable weight (e.g., a 500 g reagent container) on a glass plate on top.
9. Allow the transfer to proceed for 15 to 24 hours. It may be necessary to replace the 10X SSC and the paper towels during transfer.
10. Disassemble the transfer sandwich and mark the positions of the wells on the membrane with a pencil. Carefully peel the membrane off the gel and rinse briefly in 10X SSC.
11. Air-dry the filter and then bake for one to two hours at 80° C under vacuum. Once the membrane has been baked, the RNA will be irreversibly bound to it, and it is no longer necessary to use RNase-free reagents. Store at room temperature until needed.

Generating Strand-Specific Probes by Primer Extension

The immobilized RNA is detected by hybridization to a labeled probe in much the same way that DNA hybridizations are carried out. Any method of labeling the probe DNA or RNA may be used. However, it is often desirable to use a probe labeled on one strand only so that hybridization is limited to transcripts in one direction. This is particularly important for analysis of genes expressed in the *polh* locus because several antisense transcripts (i.e., in the opposite direction to polyhedrin transcription) occur in this region. The technique described here involves the extension, in the presence of a labeled nucleotide, of a primer bound to single-stranded DNA. The method is fast and simple and routinely yields strand-specific probes of specific activity $> 1 \times 10^8$ cpm/μg. However, the DNA clone must be in a vector that can be used to generate single-stranded DNA.

Materials

1. Single-stranded DNA: The probe DNA will need to be cloned in a plasmid with a bacteriophage f1 origin of replication, which is designed to allow the generation of single-stranded DNA [e.g., the pBluescript series of plasmids (Stratagene)]. These plasmids generally come in pairs, permitting the generation of either strand of DNA. Remember, when cloning your DNA, to ensure that you use a plasmid that will allow you to generate single-stranded DNA having *the same orientation* as the transcript to be analyzed. (This DNA will be used as template for the synthesis of the labeled strand which must be complementary to the transcript of interest.) Prepare the single-stranded DNA as described by the suppliers of the particular plasmid you are using.
2. Appropriate primer: The primer you use will depend on the particular plasmid into which your probe DNA has been cloned. Generally, there are primer binding sites at either end of the cloning sites to facilitate sequence analysis. For DNA cloned in a pBluescript plasmid (Stratagene), the primer used will either be the T3 or T7 primer. Remember that only one of these primers will bind to the single-stranded DNA you have generated.
3. 5X buffer: 200 mM Tris-HCl, pH 7.5; 100 mM $MgCl_2$; 250 mM NaCl
4. [α-^{32}P]dCTP (3000 Ci/mmol; NEN # NEG 013C)
5. dNTPs: dATP, dGTP, and dTTP may be obtained as 100 mM aqueous solutions from Sigma (# D 4788, # D 5038, and # T 9656). Prepare 10 mM solutions in 3 mM Tris-HCl, pH 7.0, 200 μM EDTA and store at – 20° C.
6. 0.1 M Dithiotreitol (DTT; Sigma # D 9779)
7. Sequenase (T7 DNA polymerase; USB # 70722). Store at –20° C.
8. 10% SDS
9. TE
10. Disposable column (e.g., Bio-Rad polyprep column # 531-1550)
11. 1 ml disposable syringe (Becton Dickinson # 9602)
12. Glass wool
13. Sephadex G50-80 (Sigma G-50-80)
    a. Swell the Sephadex beads in a large volume of TE (10 g will swell to 100 mls).
    b. Let the beads settle and remove the liquid phase.
    c. Add in fresh TE so that the volume of the resin and liquid phase are approximately equal.
    d. Autoclave and store at room temperature.
14. 15 ml conical centrifuge tubes (e.g., Falcon # 2067)

Method

1. Annealing the primer and template DNA
    a. Mix the following on ice:

0.1-0.3 pmoles of single-stranded DNA (200 ng to 500 ng of a 4 kb molecule)
1 to 3 pmoles of primer (5 ng to 15 ng of a 17-mer)
2 µl 5X buffer
ddH_2O to a volume of 10 µl

b. Incubate at 65° C for one hour.
c. Cool slowly to room temperature.

2. Incorporation of label by primer extension
   a. Dry down 50 µCi [α–^{32}P]dCTP (see note).
   b. Add the following to the dried label:
      annealed primer-template mix from Step 1
      0.5 µl of dGTP, dATP, and dTTP stocks
      2 µl 5X buffer
      1 µl 0.1 M DTT
      3 units Sequenase (diluted to a concentration of 1.5 units/µl in ice-cold TE immediately prior to use)
      Bring to a volume of 20 µl with ddH_2O.
   c. Incubate at 37° C for 15 minutes.
   d. Add 0.5 µl of the dCTP stock and incubate for 15 minutes longer at 37° C.
   e. Stop the reaction by adding 1 µl 10% SDS and 79 µl TE.
3. Preparation of Sephadex spun-column
   a. Plug a 1 ml syringe with glass wool or use a disposable column.
   b. Fill the column to the top with the Sephadex beads. The column may be packed more rapidly by applying a gentle vacuum. Keep adding more Sephadex until the resin completely fills the syringe.
   c. Place the syringe in a 15 ml conical centrifuge tube and centrifuge at 1000 × g at room temperature for five minutes.
   d. Discard the flow-through. Place a screw-cap microfuge tube under the syringe in the conical tube.
4. Removal of unincorporated nucleotides
   a. Pipet the labeling reaction mix onto the top of the column. Centrifuge at 1000 × g for 15 minutes at room temperature.
   b. Collect the flow-through (in the screw-cap microfuge tube). Take a 1 µl aliquot and count to determine the specific activity of the labeled probe. Store the probe at –20° C until needed.

Note:

50 µCi of [α–^{32}P]dCTP, at a specific activity of 3000 Ci/mmol, represents 16.5 pmol of dCTP. If you use 0.3 pmoles of template single-stranded DNA, 16.5 pmol of dCTP should allow the synthesis of approximately 150 nt of the second strand. The cold chase (Step 2-d) allows the extension of this labeled segment. If necessary, the length of the labeled segment of the second strand can be increased by using more labeled nucleotide or using less template.

Hybridization

The theory behind the hybridization of labeled probes to RNA, immobilized on a solid matrix, is essentially the same as that for hybridization to immobilized DNA (see Southern blot protocol in Chapter 13), except that the melting temperature (T_m) of an RNA-DNA hybrid is somewhat higher than the corresponding DNA-DNA hybrid (see note). However, as for Southern blot analysis, you will be using probes that are 100% identical to the target sequence and that are moderately large. The small differences likely to occur between the lengths and the G + C contents of the probes used in different experiments will not affect the T_m very dramatically. Thus, a single set of hybridization conditions that should work well in all cases can be described. Readers interested in a more extensive discussion of the parameters of hybridization of nucleic acids on a solid support should consult the review by Wahl et al. (1987).

Materials

1. 20X SSC
2. 10% SDS
3. 50X Denhardt's reagent
   OR
   1X BLOTTO: 5% (w/v) nonfat dried milk; 0.02% sodium azide (Sigma # S 2002). Store at 4° C.
   Caution: Sodium azide is poisonous. Label all solutions clearly and handle with care.
4. 10 mg/ml denatured salmon-testes DNA (Sigma # D7656). This is supplied as a sonicated, denatured solution, that has been phenol-chloroform-extracted and ethanol-precipitated.
5. Formamide (Fluka # 47670). Fluka formamide generally does not need to be deionized. Aliquot and store at –20° C.
6. 1 M NaH_2PO_4
7. 1 M Na_2HPO_4
8. 1 M sodium phosphate: Add 1 M NaH_2PO_4 to 1 M Na_2HPO_4 until the pH is 7.0.
9. Prehybridization solution: 50% formamide; 5X SSC; 50 mM sodium phosphate, pH 7.0; 5X Denhardt's or 0.05X BLOTTO; 100 μg/ml salmon-sperm DNA; 0.1% SDS
10. Hybridization solution: 50% formamide; 5X SSC; 20 mM sodium phosphate, pH 7.0; 1X Denhardt's or 0.01X BLOTTO; 0.1% SDS

Method

1. Wet the blotted membrane by floating on a solution of 5X SSC for five minutes at 42° C.
2. Seal the membrane in a heat-sealable plastic bag. Cut one corner of the bag and add 150 μl of prehybridization solution/cm^2 of membrane. Seal the bag, trying to eliminate all the air bubbles. Submerge the bag in a 42° C water bath and incubate for at least 30 minutes (it can be much longer—several hours to overnight).
3. Prepare the probe as follows:
   a. The probe will be used at a concentration of 1 ng/ml to 2 ng/ml of hybridization solution. Put the required amount of probe in a microfuge tube.
   b. Add sheared salmon-testes DNA to the probe so that the concentration in the hybridization solution will be 100 μg/ml.
   c. Denature the probe and salmon-testes DNA by incubation at 100° C for five minutes. Place immediately on ice.
4. Remove the prehybridization solution from the membrane and add the same volume of hybridization solution (preheated to 42° C) to the bag. Add the denatured probe/salmon-testes DNA mix and quickly seal the bag, removing all air bubbles. To minimize the possibility of a leak of radioactive solution, seal the bag in a second heat-sealable bag. Submerge in a 42° C water bath and incubate overnight.
5. Carefully remove the hybridization solution from the bag. Then, remove the membrane and place it immediately in a dish containing a large volume of

2X SSC, 0.1% SDS at room temperature. Incubate at room temperature with agitation for five minutes.

6. Repeat this wash step another three times, then wash twice in 0.1X SSC, 0.1% SDS at 50° C for 15 minutes each time.
7. Wrap the membrane in Saran Wrap or other plastic film. Do not allow the membrane to dry after washing. Label asymmetrically with radioactive ink (waterproof ink, with old ^{32}P added to give approximately 1000 cpm/µl).
8. Expose the membrane to X-ray film (Kodak X-OMAT AR or equivalent) at room temperature. The appropriate exposure time will have to be determined empirically. Develop the film after one to two days and then reexpose as necessary. You can reduce the exposure time by using an intensifying screen. Place the film between the membrane and the screen and expose at –70° C. If you use an intensifying screen, exposure for several hours to overnight should be adequate. The signal from the radioactive ink will allow you to line up the autoradiograph with the membrane to facilitate the identification of the bands.

Note:

The T_m of a DNA-RNA hybrid (the temperature at which 50% of the hybrid is dissociated into single strands) can be estimated from the equation

$T_m = 79.8^o + 18.5(\log[Na^+]) + 58.4(\text{mole fraction } G + C) + 11.8(\text{mole fraction } G + C)^2 - 820/L - 0.5(\% \text{ formamide})$

where L is the length in nucleotides of the hybrid (Wahl et al., 1987). This relationship is valid for hybrids greater than approximately 50 bp.

15

POST-TRANSLATIONAL MODIFICATION

REVIEW OF POST-TRANSLATIONAL MODIFICATION IN BACULOVIRUS INFECTION

One of the major reasons for choosing the baculovirus expression vector system is to allow expression of a eukaryotic gene in a eukaryotic cell, thereby taking advantage of the pathways in these cells that facilitate the folding, modification, and assembly of the protein product. These features become particularly important for proteins that are known to undergo complex post-translational modifications.

One of the most frequently asked questions concerning the baculovirus expression system is whether the heterologous protein will be folded, modified, and/or assembled in baculovirus-infected insect cells in the same manner that it is in its natural environment. This question cannot be answered unambiguously; precise answers must be determined empirically for each different protein. However, some general statements can be made concerning the likelihood and nature of certain types of post-translational modifications. In this section, several major types of post-translational modification are discussed: proteolytic processing (signal-peptide cleavage and internal proteolytic cleavage), N-terminal blocking, phosphorylation, glycosylation (N-linked and O-linked), lipid modification, α-amidation, disulfide bond formation, oligomerization and assembly. Subcellular localization of the heterologous protein is also discussed. Before discussing each of these post-translational events, however, a few general points regarding gene expression systems are noted.

Often it is not known *a priori* whether the protein of interest is subject to post-translational modification or, if it is modified, whether that modification is essential for biological activity. In some cases, the modification has no observable effect on protein activity but, in other cases, such modifications can alter the biological properties (for example, the solubility, stability, *in vivo* clearance rate, or specific activity) of the protein. More dramatic changes include failure of the protein to fold and/or assemble properly in the ER, resulting in lack of biological activity or immunogenicity.

For those concerned about the possible effects of post-translational modification on the eukaryotic protein of interest, it is best to express the gene in a eukaryotic cell (versus a prokaryotic cell). The chances of obtaining biologically active material without further manipulation appear to be generally better in the more similar cellular environment. However, it is important to be aware that there may be differences in the post-translational modification of the expressed gene product as compared to the same protein in its normal environment. This is true of any eukaryotic expression vector—baculoviruses included. (Indeed, the analysis of such differences could be instructive regarding protein structure/function relationships.) The primary advantage of using baculovirus vectors lies in the high levels of expression and easy identification of heterologous proteins.

Some types of post-translational modifications (e.g., neurohormone processing) may be species- or tissue-specific. If the desired type of post-translational modification occurs only in a specific cell type, it is quite likely that no available expression system, even one developed from the same species as the gene source, will provide the desired modification unless specific efforts have been made to retain the differentiated state of the tissue of interest. To date, few, if any, expression systems have been developed with this aim in mind.

Another factor to consider is the efficiency and heterogeneity of post-translational modification in expression systems. A system that is designed to express a heterologous gene at an extremely high rate may overwhelm the ability of the cell to modify the protein product. This often seems to be the case for the baculovirus system in which the N-glycosylation and phosphorylation systems may fail to keep pace with the very high levels of expression occurring in the last phase of the infection process. The researcher should be aware, but not unduly alarmed, that the post-translational modification of the protein products may be heterogeneous. Concern about such heterogeneity might be unwarranted because heterogeneity in post-translational modification is not uncommon for some proteins, even when synthesized in their natural cellular environment at their natural expression levels.

If heterogeneity is of known concern, several steps can be taken to maximize the ratio of fully processed versus underprocessed products in the baculovirus expression system. We recommend, under these specific circumstances, the use of promoters that are active earlier in infection, and the harvesting of proteins at earlier times during the infection (see Chapter 6).

Proteolytic Processing

Signal Peptides

A broad spectrum of eukaryotes use short (15 to 40 residue) signal peptides for the purpose of directing proteins into the endoplasmic reticulum (ER). These signal peptides, including those of mammalian origin, are usually processed normally in the baculovirus expression system. In many reported studies, proper cleavage of signal peptides in baculovirus-infected cells is inferred from the size of the product or its presence in the plasma membrane or extracellular culture fluid. At least a dozen different studies, however, have included amino acid sequence analysis of the N-terminus of the baculovirus-produced protein to confirm that the signal cleavage site is identical to that observed in the original source. Signal peptides of mammalian, plant, and yeast origin have all been shown to direct proteins into the ER and to be properly cleaved in baculovirus-infected insect cells.

Some signal peptides appear to be used and processed more efficiently than others in insect cells. The signal sequence of human tissue plasminogen activator (tPA) is reported to be rapidly recognized and cleaved (Jarvis and Summers, 1989). Signal sequences from more diverse sources, including plants, may also be efficiently recognized for cleavage and secretion (Bustos et al., 1988). In the case of plant prepropapain, however, signal cleavage may be a rate-limiting step for secretion (Vernet et al., 1990). A fivefold increase in the secretion of propapain was reported when the natural signal peptide was replaced with the signal peptide from melittin, a major component of honeybee venom (Tessier et al., 1991). (This signal sequence is used in the transfer plasmid pVT-Bac; see Chapter 7.) There is also a report that a bacterial signal sequence of the *Bacillus anthracis* protective antigen is not efficiently cleaved in the baculovirus system (Iacono-Connors et al., 1991).

In a systematic study of the effects of three different signals on two different secreted proteins and a nonsecreted protein, however, Jarvis and his colleagues found no significant effect of the signal on the level of secreted product or on the rates of secretion (D.L. Jarvis, personal communication). It is likely that a few signals (e.g., plant prepropapain) are not used efficiently and, for expression of proteins with such signals, expression can be enhanced by changing the signal. As this involves extensive modification of the existing heterologous gene sequence, we do not recommend changing signals for initial expression studies. If ratios of intracellular preprotein and extracellular mature protein or kinetics of secretion suggest that cleavage of the signal is rate-limiting, then signal replacement can be considered when optimizing the expression system.

There are two baculovirus proteins that are known to be directed to the ER during wild-type virus infection: gp64 and EGT. The glycoprotein gp64, also known as gp67, is a major structural component of the budded form of AcMNPV and is membrane bound. The predicted sequence of the gp64 protein revealed a hydrophobic N-terminus (Whitford et al., 1989). Although a likely signal cleavage site exists between two alanines at residues 38 and 39 of the published sequence, the amino acid sequence of the N-terminus of the mature gp64 protein has not yet been published.

The other ER-directed AcMNPV protein, the secreted EGT protein (O'Reilly and Miller, 1990), has a shorter hydrophobic N-terminus. N-terminal sequencing of the mature EGT protein (O'Reilly et al., 1992) reveals a cleavage site between two alanine residues at positions 18 and 19. This shorter signal should prove useful in future baculovirus expression work.

Proteolysis Including Proprotein Processing

The prosequences of precursor proteins may or may not be cleaved by the baculovirus expression system, depending on the nature of the protease involved in the process. Examples of accurate processing, inaccurate processing, and lack of processing have all been reported.

Inaccurate proteolytic processing was reported for human gastrin-releasing neuropeptide precursor (proGRP), which requires cell-specific enzymes (i.e., those found in endocrine cells) for complex proteolytic processing (Lebacq-Verheyden et al., 1988). This process is thought to include a trypsinlike protease cleavage at a pair of basic amino acids (residues 30–31), cleavage by a carboxyl B-like exopeptidase (to residue 28), and an α-amidating enzyme converting the terminal glycine to a C-terminal amide. In insect cells, the signal peptide is removed accurately and the prohormone is further cleaved into peptides that are similar in size to those found in mammals. However, the prohormone did not appear to be cleaved between the basic amino acids at residues 30 to 31. Instead, a novel cleavage reaction was observed between Met27 and Gly28. The loss of the glycine residue would obviate the ability of the cell to α-amidate the carboxy terminus of the mature neuropeptide (residues 1–27). Thus the insect cells appeared to produce biologically active gastrin neurohormone by a different route than mammalian cells; the insect-produced peptide, however, was not as active as its amidated mammalian counterpart.

There are a number of examples of prosequences that are not proteolytically cleaved and, in most of these cases, it appears that additional cell-specific or species-specific proteases are required for the cleavage. In the case of frog amidating enzyme, for example, the signal peptide of the preproprotein was cleaved properly but the proprotein was not cleaved to the mature protein (Suzuki et al., 1990). The C-terminal 138 residues of the *Schistosoma mansoni* precursor failed to be cleaved in the baculovirus expression system (Felleisen et al., 1990); it is known that this cleavage also does not occur for *in vitro* synthesized protein. The E2 peplomer glycoprotein of coronaviruses was not cleaved upon expression in insect cells (Yoden et al., 1989).

Plant prepropapain is a curious example of inefficient proprotein processing. Although the 26-residue signal peptide of prepropapain was cleaved correctly and the proprotein was secreted (Vernet et al., 1990), the secreted propapain was not further processed in the cell culture media. However, cleavage was induced by lowering the pH of the culture media in the presence of reducing agents and raising the temperature. N-terminal cleavage in the baculovirus extracellular media, however, occurred at a site a few residues from that observed in plants. This cleavage was due to a cysteine protease apparently released from lysed cells and was found only in media from virus-infected cells. The baculovirus-produced "mature" papain, however, was as active as the plant-derived papain. In this case, the baculovirus system succeeded in producing active papain where *E. coli* and yeast expression systems had failed.

There are numerous examples where proper proteolytic cleavage of a heterologous protein is observed. Many of these examples include cleavages at dibasic sites. For example, influenza virus (fowl plaque) hemaglutinin produced in insect cells is cleaved, although inefficiently, between the arginine and glycine of the sequence Lys-Lys-Arg-Lys-Lys-Arg-Gly (Kuroda et al., 1986). A number of other vertebrate glycoproteins require proteolytic processing at similar cleavage sites, and many can probably be processed properly in baculovirus expression systems.

It is likely, however, that proteolytic processing will not be efficient, particularly if synthesis is very high. For example, cleavage of the influenza virus hemaglutinin (HA) was slow and relatively inefficient (Kuroda et al., 1991). The half-time for cleavage of HA in insect cells was 90 minutes compared to 30 minutes in a vertebrate cell line, and a substantial proportion of the HA remained uncleaved. Some influenza virus HAs are not cleaved at all in the baculovirus system (Possee, 1986). The precursor of the F-protein of measles virus and the gB 150K protein were inefficiently cleaved (Vialard et al., 1990; Wells et al., 1990), as was HIV 1 gp160 (Hu et al., 1987). Propeptide cleavage at a dibasic cleavage site in proattacin was also slow and inefficient even though this is an insect-derived protein (Gunne et al., 1990). For those wishing to optimize synthesis of mature forms in these cases, coexpression of the appropriate protease gene or expression of the protein from a signal peptide/mature protein gene fusion might be considered. One might also consider production of the recombinant protein in insect larvae; fowl plaque virus HA was cleaved more efficiently in larvae than in cell culture (Kuroda et al., 1989).

N-terminal Blocking

Human aldose reductase produced in the baculovirus system contains an acetylalanine at the N-terminus (Nishimura et al., 1991). This process, which involves the removal of the N-terminal methionine and subsequent acylation of the α-NH_2 group to yield an acetylalanyl terminus, is also observed in aldose reductase purified from mammalian lens.

Phosphorylation

Many proteins that are known to be phosphorylated in their natural cell environment have also been reported to be phosphorylated in the baculovirus expression system. Examples are now too numerous to list. The following discussion focuses on some specific examples that may be characteristic of the types of phosphorylation and differences in phosphorylation patterns that might be anticipated in the baculovirus expression system. The most informative phosphorylation studies have been those in which the nature of the phosphorylation with respect to the exact residues phosphorylated have been compared in the native cell and in the baculovirus-infected insect cell.

As detailed in the discussion following, some types of phosphorylation occur and some do not. Serine and threonine phosphorylation appears to be relatively common but tends to be inefficient; this may be due to the inability of endogenous kinases to keep up with heterologous gene expression and/or the decline in kinase activities at very late times during the infection. Tyrosine phosphorylation is also observed but probably requires more specific enzymes, some of which are likely not to be present or active in the baculovirus-infected cells. A number of researchers find that the low background of endogenous protein kinase activity in baculovirus-infected cells can be used to advantage in protein phosphorylation studies.

Specific Cases of Protein Phosphorylation

The SV40 large tumor (T) antigen (SV40 T-antigen) is phosphorylated in monkey cells and in insect cells on a number of specific serine and threonine residues. Qualitatively, the phosphorylation pattern of monkey and insect T-antigens are remarkably similar; the same (multiple) phosphopeptides (modified at serine and/or threonine) are observed in immunoaffinity-purified SV40 T-antigen, following $^{32}PO_4$-labeling (Hoss et al., 1990). Quantitatively, however, differences are observed in the relative intensity of the different phosphopeptides, suggesting an under-phosphorylation of specific serine residues. The major phosphorylation sites in baculovirus-expressed T-antigen are those thought to be modified in mammalian cells by cytoplasmic protein kinases, whereas the underphosphorylated serine residues are those with rapid phosphate turnover that are phosphorylated by nuclear kinases. Most of the SV40 T-antigen, however, does localize to the nucleus of infected insect cells (O'Reilly and Miller, 1988). It may be that the under-phosphorylation is related to overexpression of the protein at very late times in the infection. The relative level of phosphorylation of SV40 T-antigen in the baculovirus expression system was found to be maximum at 24 h pi; phosphorylation did not keep pace with the bulk of T-antigen expression between 24 h pi and 48 h pi (O'Reilly and Miller, 1988). Heterogeneity of SV40 T-antigen phosphorylation is observed in mammalian cells, too, when overexpressed (Hoss et al., 1990).

The phosphorylation of the human retinoblastoma gene product (RB) is cell-cycle regulated in human cells. In the S to M phase, RB can be phosphorylated on ten tryptic peptides by the *cdc2* gene product; these sites are then dephosphorylated in G_1. RB synthesized using the baculovirus expression system is hypo-phosphorylated (Lin et al., 1991). Low levels of seven of the ten tryptic phospho-peptides were observed, and the relative intensities differed from that observed in mammalian cells. In an independent study of expression of the *Schizosaccharomyces pombe cdc2* gene in insect cells, low-level phosphorylation of this serine/threonine protein kinase was observed in the presence of the *wee1* gene product (Parker et al., 1991). Coexpression of *cdc2, wee1,* and *cyclin* A or B genes resulted in a dramatic increase in cdc2 phosphorylation. Thus, phosphorylation of cdc2 requires specific activators that either are not present in the baculovirus-infected cells or are present in very low abundance. The low level of these factors may be due to the turnoff of host cell functions late in infection, a G1-like environment in the infected cell, or the simple inability of the endogenous factors to keep up with the high level of heterologous gene expression.

Tyrosine phosphorylation of proteins has been reported for several baculovirus expression systems. In the most extensively studied system, the mammalian tyrosine kinase, $pp60^{c\text{-}src}$, is normally phosphorylated on Tyr527 in mammalian cells and auto-phosphorylated on Tyr416 *in vitro.* Baculovirus-expressed $pp60^{c\text{-}src}$ is phosphorylated on Tyr416 primarily and, to a much lesser amount, on Tyr527 (Piwnica-Worms et al., 1990). The reduced levels of Tyr527 phosphorylation are apparently due to the absence or inactivity of the appropriate tyrosine kinase. Lack of phosphorylation on Tyr527 appears to result in constitutively high enzyme activity. *In vivo,* mammalian $pp60^{c\text{-}src}$ can interact with the middle t-antigen (MTAg) of polyomavirus. When expressed in insect cells, MTAg was not phosphorylated on tyrosine residues, indicating that it did not interact with any baculovirus or insect tyrosine kinases. When coexpressed with $pp60^{c\text{-}src}$, however, MTAg formed a complex with $pp60^{c\text{-}src}$ and was phosphorylated on tyrosine residues (Piwnica-Worms et al., 1990). In another example, it was found that coexpression of $pp60^{v\text{-}src}$ with a highly conserved serine-specific protein kinase, $pp90^{rsk}$, resulted in a 100-fold increase in the specific activity of $pp90^{rsk}$ with respect to specific protein substrates (Vik et al., 1990).

Applications of Hypophosphorylation in Baculovirus-Infected Cells

Hypophosphorylation in the baculovirus expression system has been used to trace kinases responsible for specific phosphorylation events: cdc2 was found to be the kinase responsible for RB modification (Lin et al., 1991); $pp60^{v\text{-}src}$ was found to activate $pp90^{rsk}$ (Vik et al., 1990); and cyclin A or B was found to activate cdc2 via *wee1* (Parker et al., 1991). Other workers have used the baculovirus system specifically because of the low level of endogenous protein kinase activity [e.g., protein kinase C and Ca^{+2}/calmodulin-dependent protein kinase II (Brickey et al., 1990; Freisewinkel et al., 1991)].

Protein Kinase(s) Encoded by Baculoviruses

Protein kinase activity is associated with baculovirus particles (Miller et al., 1983a; Wilson and Consigli, 1985). Wilson and Consigli (1985) reported that the protein kinase activity associated with viral capsids is dependent on the presence of Mn^{+2}. Although a protein kinase gene has not yet been reported in the AcMNPV genome, it is likely that one will be found due to the finding of a protein kinase homolog in the genome of the baculovirus, *Helicoverpa zea* nuclear polyhedrosis virus (W.C. Rice, 1991 meeting of the American Society of Virology). The specificity of baculovirus protein kinase(s) remains to be established, although the kinase described by Rice appears to be related to protein kinase C.

N-glycosylation

Asparagine-linked glycosylation (N-glycosylation) is observed for many proteins targeted to the endoplasmic reticulum (ER) and containing the sequence Asn-X-Ser/Thr in an appropriate context for recognition by oligosaccharyltransferases found in the ER and Golgi. The first step in N-glycosylation is the cotranslational transfer of $Glc_3Man_9GlcNAc_2$ from a lipid carrier to the nascent protein in the lumen of the rough ER. Although this step appears to be conserved among lower eukaryotes, plants, invertebrates, and vertebrates, there are notable differences in the subsequent trimming and modification of this structure observed in different species. Reviews on the basic structure and assembly of N-linked oligosaccharides (Kornfeld and Kornfeld, 1985), as well as the function of and changes in the nature of protein glycosylation (Goochee and Monica, 1990), are recommended for readers interested in these subjects.

For the purposes of this discussion, high mannose oligosaccharides are those having three to nine mannose residues attached to the two N-acetylglucosamine residues linked to asparagine (i.e., $Man_{3\text{-}9}GlcNAc_2$-Asn). Complex oligosaccharides are those containing GlcNAc, galactose or sialic-acid additions to a $Man_3GlcNAc_2$ core, while hybrid oligosaccharides contain at least one of these additions on one branch of the core and mannose residues on the other branch of the core (Kornfeld and Kornfeld, 1985).

The majority of membrane-bound and extracellular proteins of animals are N-glycosylated, and the biological significance of such glycosylation is just beginning to be understood. Questions about how insect N-glycosylation may differ from mammalian N-glycosylation, and how any such differences might affect protein properties are impossible to answer in a general way. The role of glycosylation in the function of many proteins is unknown, and the studies that *have* been done show that roles vary among different proteins. The following discussion attempts to summarize what is known currently about the differences between N-glycosylation of vertebrates and baculovirus-infected insect cells.

Proteins that are N-glycosylated in vertebrate cells are generally also glycosylated in insect cells. As discussed earlier, signal sequences usually function in insect cells to direct the protein to the ER. Like most eukaryotic ERs, insect ER enzymes have the capacity to attach at least a $Man_9GlcNAc_2$ to the same sites (asparagines in the sequence context Asn-X-Ser/Thr) recognized by vertebrate enzymes. Usually, the $Man_9GlcNAc_2$ moiety is trimmed to shorter oligosaccharide structures, such as $Man_3GlcNAc_2$, in both vertebrate and insect cells. In vertebrates, these shorter core structures serve as the framework for complex oligosaccharide synthesis involving further GlcNAc, Gal, or sialic-acid additions. In insect cells, this additional, complex oligosaccharide synthesis does not appear to occur in many cases.

In the vast majority of studies, glycosylation of baculovirus-produced proteins has been characterized only by comparison of protein sizes following: (1) treatment of infected cells with tunicamycin; and/or (2) treatment of the expressed proteins with endoglucosaminidase H and/or endo F. (Endo F is often a mixture of endo-β-N acetylglucosaminidase F and glycopeptidase F from *F. meningosepticum*. It is also known as PNGase F or N-glycanase.) Tunicamycin treatment inhibits lipid-linked $Glc_3Man_9GlcNAc_2$ addition in both insect and vertebrate cells and thus blocks the first step of N-glycosylation. As a result, the proteins produced in the presence of tunicamycin are generally smaller and of a single size. (Exceptions are found if proteolytic cleavage, for example, depends on glycosylation; this has been documented in several cases.) Tunicamycin treatment can thus provide information on whether basic N-glycosylation is occurring and the size of the unglycosylated form of protein. Endo H cleaves high-mannose oligosaccharides between the two GlcNAcs in the sequence Asn-GlcNAc-GlcNAc-X, but does not cleave proteins with oligosaccharide additions that have been processed to $Man_3GlcNAc_2$ (Hsieh and Robbins, 1984), or processed to the complex form. Endo F can cleave processed oligosaccharides, including complex forms, at the protein attachment point. Changes in the size of proteins following endo H or endo F cleavage therefore provide a measure of those proteins that are still in the high-mannose form (Endo H^S, Endo F^S) and those that have been either trimmed to the $Man_3GlcNAc_2$ core or further modified to complex oligosaccharides (Endo H^R, Endo F^S).

In most cases reported in the literature, the size of baculovirus-produced N-glycoproteins is reduced by tunicamycin. Depending on the protein, tunicamycin treatment may also block or greatly reduce secretion, indicating a dependence of transport through the ER and/or secretion on N-glycosylation, which in turn may affect folding, oligomerization, and so on. In most reported cases using endo H and endo F, the N-glycoproteins are initially sensitive to endo H and endo F but become resistant with time to endo H [e.g., influenza hemaglutinin (HA) (Kuroda et al., 1990)]. In many cases [e.g., tPA (Jarvis and Summers, 1989)], significant quantities of endo H- sensitive structures remain, perhaps due to inefficient removal of mannose residues down to the basic $Man_3GlcNAc_2$ core.

Influenza virus HA is one of the few proteins for which the chemical structures of the oligosaccharides of baculovirus-produced and vertebrate-produced proteins were characterized and compared (Kuroda et al., 1990). Baculovirus-produced HA protein contains both endo H-sensitive and endo H-resistant oligosaccharides. The proportion of endo H-sensitive oligosaccharides decreases with increasing length of the chase period following radioactive labeling with [2-^{3}H]mannose. These endo H-sensitive oligosaccharides consisted of $Man_{5-9}GlcNAc_2$ structures. The endo H-resistant glycans consisted of two basic structures: $Man_3GlcNAc_2$ with and without fucose attached to the protein proximal GlcNAc. Comparison of each of the seven positions of glycosylation on the HA protein, produced either in chicken embryo cells or in the baculovirus expression system, is revealing (Kuroda et al., 1990), particularly when viewed in relation to the three-dimensional structure determined by X-ray crystallography. Those positions (i.e., asparagines 28 and 406) that have high-mannose sidechains (e.g., $Man_{6-9}GlcNAc_2$) in chicken cells also appear to be incompletely trimmed ($Man_{5-9}GlcNAc_2$) in baculovirus-produced proteins. The other five positions (i.e., asparagines 12, 123, 149, 231, and 478) that appear to have complex oligosaccharides in chicken cells are primarily $Man_3GlcNAc_2$ (+/− fucose) in baculovirus-produced hemaglutinin. (The fucose modification on $GlcNAc_2$ is also observed in chicken cells; it is not unique to insect cells.)

The predominant view of baculovirus N-glycosylation from the HA study is that: (1) the insect system recognizes and modifies the same sites as the vertebrate system; (2) the modifications to the lipid-transferred oligosaccharide in insect cells involve primarily mannose trimming with some fucosyl transfer, but no complex additions involving galactose or sialic acid. This finding is remarkably consistent with an earlier in-depth study of glycosylation of a Sindbis virus glycoprotein in vertebrate and mosquito cells (Hsieh and Robbins, 1984). The observation is also consistent with earlier observations (Butters and Hughes, 1981; Butters et al., 1981) that mosquito cells lack galactose and sialic acid transferases required for complex glycosylation. The oligosaccharide modification in insect-produced HA protein is relatively homogeneous compared to the complex glycosylation found in mammalian-cell-produced proteins. Such homogeneous modification may be useful in future structural and functional studies.

A different result, however, was obtained in structural studies on the oligosaccharides of human plasminogen derived from the baculovirus expression system (Davidson et al., 1990; Davidson and Castellino, 1991a; Davidson and Castellino, 1991c). Human plasminogen has a single N-glycosylation site, Asn289. Microheterogeneity in the oligosaccharide sidechain at this site is observed in both mammalian and insect cells (Davidson and Castellino, 1991b). At least 16 different oligosaccharide structures have been characterized at this one site in both insect and Chinese hamster ovary cells. Most remarkably, some of the structures found on baculovirus-produced plasminogen contain galactose and sialic acid, and the structures, as determined by high pH anion exchange chromatography, include biantennary, triantennary, and tetraantennary complex classes of oligosaccharides with varying extents of outer-arm completion (Davidson and Castellino, 1991a). No hybrid oligosaccharides were observed in plasminogen isolated from either mammalian or baculovirus systems.

These workers also note that the proportion of complex oligosaccharides observed in plasminogen secreted from infected insect cells increases during the very late phase (20 h pi–96 h pi). This might reflect a time dependence for fully glycosylated proteins to move through the secretory pathway; movement through the secretory pathway was considered rate-limiting for tPA secretion previously (Jarvis and Summers, 1989). Davidson and Castellino suggest, alternatively, that it might reflect an induction of the glycosylation pathway, presumably due to changes in the cell environment; the cell environment is known to influence the nature of protein N-glycosylation in other cell systems (Goochee and Monica, 1990). Because most host-cell functions are shut off at these very late times, an unmasking rather than an induction would seem more likely if this possibility is considered. The low yields of plasminogen in cell fluid collected from 20 h pi to 60 h pi favor the hypothesis of very slow movement through the secretory pathway. The efficiency of glycosylation of plasminogen remains to be determined.

For the most part, the observation of complex N-glycosylation of human plasminogen is unexpected and novel. In the vast majority of other reports concerning baculovirus expression, the existing evidence suggests shorter, noncomplex oligosaccharides, as confirmed for influenza HA (Kuroda et al., 1990) and for interleukin-3 by chromatographic and mass spectrometric methods (Svoboda et al., 1991). The latter study, however, did suggest that a low level of neuraminic acid might be present in a portion of the heterogeneous oligosaccharides found on IL-3. Although Davidson and Castellino used SF-21AE cells in their studies, the modifications they observed are also found in other lepidopteran cells (Davidson and Castellino, 1991c). It is possible that human plasminogen contains unique and necessary signals to enter the insect Golgi apparatus for further processing. Thus, the ability of insect cells to assemble complex oligosaccharides on the basic oligosaccharide core may be dependent on the nature of the protein. The work on plasminogen points to the necessity of structurally analyzing endo H-resistant oligosaccharides from baculovirus-expressed proteins for the possibility of complex glycosylation. Investigators might also consider an approach that was employed for human EGF receptor (Greenfield et al., 1988): protein sizes from insect and mammalian cells were compared following treatment with tunicamycin or monensin, an inhibitor of Golgi transport and terminal glycosylation. Monensin, however, may interfere also with some trimming reactions.

Several reports have noted that the proportion of proteins that are N-glycosylated and/or secreted decreases from 24 h pi through 72 h pi, suggesting that the function of the secretory pathway is declining during this period (e.g., Jarvis and Summers, 1989; Murphy et al., 1990a). This would be consistent with the biology of the virus. There are approximately 10 glycoproteins made specifically during wt baculovirus replication (Kelly and Lescott, 1983; Stiles and Wood, 1983). The major N-glycosylated protein is gp64 (also known as gp67), a structural protein of the membrane of the budded virus that is maximally synthesized between 12 h pi and 24 h pi (Carstens et al., 1979; Goldstein and McIntosh, 1980; Kelly and Lescott, 1983; Stiles and Wood, 1983). During the very late phase, occluded virus assembly occurs in the nucleus and the need for the secretory pathway would be expected to decline.

O-glycosylation

O-glycosylation involves the post-translational addition of sugars to specific hydroxylated amino acids (e.g., serine or threonine residues). In animal cells, the addition of N-acetylgalactosamine to serine or threonine residues is a common pathway for mucins and many other membrane-associated or secreted proteins, but other types of O-glycosylation are also observed, such as O-xylosyl-Ser(Thr) glycosylation of proteoglycans, O-galactosyl-hydroxylysine glycosylation of collagen, and O-glucosaminyl-Ser glycosylation of nuclear pore and cytoplasmic proteins [reviewed by (Carraway and Hull, 1989; Hart et al., 1989)]. O-glycosylation may involve only a single sugar or an oligosaccharide, which is built one residue at a time by transfer from a nucleotide derivative. For plasma-membrane-associated or secreted proteins, it is thought that the first one or two sugars are transferred in the ER and that additional sugars may be added, depending on the protein, in the Golgi or trans-Golgi network.

O-glycosylation in baculovirus-infected cells was originally described for a 42 kDa virus structural protein based on: (1) labeling of the protein with N-[1-^{3}H]-GlcNAc but not with [$^{2-3}$H]mannose; and (2) the resistance of the 42 kDa GlcNAc labeling to tunicamycin. This protein is found in the nucleus of infected cells, as it is a structural protein of the occluded form only (Stiles and Wood, 1983). Similarly,

Whitford and Faulkner have recently determined that a 41 kDa occluded-virus structural protein is O-glycosylated with N-acetylglucosamine (Whitford and Faulkner, personal communication).

The ability of baculovirus-infected cells to O-glycosylate heterologous proteins was thoroughly documented for a pseudorabies virus (PRV) glycoprotein, and the nature of the glycosylation of the protein in insect cells was compared to that occurring in mammalian cells (Thomsen et al., 1990). PRV protein, isolated from mammalian cells, contained only the disaccharide Galβ1-3GalNAc, with or without one or two sialic residues attached. PRV protein from the baculovirus expression system, however, contained primarily the monosaccharide, GalNAc, and lower amounts (12%–26%) of the disaccharide Galβ1-3GalNAc. No sialic acid was found on the insect cell-derived structures. Measurement of glycosyltransferases in insect and mammalian cells revealed comparative levels of UDP-GalNAc:polypeptide, N-acetylgalactosaminyltransferase activity, but markedly lower levels of UDP-Gal:N-acetylgalactosamine, β1-3 galactosyltransferase activity in insect cells. In another study, human β-choriogonadotropin was found to have both O-linked and N-linked glycosylation (Chen et al., 1991b); the O-linked residues were reported to be dissaccharides GalN-Gal, although structural determination was not complete.

Thus, O-glycosylation clearly takes place in the baculovirus-infected cells. This O-glycosylation may be similar, but not necessarily identical, to that observed in mammalian cells. The nature of the O-linked sugar may depend on the nature of the protein or its subcellular location in the insect cells.

Lipid Modification: Fatty Acid Acylation, Polyisoprenylation

At least two virus structural proteins of the budded form of AcMNPV are acylated with a fatty acid, as determined by radiolabeling with [^{3}H]palmitic acid and/or [^{3}H]myristic acid (Roberts and Faulkner, 1989). The nature of the fatty acid attached to one of these proteins, the gp64/gp67 envelope glycoprotein, was characterized in detail and found to be a nonthio-type ester linkage with palmitic acid. Palmitic acid was the primary modification regardless of whether the radiolabel was supplied as myristic acid or palmitic acid (Roberts and Faulkner, 1989). Thus, myristic acid appeared to be metabolized to palmitic acid before use as an acyl donor. Chain elongation is a common metabolic change to exogenously supplied fatty acids.

Pulse-labeling of wt AcMNPV-infected cells at 48 h pi with [^{3}H]myristic acid incorporates radioisotope into at least three proteins of ca. 67 kDa, 22 kDa, and 15 kDa (Belsham et al., 1991). Furthermore, a number of investigators have reported that heterologous gene products are myristoylated in the baculovirus expression system (e.g., Lanford et al., 1989; Luo et al., 1990; Piwnica-Worms et al., 1990; Belsham et al., 1991; Morikawa et al., 1991). In the study of picornavirus capsid protein, incorporation of [^{3}H]myristic acid was observed when the protein was produced with the leader peptide or with an N-terminal methionine residue adjacent to the consensus sequence (G-X-X-S/T) for myristoylation (Belsham et al., 1991). Because methionine is not a substrate for myristoylation, it was assumed that the methionine as well as the leader were removed properly in the insect system before myristoylation. All the reports describing myristoylation in the baculovirus expression system, however, have demonstrated only that [^{3}H]myristic acid is incorporated into the protein. Considering the gp64 research (Roberts and Faulkner, 1989) described earlier, the nature of the incorporated fatty acid in these studies should have been analyzed chemically.

Several studies have reported palmitoylation of heterologous gene products, particularly CAAX-box-containing proteins that require prior isoprenylation. Only 10% to 20% of the Ha-ras p21 (a CAAX-box-containing protein) produced in the baculovirus expression system is post-translationally modified (Lowe et al., 1990). But that 10% to 20% of the product, the membrane-bound fraction, is metabolically labeled with [^{3}H]mevalonate as well as [^{3}H]palmitic acid; the radiolabel from [^{3}H]mevalonate was not labile in hydroxylamine, while that derived from [^{3}H]-palmitate was labile as expected for appropriate chemical modifications of this nature. The nature of the mevalonate labeling is consistent with a polyisoprene unit. No mevalonate was observed in wt virus-infected cells although significant incorporation was observed in mock-infected insect cells.

The nature of isoprenylation in baculovirus-infected cells was chemically determined for H-ras and Rap1A (Krev-1), a CAAX-containing protein related to H-ras (Buss et al., 1991). Rap1A was modified through a thio-ester linkage with a C_{20} (geranylgeranyl) isoprenoid whereas H-ras was modified with a C_{15} (farnesyl) group. These modifications were the same in baculovirus-produced proteins as in mammalian-derived proteins, indicating a remarkable evolutionary conservation of this post-translational modification system. The isoprenylation reaction was inefficient and did not keep pace with the baculovirus expression of Rap1A or ras. Addition of mevalonate to the growth medium did not increase the proportion of processed protein (Buss et al., 1991). H-ras is palmitoylated and Rap1A is not palmitoylated when produced in insect cells and in mammalian cells.

In mammalian cells, methylation accompanies isoprenylation at the cysteine of CAAX-box-containing proteins following trimming of the AAX from the carboxyl end. Methylation of the α-carboxyl end of both Rap1A and H-ras was detected by monitoring transfer of the radiolabeled methyl group of [^{3}H]methylmethionine into protein and volatilization of the incorporated label using NaOH (Buss et al., 1991). The electrophoretic mobility of the methylated protein indicated that trimming of AAX residues had occurred before methylation, presumably on the resulting C-terminal cysteine. There is evidence to suggest that methylation also occurs on the Kirsten-ras p21 protein produced in insect cells (Lowe et al., 1991).

In a study of a CAAX-box-containing protein from *Xenopus* (Kloc et al., 1991), palmitoylation was demonstrated by monitoring [^{3}H]palmitic acid incorporation, followed by methanolysis and thin-layer chromatography, to show the release of methyl palmitate from the protein. These workers also demonstrated that palmitoylation of this protein was dependent on prior isoprenylation, as is the case with other CAAX-box-containing proteins, by inhibiting palmitoylation with lovastatin, an inhibitor of endogenous mevalonic acid synthesis. They also noted that a virus-specific protein (reported to be ca. 95 kDa) incorporated radiolabeled palmitic acid by an isoprenyl-independent pathway.

α-amidation

A number of proteins, particularly small peptides, are amidated at their carboxyl terminus. Such peptides are usually synthesized as a larger precursor and then processed to a peptide with a C-terminal glycine residue. Conversion of the C-terminus to an α-amide group occurs by an oxidative enzymatic process that probably involves an initial hydroxylation at the α-carbon of the glycine.

Determination of the mechanism for the α-amidation reaction was actually facilitated using a baculovirus expression system to express the gene encoding frog peptidylglycine α-hydroxylating monooxygenase (Suzuki et al., 1990). Production of this enzyme, using the baculovirus expression system, allowed conversion of a model peptide substrate to the hydroxylated form. An endogenous enzyme found in the infected insect-cell culture fluid, separable from the expressed monooxygenase, could convert the hydroxylated peptide to the amidated product at physiological pHs. This work demonstrated that two different enzymes are involved in the α-amidation reaction.

Based on existing data for the baculovirus expression system, it is difficult to predict whether a protein will be properly α-amidated. Amidation of the C-terminus was reported for cecropin A, an antibacterial insect peptide, when produced in recombinant baculovirus-infected *Trichoplusia ni* larvae and pupae of *Hyalophora cecropia* (Hellers et al., 1991). Amidated products constituted 20% to 30% of the total product in *T. ni* larvae and 70% in the pupal expression system. Amidation was not observed for sarcotoxin IA, another insect antibacterial peptide, when expressed in *B. mori* cells (Yamada et al., 1990). The terminal residue of the produced sarcotoxin was glycine and should have been a suitable substrate for α-amidation. The lack of amidation suggests that these cells were inefficient or deficient in one or both of the enzymes required for α-amidation.

Thus far, it seems most likely that a larval- or pupal-based expression system would offer the best possibility of obtaining amidated products. However, it also appears likely that if an α-hydroxylating monooxygenase is coproduced by expressing, for example, the frog gene, α-amidation might be achieved in cell culture.

Subcellular Localization

For the most part, proteins that are synthesized using the baculovirus expression system are directed to the appropriate subcellular location. (Chloroplast localization would be an obvious exception.) As noted earlier, signal sequences from many sources generally function in insect cells to direct proteins to the ER. Proteins have been found to be membrane-bound or secreted as expected. Most proteins with internal nuclear localization signals have been found to be targeted to the insect nucleus (e.g., Miyamoto et al., 1985; O'Reilly and Miller, 1988). The nuclear localization signal of AcMNPV polyhedrin has been defined (Jarvis et al., 1991) and is similar to signals described for mammalian and amphibian systems.

One demonstration of the subcellular localization features of the baculovirus expression system is the study of rat glucocorticoid-receptor overexpression (Alnemri et al., 1991a). Infected insect cells produced approximately 2×10^6 receptor molecules per cell, approximately 30 times more than that expressed normally in hepatocytes. The receptor was cytoplasmic in the absence of ligand but, upon binding a glucocorticoid agonist, it translocated from the cytoplasm to the nucleus. In its cytoplasmic form, it had a Stokes radius identical to the product from rat liver; its high molecular weight ($M_r > 300{,}000$) was characteristic of a complex that, in rat liver, is thought to include hsp90. The hsp90 protein appears to be released from the receptor upon ligand addition and nuclear localization.

An example of the usefulness of proper cellular localization is the demonstration of potassium channel function in baculovirus-infected insect cells (Armstrong and Miller, 1990; Klaiber et al., 1990). Expression of the *Drosophila* shaker gene resulted in the formation of functional potassium-ion channels, the activity of which could be monitored using patch clamp techniques. Uninfected SF cell lines have a relatively low background from an electrophysiological view and are now being used extensively for monitoring other electrophysiologically active proteins [e.g., anion conductance of cystic fibrosis gene product (Kartner et al., 1991)].

There is a report of inappropriate cellular localization using the baculovirus expression system. Human β-galactosidase, which is normally directed to the lysosome in human cells, was found in the cytoplasm and secreted into the media when expressed in insect cells (Itoh et al., 1990). Expression levels were 300 times higher in the baculovirus-infected cells than in human cells, and it may be that the lysosomal localization pathway was overwhelmed by the high level of expression.

Folding, Disulfide Bond Formation, Dimerization, Oligomerization, and Larger Assemblages

There are now numerous reports in which the baculovirus expression system has been used to successfully produce a biologically active protein where a prokaryotic system has failed to do so. Usually, the failure of the prokaryotic system is attributed to lack of post-translational modification and/or aggregation of the protein into insoluble complexes. Although there are a few reports of insoluble complexes in insect cells, insoluble aggregates are not usually produced in the baculovirus system, and most proteins that do form aggregates are known to assemble or form aggregates in their natural setting. Insect cells, like other eukaryotes, have a secretory pathway including the ER, Golgi, and molecular chaperones, that facilitates the appropriate folding, disulfide bond formation, and oligomerization that are critical to the biological activity of many eukaryotic proteins.

Appropriate formation of both intrachain and interchain disulfide bonds has been reported for a number of proteins. Reports of the formation of secreted, functional immunoglobulin heterodimers using baculovirus coexpression vectors are excellent examples of the capability of the baculovirus expression system (Hasemann and Capra, 1990; zu Putlitz et al., 1990). Other examples include disulfide-linked dimer formation of v-sis/platelet-derived growth factor B protein (Giese et al., 1989) and the formation of disulfide-bridged homodimers of human interleukin-5 (Ingley et al., 1991).

Although there are a number of examples of oligomer formation, the trimerization of influenza hemagglutinin is reported to be approximately three times slower in insect cells than in vertebrate cells (Kuroda et al., 1991). Trimerization of hemagglutinin is an ordered process occurring in the ER of vertebrates, and the delay in trimerization may account, in part, for the delay in movement of this protein through the ER.

Baculovirus expression systems allow not only the formation of homodimers and heterodimers but also the assembly of much more complex structures such as heterologous virus particles. Examples of virus assembly include polyomavirus capsid assembly (Montross et al., 1991); parvovirus capsid assembly (Brown et al., 1991); hepatitis B core assembly (Hilditch et al., 1990); rotavirus core-particle formation (Labbe et al., 1991); hollow fiber formation by caulimovirus gene 1 product (Vlak et al., 1990); viruslike particles of retroviruses (Gheysen et al., 1989; Rasmussen et al., 1990); and Newcastle disease viruslike envelopes (Nagy et al., 1991). Corelike particles of bluetongue virus (BTV) were produced when the two major core proteins were coexpressed using a dual-gene expression vector (French and Roy, 1990). Using three different viruses, two of which expressed two BTV genes, a total of five BTV genes were expressed simultaneously, resulting in the assembly of double-shelled viruslike particles of BTV (Loudon and Roy, 1991).

The ability to coexpress two or more genes to achieve complex formation and/or biologically relevant protein structures (e.g., St. Angelo et al., 1987; O'Reilly and Miller, 1988; Hasemann and Capra, 1990) is a very useful feature of the baculovirus expression system that will probably be more extensively exploited in future expression work. Coexpression of two genes can be achieved either by coinfecting with two viruses, each expressing a single gene, or constructing a single virus that expresses two or more heterologous genes (see Chapter 6 for further discussion).

Substitution of Unusual Analogs into Proteins Produced by the Baculovirus Expression System

It is likely that the baculovirus system will find many novel applications for biochemistry research. Two unique applications that may stimulate further thinking are selenomethionyl incorporation into baculovirus-expressed proteins and substitution of heme analogs into heme-containing proteins. Selenomethionine can be useful as an isomorphous replacement to resolve the phase problem in X-ray diffraction analysis of protein structure. In high concentration, however, selenomethionine is toxic to prokaryotic and eukaryotic cells. Using human choriogonadaotropin as a model system, it was found that over 80% of the methionine residues can be replaced with selenomethionine if selenomethionine is used in place of methionine in the medium of baculovirus-infected cells during the very late phase of expression (Chen and Bahl, 1991b and c).

In another application of the expression system, the synthesis of active rat P450 IIA1, a membrane-bound hemoprotein, was found to be partially dependent on exogenously added protoheme (Asseffa et al., 1989). It was proposed that the baculovirus system will provide a novel means to produce high levels of cytochrome P450s with modified heme moieties.

METHODS FOR THE ANALYSIS OF POST-TRANSLATIONAL MODIFICATIONS OF HETEROLOGOUS GENE PRODUCTS

It will be apparent from the preceding discussion that the thorough characterization of the post-translational modification of a recombinant protein may require considerable effort. However, a lot of information can be obtained through simple modifications of some of the protein analysis techniques described in Chapter 14. The following is a discussion of how you may evaluate the glycosylation, phosphorylation, acylation, and proteolytic processing of a recombinant protein in the baculovirus expression system. All the protocols incorporate methods for the characterization of protein synthesis described in Chapter 14, including metabolic labeling, immunoprecipitation, SDS-PAGE, and immunoblotting. These procedures are only summarized in this chapter, and the reader should consult the appropriate sections of Chapter 14 for complete protocols and materials lists.

Glycosylation

Three simple methods that supply information on the extent of N-glycosylation of a protein are described here. These are labeling with radioactive sugars, treatment of the protein with specific glycosidases, and treatment of the cells with inhibitors of glycosylation. The investigation of O-glycosylation using labeling techniques in conjunction with glycosylation inhibitors is also discussed briefly.

Metabolic Labeling with Radioactive Sugars

A wide range of radiolabeled sugars is commercially available. Essentially, any of these can be used in metabolic labeling experiments. The sugar you use will depend on your particular application. The following protocol, using [^{3}H]mannose, is suitable for examination of N-glycosylation because mannose residues occur on all N-glycoproteins. For examination of O-glycosylation, the infected cells may be labeled, in the presence or absence of tunicamycin, with D-[1,6-^{3}H]glucosamine hydrochloride (30-60 Ci/mmol; NEN # NET-557A). Glucosamine is a precursor of N-acetylgalactosamine, therefore the label will be incorporated into O-glycoproteins. O-glycosylation is not inhibited by tunicamycin, and specific labeling in the presence of tunicamycin is good evidence that the protein is modified by O-glycosylation.

CONTROLS: Controls should include a time course performed with wt virus (and/or parent virus if different), as well as one dish of mock-infected cells. If you analyze the samples by immunoprecipitation or immunoblotting, remember to use the appropriate control antibodies (see the protocols in Chapter 14). A common problem with metabolic labeling with radioactive sugars is that they are readily converted to radiolabeled amino acids, which are then incorporated into proteins. To control for this problem, check to make sure other (nonglycosylated) proteins are not labeled, and support your conclusions with data from another experimental approach such as tunicamycin treatment or endoglycosidase cleavage.

Materials

1. SF cells grown to approximately 1×10^6 cells/ml
2. 35 mm tissue-culture dishes (see Chapter 11)
3. Recently titered stocks of your recombinant and control viruses
4. Complete tissue-culture medium. (See Chapter 11 for details concerning various media.)
5. Glucose-deficient tissue-culture medium. Prepare tissue-culture medium as described in Chapter 11, omitting glucose.
6. D-[2-^{3}H]mannose (20-30 Ci/mmol in aqueous solution; NEN # NET-570A)
7. PBS, pH 6.2

Method

1. Seed and infect the cells as described for the standard time course experiment in Chapter 14.
2. Two hours before each time point, remove the medium from the plates and replace with the same volume of glucose-deficient medium. Incubate at 27° C for one hour.
3. One hour before each time point, remove the glucose-deficient medium from the cells. Replace with 0.5 mls glucose-deficient medium containing 50 µCi [^{3}H]mannose. Incubate at 27° C or room temperature for one hour with gentle rocking.
4. At the time point, remove the tissue-culture fluid containing the radiolabel and dispose of appropriately. Rinse the cells *gently* three times with 2 mls each of PBS. Collect the secreted proteins in PBS or lyse the cells immediately as described in Chapter 14. Add protease inhibitors if necessary. Analyze the samples by immunoprecipitation and/or SDS-PAGE and fluorography. Depending on the protein being analyzed (e.g., proteins with a single N-linked oligosaccharide), the incorporation of radiolabeled sugar may not be very efficient.

Cleavage with Glycosidases

Cleavage with endoglycosidases F and H is described here. The specificities of these enzymes were addressed in detail in the preceding introduction. In both cases, cleavage is most conveniently carried out on metabolically labeled, immunoprecipitated samples. Note that a number of O-linked oligosaccharide-specific endoglycosidases are now available, which may facilitate the characterization of O-glycoproteins (see, for example, Thomsen et al., 1990).

CONTROLS: Keep an aliquot of each sample as an uncleaved control. This aliquot should be treated in an identical manner, with the enzyme omitted.

Materials

1. Immunoprecipitates of appropriate radiolabeled samples and controls
2. Elution buffer: 50 mM Tris-HCl, pH 6.8; 0.5% (w/v) SDS; 0.1 M β-mercaptoethanol
3. Endo F buffer: 40 mM sodium phosphate, pH 7.2; 4% (v/v) NP40; 20 mM EDTA
4. Endo F (endoglycosidase F/N-glycosidase F mixture; Boehringer Mannheim # 878 740). Store at 4° C.
5. Endo H buffer: 200 mM sodium citrate, pH 5.5
6. Endo H (endoglycosidase H; Boehringer Mannheim # 1088 726). Store at 4° C.
7. PMSF (Fluka # 78830); 10 mg/ml in isopropanol. Store at room temperature. Make a 2 mg/ml working stock. PMSF is not stable in aqueous solution, so it should only be added to the sample just before the enzyme is added.
   Caution: PMSF is very harmful and may be fatal if inhaled, swallowed, or if it contacts skin. Wear protective clothing and a mask when handling PMSF.
8. 2X SDS gel-loading buffer

Method

1. Infect, label, and immunoprecipitate the samples with the appropriate antibodies. Initially, try using 10 µl of infected cell lysates or 100 µl of extracellular fluid for immunoprecipitations.
2. Resuspend the washed SAC or protein A pellets in 30 µl elution buffer and boil for five minutes.
3. Centrifuge in a microfuge for five minutes and divide the supernatant into three aliquots.
4. Set up the cleavage reactions as follows:
   Endo F cleavage:
   10 µl sample
   5 µl Endo F buffer
   1 µl 2 mg/ml PMSF
   5 µl (250 mU) Endo F
   Endo H cleavage:
   10 µl sample
   5 µl Endo H buffer
   1 µl 2 mg/ml PMSF
   5 µl (5 mU) Endo H
   For the untreated control, dilute 10 µl of sample with 10 µl of either buffer, and add 1 µl 2mg/ml PMSF.
5. Incubate at 37° C for 24 hours. Stop the reaction by adding an equal volume of 2X SDS gel-loading buffer and boil for five minutes. Analyze by SDS-PAGE. Removal of sugar residues should result in a change in mobility of the protein of interest. Alternatively, if the protein was labeled with [^{3}H]mannose, cleavage should result in the loss of the label.

Inhibitors of Glycosylation

The *Streptomyces* antibiotic tunicamycin is commonly used as a tool to study protein glycosylation because it inhibits an early step in the N-glycosylation of proteins. (See introduction for discussion of mode of action.) However, it does not inhibit O-glycosylation. Tunicamycin treatment in conjunction with the metabolic labeling of infected cells is described here. However, it may be used equally in combination with other methods of protein analysis, such as immunoblotting. Note that tunicamycin treatment may partially inhibit protein synthesis in treated cells so that the total uptake of radiolabeled amino acid may be reduced. In addition, since N-glycosylation facilitates the secretion of certain proteins, tunicamycin treatment may cause a reduction in the secretion of extracellular proteins.

CONTROLS: Perform the experiment in duplicate so you may compare identical samples with and without tunicamycin treatment.

Materials

1. SF cells grown to approximately 1×10^6 cells/ml
2. 35 mm tissue-culture dishes (see Chapter 11)
3. Recently titered stocks of your recombinant and control viruses
4. Complete tissue-culture medium. (See Chapter 11 for details concerning various media.)
5. Complete medium and methionine-deficient tissue-culture medium with tunicamycin. Add tunicamycin (Boehringer Mannheim # 1243 080) to a final concentration of 5 μg/ml just before use. Sterilize by filtration through a 0.2 μm filter (Gelman Acrodisk # 4192).
6. Tran^{35}S-label (> 1000 Ci/mmol in aqueous solution, ICN # 51006). Store in small aliquots under nitrogen at –80° C.
7. PBS, pH 6.2
8. PBS plus tunicamycin (for secreted proteins only): Add tunicamycin to 5 μg/ml just before use.

Method

1. Infect SF cells with the appropriate viruses.
2. At least six hours before labeling, remove the medium and refeed with complete medium plus tunicamycin.
3. One hour before labeling, remove the medium from the cells and replace with the same volume of methionine-deficient medium plus tunicamycin. Incubate at 27° C for one hour.
4. Replace the medium with 0.5 mls methionine-deficient medium plus tunicamycin containing 25 μCi Tran^{35}S-label. Incubate at 27° C or room temperature for one hour with gentle rocking.
5. Remove the tissue-culture fluid containing the radiolabel and dispose of appropriately. Rinse the cells *gently* three times with 2 mls each of PBS. Collect secreted proteins following an additional incubation in PBS plus tunicamycin. Otherwise, lyse the cells immediately, adding protease inhibitors if required. These samples may now be analyzed by SDS-PAGE or immunoprecipitation. Other proteins may coimmunoprecipitate with your protein of interest from the tunicamycin-treated samples. Some of these may be chaperonins that are bound to malfolded protein molecules (Jarvis et al., 1990b).

Phosphorylation

This section describes the analysis of the state of phosphorylation of a protein by metabolic labeling with $^{32}P_i$ during a time course experiment. The optimal length of time to label the cells is difficult to predict and should be determined empirically for each protein. We recommend trying both a one hour pulse-label, and long-term labeling for 12 hours in initial experiments.

CONTROLS: Include a time course performed with wt virus (and/or parent virus if different), as well as one dish of mock-infected cells. If you analyze the samples by immunoprecipitation or immunoblotting, remember to use the appropriate control antibodies (see protocols in Chapter 14).

Materials

1. SF cells grown to approximately 1×10^6 cells/ml
2. 35 mm tissue-culture dishes (see Chapter 11)
3. Recently titered stocks of your recombinant and control viruses
4. Complete tissue-culture medium. (See Chapter 11 for details concerning various media.)
5. Phosphate-deficient tissue-culture medium. Prepare tissue-culture medium as described in Chapter 11, omitting sodium phosphate and tryptose, lactalbumin hydrolysate, or yeastolate as applicable. Add 7.5 mM Hepes, pH 6.2, to replace the sodium phosphate as a buffer. Adjust the pH of the medium to 6.2.
6. $^{32}P_i$ (8500-9100 Ci/mmol, in aqueous solution; NEN # NEX-053)
7. PBS, pH 6.2
8. NP40 Lysis buffer plus EDTA: 1% NP40; 150 mM NaCl; 50 mM Tris-HCl, pH 8.0; 5 mM EDTA. Store at 4° C. The EDTA inhibits endogenous kinases that might phosphorylate proteins after lysis.
9. 2X SDS gel-loading buffer

Method

1. Seed the cells and infect as described for the standard time course experiment in Chapter 14.
2. Incubate the cells in 2 mls phosphate-free medium for one hour prior to labeling.
3. Remove the phosphate-free medium. For short-term labeling, incubate the cells in the presence of 100 μCi $^{32}P_i$ in 0.5 mls of phosphate-free medium for one hour at room temperature or 27° C. For long-term labeling, 25 μCi of $^{32}P_i$ should be sufficient. However, it is advisable to increase the volume of phosphate-free medium to 1 ml. There is no need to supplement the phosphate-free medium with serum. Incubate at 27° C for 12 hours.
4. After labeling, rinse the cells *gently* three times with 2 mls of PBS. If you do not want to analyze secreted proteins, the cells may be lysed at this point. To examine both intracellular and extracellular fractions, continue as follows.
5. Incubate the cells in 0.5 mls PBS for one hour at room temperature with gentle rocking.
6. Collect the PBS, spin for five minutes in a microfuge at 4° C, and store the supernatant (the extracellular fraction) at –20° or –80° C.
7. Lyse the cells using 1X SDS gel-loading buffer or the NP40-based lysis buffer plus EDTA as required. Add protease inhibitors if necessary. These samples may now be analyzed by SDS-PAGE or immunoprecipitation. There is no need to fluorograph the gels. The signal may be enhanced with a screen if necessary.

Fatty Acid Acylation

Fatty acid acylation may also be evaluated by labeling prior to analysis by immunoprecipitation or SDS-PAGE. This section describes the characterization of myristoylation and palmitoylation of baculovirus-produced proteins during a time course experiment.

> CONTROLS: Include a time course performed with wt virus (and/or parent virus if different), as well as one dish of mock-infected cells. If you analyze the samples by immunoprecipitation or immunoblotting, remember to use the appropriate control antibodies (see protocols in Chapter 14). Run duplicate gels so that one gel may be subjected to hydroxylamine or alkali treatment.

Materials

1. SF cells grown to approximately 1×10^6 cells/ml
2. 35 mm tissue-culture dishes (see Chapter 11)
3. Recently titered stocks of your recombinant and control viruses
4. Complete and incomplete tissue-culture medium. If you are using serum-free medium, omit any lipid supplement for the labeling medium. (See Chapter 11 for details on various media.)
5. [9,10-^{3}H(N)]myristic acid (10-60 Ci/mmol, NEN # NET-830). Lyophilize the label and resuspend in the same volume of DMSO prior to use.
6. [9,10-^{3}H(N)]palmytic acid (30-60 Ci/mmol, NEN # NET-043). Lyophilize the label and resuspend in the same volume of DMSO prior to use.
7. PBS, pH 6.2
8. 1.0 M hydroxylamine (Sigma # H 2391)
9. 1.0 M Tris-HCl, pH 10.0

Method

1. Seed the cells and infect as described for the standard time course experiment in Chapter 14.
2. Incubate the cells in 2 mls incomplete tissue-culture medium (or serum-free medium without lipids) for one hour prior to labeling.
3. Remove the incomplete medium. Label with 500 μCi of either label in 0.5 mls of incomplete medium (or serum-free medium without lipids) for one to four hours at 27° C.
4. After labeling, rinse the cells *gently* three times with 2 mls each of PBS. Collect the secreted proteins in PBS, or lyse the cells immediately as before. Include protease inhibitors if necessary. Analyze the samples by immunoprecipitation or SDS-PAGE. We recommend running all samples on duplicate gels.
5. Following fixation of the SDS gels, soak one gel of each pair in 1 M hydroxylamine or 1 M Tris, pH 10.0, for 16 hours at room temperature. Then rinse the gels thoroughly in water and fluorograph as normal. These treatments should remove palmitate, but not myristate, from proteins. Thus, loss of the label from a band following treatment suggests that the protein was palmitoylated.

Proteolytic Processing

Pulse-chase labeling of infected cells can provide a significant amount of information about the proteolytic processing of a protein, especially when used in combination with immunoprecipitation analysis. The principle of the method is to metabolically label the infected cells for a brief period of time. Only proteins that are being synthesized during this period will incorporate the label. The label is then removed and the cells incubated for various periods of time without the metabolic label. Generally, the samples are "chased" by incubation in medium containing an excess of the unlabeled amino acid, to ensure that there is a rapid and complete cessation of uptake of the label. An inhibitor of protein synthesis, such as cycloheximide, can also be included in this chase medium. Monitoring the loss of radioactivity from a particular protein during this chase period provides information on the rate of turnover of the protein. In addition, it is often possible to identify bands that transiently or continuously accumulate label during the chase. These are likely to represent processing intermediates and mature forms of the pulse-labeled protein. When the information obtained is combined with data concerning the steady-state levels of proteins during infection (from SDS-PAGE and immunoblotting experiments), it is possible to describe, in some detail, the proteolytic processing of a given protein.

As discussed in the labeling protocol in Chapter 14, the choice of radiolabeled amino acid is critical to the success of the experiment. Remember that any fragments of the protein that lack the particular amino acid used will not be detected. The procedure described here is pulse-chase labeling with Tran^{35}S-Pabel.

The appropriate time after infection to carry out the pulse-labeling can be determined by first carrying out a time course experiment as described in Chapter 14. The data obtained will indicate the time at which synthesis (incorporation of radiolabel) of the required protein is maximal. For many applications, approximately 36 h pi is appropriate.

The time lengths for the chase should be determined empirically, as suitable chase periods depend on the stability of the protein of interest. The following pulse-chase protocol is based on that of Jarvis and Summers (1989).

CONTROLS: It is useful to include pulse-labeled samples derived from cells infected for the same length of time with wt virus (and/or the parent virus if different), as well as pulse-labeled samples from mock-infected cells. For immunoprecipitation analysis, carry out the control immunoprecipitations on pulse-labeled samples only. When analyzing the samples by immunoprecipitation, it may be worthwhile to also analyze nonimmunoprecipitated samples. This should reveal many proteins of varying stabilities in the sample, and can help confirm that a good chase was achieved.

Materials

1. SF cells grown to approximately 1×10^6 cells/ml
2. Tissue-culture dishes (see Chapter 11)
3. Recently titered stocks of your recombinant and control viruses
4. Complete tissue-culture medium. (See Chapter 11 for details on various media.)
5. Methionine-deficient tissue-culture medium
6. Chase medium: incomplete tissue-culture medium supplemented with 1 mg/ml methionine and 10 μg/ml cycloheximide (Sigma # C 6255)
7. Tran^{35}S-label (> 1000 Ci/mmol in aqueous solution, ICN # 51006). Store in small aliquots under nitrogen at –80° C.
8. PBS, pH 6.2
9. NP40 lysis buffer
10. 2X SDS gel-loading buffer

Method

1. Seed and infect the cells as described for the standard time course experiment in Chapter 14. Infect $\approx 1 \times 10^6$ cells for each chase period you wish to analyze. The cells may be in one or more plates.
2. One hour before the chosen labeling time, remove the medium from the plates and replace with the same volume of methionine-deficient medium. Incubate at 27° C for one hour.
3. Remove the cells from the plate(s) and centrifuge at $1000 \times g$ for two minutes in a 15 ml conical centrifuge tube. Gently resuspend the cells in 0.5 mls methionine-deficient medium containing 250 μCi Tran^{35}S-label. Incubate at 27° C for 5 to 15 minutes in a shaking incubator at 100 RPM to 120 RPM.
4. At the end of the labeling period, *immediately* dilute the cells tenfold in *warm* (27°C) chase medium. Pellet the cells and resuspend in 1 ml chase medium for each chase period to be analyzed. For only the pulsed-labeled samples (no chase), withdraw 1 ml at this point, and place on ice. Rinse by pelleting

and resuspending three times with *cold* PBS and lyse as before. Protease inhibitors should always be included in pulse-chase experiments. Incubate the remainder of the sample at 27° C on a shaker at 100 RPM to 120 RPM.

5. At the end of each chase period, take out 1 ml and rinse and lyse the cells (with protease inhibitors) as before. Analyze by immunoprecipitation and SDS-PAGE. The rate of turnover of a particular protein can be evaluated by measuring the loss in radioactivity after the different chase times, relative to a sample that was only pulse-labeled. Processing intermediates will appear as bands that become labeled transiently during the chase period, whereas end products will accumulate label during the chase period.

IV

METHODS FOR SCALE-UP OF PROTEIN PRODUCTION AND USE OF INSECT LARVAE

Once a recombinant baculovirus vector has been generated, many investigators wish to scale up protein production to have sufficient quantities for further research. To produce larger quantities of protein using baculovirus vectors, there are three basic options: (1) combining the products of many small-scale cultures; (2) infecting large-volume cell cultures; or (3) producing the protein in insect larvae. Note that the options available to you may depend on the particular virus and host species you used. For example, the large-volume culture of *Bombyx mori* cells has not been reported yet, and scale-up with a BmNPV-based vector is generally achieved using either multiple small-scale cultures or by infection of *B. mori* larvae.

This part of the manual begins with a comparison of the relative merits of scale-up in cell culture versus insect larvae and with a general discussion on scale-up of protein production in cell culture. Chapters 16 and 17 then provide two protocols for the large-scale culture and infection of SF cells. These chapters were contributed by David L. Clemm and Cheryl Isaac Murphy, respectively, both of whom are industrial scientists with extensive experience in large-scale baculovirus-directed protein production. Finally, Chapter 18 describes the maintenance and infection of larval hosts of both AcMNPV and BmNPV and includes protocols for harvesting recombinant proteins produced in the insect.

PRODUCTION IN CELL CULTURE VERSUS INSECT LARVAE

Scale-up of recombinant protein production in cell cultures has the following advantages: (1) The protein is produced in the same system used for the initial generation of the recombinant virus and for the characterization of the expressed gene product. This minimizes the possibility that, after scale-up, the protein will display any unexpected or undesirable properties that had not been previously detected. (2) Cell-culture-derived protein is easier to purify than protein derived from insect larvae. This is especially true if the cells are cultured in serum-free medium. (3) There may also be more uniformity in post-translational modification of the protein because only a single cell type is supporting synthesis.

The principal disadvantage of the scale-up of recombinant protein production in cell culture is cost. Commercially prepared cell-culture media are expensive, and costs may become prohibitive if very large volumes are required. In addition, large-volume cell culture requires a considerable investment in specialized equipment (e.g., bioreactors) not commonly found in molecular biology laboratories. Although cell-culture-derived proteins are likely to be modified in a uniform manner, certain types of post-translational modification are not carried out efficiently in cell culture (see Chapters 6 and 15). Finally, as mentioned already, the large-scale culture of *B. mori* cells has not been documented yet, so this approach is not available to researchers using a BmNPV-based vector.

The principal advantages offered by larval production include: (1) higher efficiency of certain types of post-translational modifications; and (2) less costly production. One of the major tissues of final-instar lepidopteran larvae is the "fat body," which plays a central role in synthesizing and secreting proteins—including lipoproteins and glycoproteins—into the hemolymph of insect larvae. Thus, larvae can be viewed as natural protein manufacturing and storage plants (Maeda, 1989a). Proteins may be more efficiently post-translationally modified in specialized larval cells, particularly with regard to glycosylation, signal cleavage, fatty acylation, and carboxy-terminal amidation.

The disadvantages of producing proteins in insect larvae may include: (1) a wider variability in post-translational modifications due to expression in a variety of cell types; (2) the need to tend living insects; and (3) the need to pay more attention to downstream protein purification. Downstream purification may be hampered by insect parts, the contents of ruptured insect guts, and contaminating proteins including proteases. If the product is secreted into the hemolymph and the insect is large enough to be bled, the product should be free of insect parts and gut contents. Even so, the product will probably constitute 2% or less of the total hemolymph protein. For proteins that are not secreted, insect gut contamination can also be avoided, if necessary, by removing the midguts before collecting the carcasses. This is somewhat tedious but is not as onerous as it may sound to the novice. Tending insects is not difficult although it may be perceived by some as a negative feature. Eggs can be purchased or obtained from a variety of sources, and the researcher needs only to hatch the eggs and feed the larvae until they are large enough to infect for protein production (see Chapter 18).

For those considering larval production, a comparison of protein yields from cell culture versus insect larvae may be helpful. For the production of polyhedra, we estimate that it takes approximately 100 *Spodoptera frugiperda* or *Trichoplusia ni* larvae (ca. 500 mg each) to equal the production from 1 liter of cells in suspension (ca. 2×10^9 cells). This ratio also appears to approximate recombinant protein yields. In one study (Medin et al., 1990), 22 larvae yielded 8 mg to 9 mg of homogeneous enzyme with an approximately 50% loss of activity during purification. The yield was estimated to be approximately 1 mg per larvae, whereas the yield of the same protein in cell culture is estimated to be approximately 200 mg/l. Thus, 100 to 200 larvae provide a yield equivalent to 1 liter of cells. Readers are reminded that most proteins may not be expressed to as high a level as in this example, and that each recombinant protein may behave differently in its production and stability characteristics in the two systems. Use of larger larvae, such as *B. mori,* may reduce the number of larvae required.

In summary, some proteins may be produced more cheaply in insect larvae than in cell culture. Production of some secreted proteins in larvae may be particularly advantageous because of the more efficient post-translational modification and because of secretion into the hemolymph. However, these advantages may be counterbalanced by greater heterogeneity of the product and more difficulty in downstream purification compared to protein production in cell culture. The larval expression system should not be overlooked if large-scale production is required, expression in cell culture is relatively low, and bioreactors for large-scale production are not readily available.

SCALE-UP OF PROTEIN PRODUCTION IN CELL CULTURE

The options available for the scale-up of protein production in cell culture are either to infect multiple small-scale cultures (using the methods described in the previous chapters) or to grow and infect the cells in large volumes. The first of these options is only practical when relatively small numbers of cells are required. Producing recombinant proteins in spinner flasks is very labor intensive, and yields cannot be optimized due to the limitations in supplying sufficient oxygen. Thus, for large-scale applications, economical and efficient methods for the growth and infection of insect cells in large volumes are needed.

The major limitation to the scale-up of insect cell cultures is the provision of sufficient oxygen without damaging the cells. Insect cells have a relatively high oxygen demand (Weiss et al., 1982). On a small scale, such as spinner flasks, the surface area-to-volume ratio is large enough so that sufficient oxygen is available by diffusion alone. However, as the volume of a vessel is increased, the surface area-to-volume ratio decreases, and the cell density that can be supported also decreases due to oxygen limitation. For example, the oxygen transfer capacity for a traditional nonsparged 100 ml spinner is approximately six times that of a 3 l spinner flask. It is calculated that the oxygen transfer capacity is adequate to support 4.8×10^6 cells/ml in a 100 ml spinner without limiting the growth rate, but only 0.7×10^6 cells/ml in a 3 l spinner (Maiorella et al., 1988).

Aeration by sparging is the usual means of supplying oxygen for large-scale cultures. However, sparging can damage insect cells (Tramper and Vlak, 1986; Tramper et al., 1986). Cell damage appears to involve bubbles bursting at the medium surface and turbulence in the region of the sparger (Tramper et al., 1987; Handa-Corrigan et al., 1989; Murhammer and Goochee, 1990a). There is also evidence that insect cells are relatively sensitive to shear stress. Whereas mammalian cells have been shown to withstand stresses of 2 to 4 Newtons (N)/m^2 (Diamond et al., 1989), shear stress experiments using SF cells in the exponential phase of growth have shown that critical shear stress occurs in the range of 0.6 to 1.0 N/m^2 (Tramper and Vlak, 1986; Tramper et al., 1986; Goldblum et al., 1990; Murhammer and Goochee, 1990a). Virus-infected cells are even more sensitive, because they swell from 1.5 to 2 times their original size, with a significant stretching of the cell membrane apparently due to the accumulation of virus within the cell (Murhammer and Goochee, 1988). It is not clear how the ability of a cell to withstand laminar shear in a viscometer relates to the ability of the cell to tolerate complex hydrodynamic stresses in a bioreactor.

The dual problem of providing sufficient agitation and oxygen without damaging the cells has been largely overcome by the addition of protective agents to the tissue-culture medium. The most commonly used is Pluronic F-68, a block copolymer of polyoxypropylene and polyoxyethylene. The strong protective activity of Pluronic F-68 during sparging has been well documented (Maiorella et al., 1988; Murhammer and Goochee, 1988; Murhammer and Goochee, 1990a; Murhammer and Goochee, 1990b; Jobses et al., 1991). In addition, it has been shown that Pluronic F-68 can protect cells from agitation damage (Michaels et al., 1991). It is hypothesized that the hydrophobic portion of the molecule interacts with the cell membrane, while the polyoxyethylene oxygen may form hydrogen bonds with water molecules to generate a hydration sheath, which provides the protection from laminar shear stress and cell-bubble interactions (Murhammer and Goochee, 1990a; Murhammer and Goochee, 1990b).

In Chapters 16 and 17, two commonly used methods for the large-scale growth and infection of insect cells in a batch fashion are provided. The discussion is confined to methods that can be fairly easily implemented in the average laboratory. Very large-scale applications (> 100 l) are not addressed. The methods described deal with growth in agitated, sparged bioreactors or in airlift bioreactors. These methods differ in how the cells are agitated. In the former, the cells are stirred mechanically; in the latter, agitation is provided by the sparging process.

For interested readers, we note also that commercial-contract production of baculovirus-expressed proteins is available through Vista Biologicals, 2120-C Las Palmas, Carlsbad, California, 92009 (telephone: 619-438-5058). Available services include process optimization, production capabilities of up to 100 l, and downstream processing.

16

SCALE-UP OF PROTEIN PRODUCTION IN A STIRRED BIOREACTOR

David L. Clemm
Ligand Pharmaceuticals, Inc., 9393 Towne Centre Drive, Suite 100
San Diego, CA 92121

This chapter provides an overview of methods and conditions for the setup and operation of a 1- to 100-liter agitated, sparged bioreactor for the cultivation and infection of insect cells to produce recombinant proteins in a batch fashion. A diagram of the bioreactor we use is presented in Figure 16-1 in the section on fermenter preparation. Cultivation of insect cells in an agitated, sparged bioreactor allows for carefully controlled growth and optimization of conditions for improved yields. It is an easy, reliable, and reproducible means to produce milligram quantities of recombinant protein.

It is well documented that sparging can damage insect cells (Tramper and Vlak, 1986; Tramper et al., 1986). Sparger design and bubble diameter are the main factors determining cell damage (Murhammer and Goochee, 1990a). Bubbles smaller than 2 mm in diameter have been shown to be cytocidal (Lynn and Hink, 1977; Jobses et al., 1991). In the method described here, the growth medium is supplemented with 0.1% to 0.2% (w/v) Pluronic F-68, which provides good protection for insect cells against both mechanical stress and damage due to sparging. Cell damage due to sparging is further mitigated by the use of a minimal air flow [0.02–0.07 vvm (volume/volume/minute); e.g., 200–700 mls/min in a fermenter with a 10 l working volume] and an appropriately large sparger orifice (≈1 mm), which produces bubbles larger than 2 mm. By comparison, the environment in an airlift bioreactor is more severe due to differences in the gas distributor and the demands of increased gas velocity and volume necessary to provide lift and agitation (Murhammer and Goochee, 1990a; Murhammer and Goochee, 1990b).

In general, conditions for the optimization of production must be determined for each recombinant protein. However, the guidelines that follow should provide good starting parameters. Numerous factors, including the type of cell line, medium composition, dissolved oxygen (DO) concentration, multiplicity of infection (MOI), cell density and stage of growth, and time of harvest, influence the yields of recombinant proteins. Generally, for a given recombinant protein, most parameters should first be optimized on a small scale in spinner flasks.

Selection of a cell line and choice of a serum-containing or serum-free medium must be determined empirically. Numerous cell lines are available for production of recombinant proteins. There is, however, a wide variation in the levels of expression among these lines. A recent comparison of 23 cell lines for the production of three different recombinant proteins highlights these differences (Hink et al., 1991). Several cell lines had higher levels of production of a given protein than the commonly used SF-9 cells. For example, for expression of human plasminogen, a

Mammestra brassica cell line was optimal, and four other cell lines, including an *S. frugiperda* line, were better than SF-9 cells. However, no single cell line was consistently better for all three proteins tested. Thus, the optimal cell line will change for each application. Factors to consider in choosing a cell line include product yield, growth rates and cell densities in suspension culture, ability to grow in serum-free medium, and the degree to which the cells support post-translational modification of the required protein.

The choice of medium depends on several factors including cost, product yield, ease of purification, and potential protein degradation. Serum-free medium has the advantages of lower cost, potentially higher yields due to higher cell densities (up to 4–8 $\times 10^6$ cells/ml versus 2–3 $\times 10^6$ in Grace's), and easier purification of secreted products due to the low protein content. However, some recombinant proteins are not expressed as well in serum-free medium (unpublished observations). In addition, serum contains protease inhibitors and large amounts of protein, which may protect sensitive, secreted proteins from proteolysis. Another consideration when using serum-free medium is the "passage effect." Cells that have been passaged in serum-free medium for long periods of time show reduced yields and require higher MOIs to achieve the same level of expression as recently adapted cells (unpublished observations). To avoid this problem, cells should be maintained in serum-containing medium, adapted to serum-free medium, and passaged only 40 to 50 times. Low-passage-number adapted cells (pass 12-15) can be frozen down and broken out as needed for scale-up.

The following protocol for the operation of a stirred, sparged bioreactor is divided into sections describing the preparation of the fermenter, recommended operating conditions, inoculation and start-up, infection, and harvesting.

FERMENTER PREPARATION

Figure 16-1 shows a diagram of the 15-liter Applikon bioreactor currently used in our laboratory.

Materials

1. Bioreactor (e.g., 7 l cell culture fermenter, Applikon part # Z611000007; or 15 l cell culture fermenter, Applikon part # Z611000010)
2. Dissolved oxygen probe (Ingold autoclavable polarographic DO electrode, part # 32-275-6702)
3. pH probe (Ingold autoclavable double-junction pH electrode, part # 456-35-90-K9)
4. Air-outlet condenser (Applikon part # Z81308L010)
5. Air-vent filters (Gelman Sciences product # 4210; Fisher # 09-730-125)
6. Quick dis-connect couplings (Cole-Palmer # L-06364). Autoclavable quick dis-connect couplings allow for easy addition of medium and cells. Female connectors attached to inlet ports, and male connectors, attached by silicone tubing to an appropriate size aspirator bottle, allow for convenient sterile transfer.
7. Silicone tubing (Cole-Palmer # L-06411)
8. Tissue-culture-grade detergent (e.g., Linbro; Flow # 76-670-94)
9. PBS pH 6.2
10. Clamps (Cole-Palmer # L-06832-99)
11. Electrical ties (Cole-Palmer # L-06830)

Methods

1. Wash probes, vessel, headplate, and fermenter parts with tissue-culture-grade detergent.

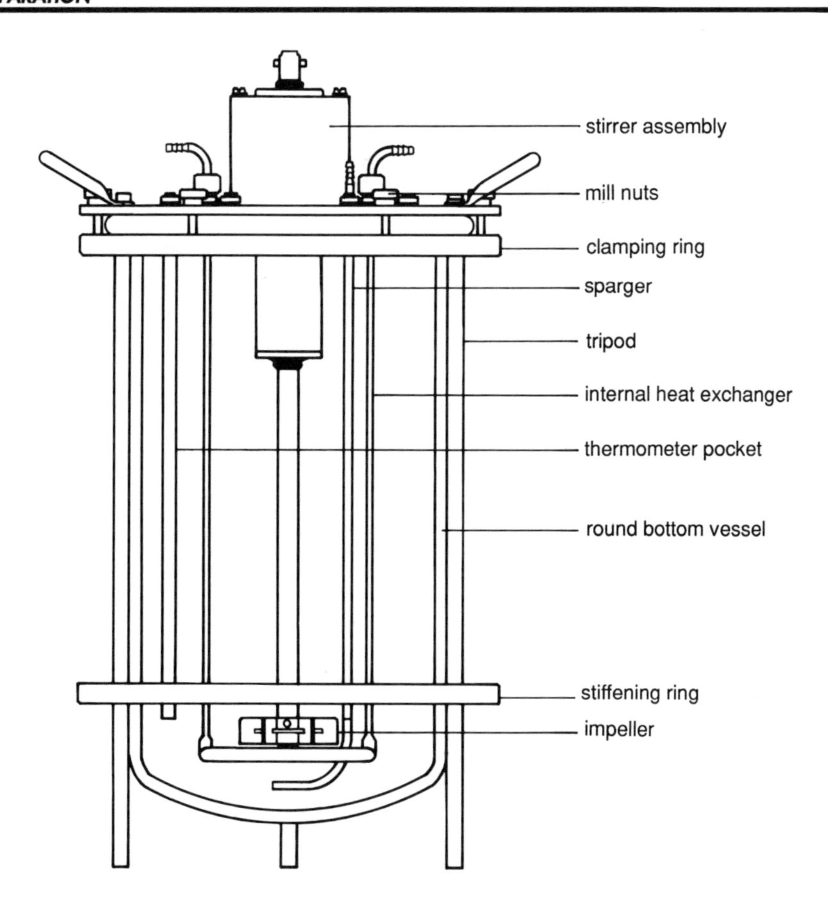

FIGURE 16-1 *Diagram of a 15-liter agitated sparged bioreactor.*

2. Rinse all parts thoroughly with tap water (five minutes), then with triple-distilled, deionized water (ddH_2O) or water of similar quality (two to three minutes). Flush individual sampling tubes, ports, sparger, silicone tubing, and connectors with ddH_2O.
3. Check all O-rings, lines, and connectors for cracks or damage and replace as needed. Secure all lines with clamps or electrical ties.
4. Fill vessel about half full with PBS or ddH_2O.
5. Attach headplate making sure O-ring seals properly, and secure uniformly.
6. Check membrane and electrolyte solution of the DO probe. Insert into a 27 mm port (Figure 16-2A) and secure.
7. Calibrate and test the pH probe. Add electrode buffer if required. Calibrate at pH 7.0 and pH 4.0. Insert into a 27 mm port (Figure 16-2A) and secure pH probe. Tape electrode plug in place.
8. Attach the air-outlet condenser to a 27 mm port.
9. Attach new air-vent filters to sparger inlet, condenser outlet, and sampling apparatus (see Figure 16-2B). Orient filters according to direction of flow. Clamp off filters to avoid any backflow of liquid during sterilization.
10. Wrap all quick dis-connect couplings with aluminum foil.
11. Secure sampling apparatus (see Fig 16-2B), cap DO and pH probes, and tighten all blind plugs except one, to allow for pressure equalization during autoclaving.
12. Autoclave for 50 to 70 minutes at 121° C and 15 psi to 18 psi.
13. Remove vessel from autoclave and let cool. Remove clamps from filter vents and tighten all plugs.

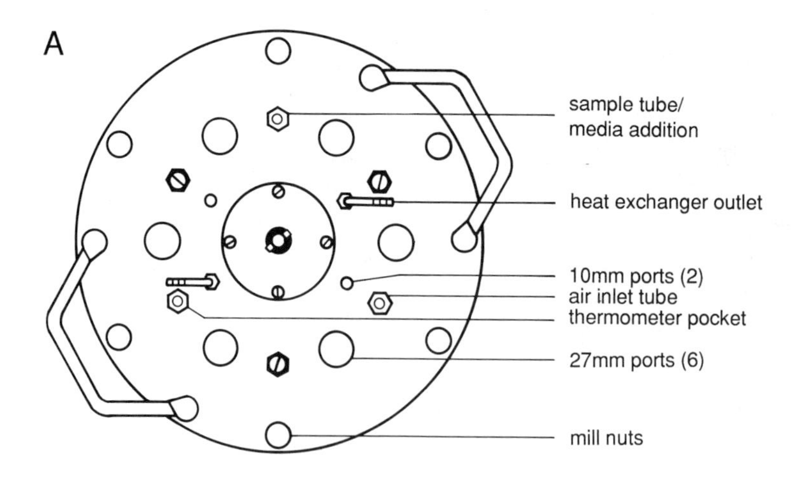

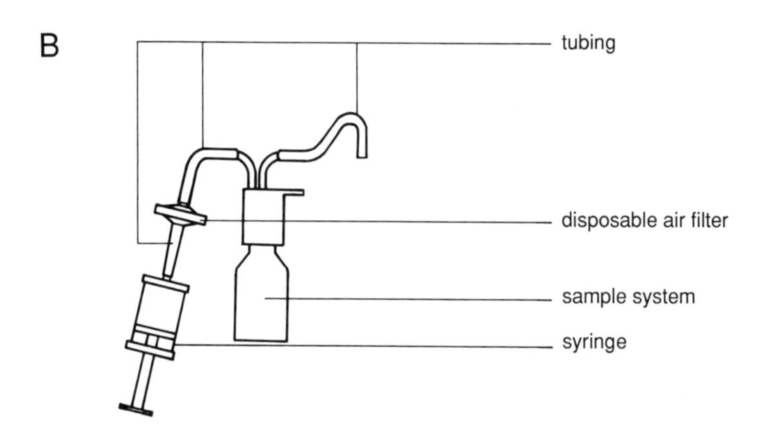

FIGURE 16-2 *Agitated sparged bioreactor. **A**. Headplate. **B**. Sampling apparatus.*

14. If the DO probe is polarographic, attach lead and polarize electrode for at least one hour prior to calibration.
15. Calibrate DO probe.
    a. Adjust vessel temperature to 27° C.
    b. Sparge air for 20 to 30 minutes to saturate water/PBS.
    c. Set 100% set point.
    d. Set 0% set point electronically, or sparge nitrogen or argon for 20 to 30 minutes and set 0% set point.

RECOMMENDED OPERATING CONDITIONS

1. Temperature: 27° C. Maintain temperature according to the vessel configuration. Common methods include a heat blanket or a refrigerated, circulating waterbath connected to a water jacket or internal heat exchanger. In the fermenter shown in Figure 16-1, an internal heat exchanger is used.
2. pH: 6.1 to 6.3. No pH adjustment is required as most insect media have a high buffering capacity.
3. Stirrer speed: 75 RPM to 200 RPM. Use the lowest agitation speed possible to keep the cells in suspension, prevent clumping, and provide adequate mixing. We recommend using a marine impeller, which provides lift without creating

excess turbulence. Stirring at 110 RPM in a vessel with a 5 l working volume provides good agitation. In larger vessels, even slower rates are sufficient. Stirring at 75 RPM to 85 RPM in a 100 l vessel will provide good mixing.
4. Dissolved oxygen: 50% of saturated air. A DO concentration of 50% of saturated air is sufficient to maintain cell growth and supply adequate oxygen for protein production during viral infection. Levels above 50% do not enhance cell growth (Hink, 1980; Murhammer, 1989).
5. Oxygen sparge rate: 0.02 to 0.07 vvm O_2, (extra-dry grade). Since insect cells have such a high oxygen demand, sparging with air will not provide enough oxygen to keep up with the demand as the population density increases. In a vessel with a 5 l working volume, cell densities over 1×10^6 cells/ml require sparging pure O_2. Sparging pure O_2 from the beginning of a run mitigates problems of foaming and cell damage associated with sparging large volumes of air. Oxygen is sterilized by passage through a 0.22 μm filter prior to introduction into the bioreactor.

INOCULATION AND STARTUP

The cells to be used as the inoculum for use in scale-up should be maintained in spinner flasks. They should be in exponential phase and have a viability of at least 95%. [Because the presence of Pluronic F-68 alters trypan-blue viability counts (Murhammer and Goochee, 1988), visually inspect cells to determine percent viability. Dead cells will appear shriveled and more refractive than live cells.] Generation of the cell inoculum should take into account the working volume of the vessel and the desired starting density. Initial densities of less than 3×10^5 cells/ml result in long lag times. Starting densities of $4–6 \times 10^5$ cells/ml allow the cells to move rapidly into exponential growth, thus shortening the run time. For large vessels (100 l), start off by inoculating cells into 40% of the final working volume. When cell densities approach mid- to late exponential phase, add additional medium up to 80% of the final working volume. The next day when the population density has again doubled, add medium up to the final working volume. This procedure always keeps the cells in exponential growth and reduces the time in culture prior to infection.

Cells should be counted immediately after they have been seeded and diluted to the final working volume, then counted daily to determine their viability and growth kinetics. Cell growth will usually lag for about one day after seeding, but should approach a population doubling time of 20 to 28 hours and maintain 95% viability after that.

Prior to starting a viral infection, it is important to determine the growth kinetics and maximum cell density of a cell line in either serum-containing or serum-free medium in the vessel to be used for protein production. This information then can be used to determine the best cell density and time of infection to optimize yields.

The old medium in which the cells were grown for the generation of the inoculum can reduce yields. Pelleting the cells and resuspending in fresh medium can increase the yield. If the inoculation volume is greater than one-tenth of the working volume, the cells to be transferred should be gently spun down (200–400 × g for 10 minutes) and resuspended in fresh, prewarmed medium just prior to transfer. This is especially important for optimization of yields in vessels with less than a 10 l working volume.

Materials

1. Complete tissue-culture medium. See Chapter 11 for details of the various media available. Note that serum-free media generally are already supplemented with Pluronic F-68. Otherwise, supplement the medium with 0.1% to

0.2% (w/v) Pluronic F-68 (JRH Bioscience # 59-91577P; GIBCO # 670-4040 AG). Addition of antifoam is generally not required due to the relatively low sparge rate. Antibiotics are usually added to reduce the chance of bacterial contamination. Supplement with penicillin-streptomycin to a final concentration of 50 units/ml (100 μg/ml) (JRH Bioscience # 59-60277) or gentamycin to 50 μg/ml (JRH Bioscience # 59-60577).

2. Sterile, calibrated bioreactor (see preceding for preparation)
3. Controller (Applikon, ADI 1030 Biocontroller, part # Z510300020)
4. Chart recorder (Yokogawa Model 4156 100 mm Micro Recorder; Cole-Palmer # L-08386-56)
5. Exponential phase (approximately 1×10^6 cells/ml) cell inoculum. See the introduction to this section for guidelines concerning the volume of cells to prepare, etc.
6. Sterile transfer bottle (aspirator bottle, Fisher # 02-972)
7. Compressed air
8. Compressed oxygen with two-stage regulator. Use extra-dry grade.
9. Rotameter (gas-flow meter) (Applikon, O_2 rotameter, part # Z3RM000000)
10. 70% ethanol

Methods

1. Remove water/PBS from vessel.
   a. Clamp off exhaust vent of air-outlet condenser and pressurize vessel with compressed air to 5 psi to 10 psi.
   b. Blow out liquid through tubing attached to a sampling tube that extends to the bottom of the vessel.
   c. Remove clamp and air line.
2. Add cells and complete medium to the transfer bottle in a sterile hood.
   a. Using sterile technique, attach quick dis-connect fittings to vessel.
   b. Gravity-feed medium and cells into bioreactor.
   c. Disconnect couplings and rinse with 70% ethanol.
3. Turn on controller and adjust operating conditions (see recommended operating conditions earlier in this chapter).
   a. Turn on stirrer motor and adjust speed.
   b. Attach cooling lines to exhaust condenser and circulate cooling tap water.
   c. Attach O_2 feed line.
   d. Turn on compressed oxygen valve and adjust regulator output to 10 psi.
   e. Adjust rotameter for appropriate O_2 flow rate.
   f. Check temperature and dissolved oxygen set points.
   g. Turn on chart recorder to monitor pH, DO, and temperature.

VIRAL INFECTION

To optimize yields, several factors, including cell density, stage of growth, and MOI, must be considered together. The relationship of the final product concentration to the MOI is highly dependent on the growth phase of the cells. Maiorella et al. (1988) and Murhammer and Goochee (1988) have found that product yields are relatively insensitive to MOIs in the range of 0.5 to 10 when cells are in early to mid-exponential growth phase. Similarly, Licari and Bailey (1991) report that product yields are relatively steady for MOI values from 0.1 to 100 with cultures infected during the early exponential phase. However, during late exponential growth phase, both cell density and product formation are strongly dependent on MOI. The higher the MOI used, the more rapid the decrease in cell viability and the greater the product yield. Cultures infected during the early exponential growth phase show a similar relationship between cell viability and MOI but do not show the effect on product yield.

The final product yield for cells infected in the early to mid-exponential phase is generally greater than that obtained for cells infected in the late exponential phase for MOIs up to 10. However, the maximum yield of recombinant protein occurs in cells infected in late exponential phase with an MOI of 100 (Licari and Bailey, 1991). On a large scale, it is not feasible to infect cultures with an MOI of 100 due to the large volume of virus stock necessary. Therefore, it is best to infect cells in mid-exponential phase with a low MOI (2 or less) to obtain optimum yields. By using a low MOI and relying on the secondary infection of uninfected cells, the amount of virus stock needed is substantially reduced.

The virus stock used to infect large-scale cultures should be well characterized, low passage number (less than passage 6), and high titer. High-passage virus stocks give reduced yields due to defective interfering particles (see Chapter 1). Infect mid-exponential phase suspension cultures grown in serum-free medium at a low MOI [0.1 plaque-forming units (pfu)/cell]. Grow infected cells for five days at 27° C, harvest the supernatant by centrifuging the cells, and determine the titer by plaque assay (see Chapter 12). Cells grown in serum-free medium should give titers of 1×10^9 pfu/ml versus $1\text{–}3 \times 10^8$ pfu/ml from cells grown in Grace's medium. See Chapter 12 for protocols for the preparation and titration of high-titer virus stocks.

Materials

1. High-titer recombinant virus stock
2. Bioreactor with mid-exponential growth phase cells greater than 95% viable ($\approx 1.0\text{–}1.5 \times 10^6$ cells/ml grown in serum-containing medium) ($\approx 2\text{–}3 \times 10^6$ cells/ml grown in serum-free medium)
3. Syringe or sterile transfer bottle, depending on volume to be transferred
4. 70% ethanol

Methods

1. Count cells and determine the percent viability. Always flush sampling lines at least once before taking a sample to count.
2. Calculate the volume of virus stock to add, based on cell count, working volume, and MOI. For example, calculate the number of mls of a virus stock with a titer of 1×10^9 pfu/ml needed to infect a 10 l working volume at 1.5×10^6 cells/ml with an MOI of 2.
   10 l = 10,000 mls $\times 1.5 \times 10^6$ cells/ml = 1.5×10^{10} cells;
   1.5×10^{10} cells $\times$ 2 (MOI) = 3×10^{10} pfu/10 l;
   3×10^{10} pfu/1×10^9 pfu/ml = 30 mls virus stock
3. Add virus stock to syringe or transfer bottle in sterile hood.
4. Using sterile technique, add virus stock to the bioreactor through alcohol-rinsed septum or quick dis-connect fitting. Dis-connect coupling and rinse with 70% alcohol.

HARVESTING

The optimal time to harvest the infected cells or conditioned medium must be determined for each recombinant virus based on time course experiments. These experiments should be done in the fermenter used for production by determining the amount of recombinant protein present at various times pi. Factors to consider when determining the best time to harvest should include the highly regulated temporal nature of heterologous protein synthesis using baculovirus promoters, the secondary infection of the remaining uninfected cells approximately 24 h pi, and intracellular as well as extracellular protein degradation.

Studies of β-galactosidase produced in SF-9 cells show that intracellular degradation at early times pi is primarily the response of a healthy, viable cell to the stressful, highly disrupted state created by the viral infection. However, as cellular processes are compromised by viral activities, the degradation rate becomes negligible. Although intracellular degradation of recombinant protein occurs, degradation activity declines as infection proceeds and is insignificant late in infection when synthesis is intense (Licari and Bailey, 1991).

If a recombinant protein is secreted into the medium, proteolysis may occur late in infection when cells begin to lyse and release proteolytic enzymes. This problem can be mitigated by determining when maximal protein synthesis occurs, using pulse-label experiments (see Chapter 14), and then harvesting prior to the decline in synthesis and before cell lysis begins. Generally, proteolysis is not a problem if harvesting occurs between 48 h pi and 72 h pi.

A variety of protease inhibitors are commercially available. See Chapter 14 for a discussion of the more commonly used ones.

Harvesting by centrifugation is effective for separating infected cells from the conditioned medium for volumes less than 10 l. However, for larger scale cultures, cells should first be concentrated using a tangential flow filter, which separates the culture into a cell-enriched fraction and a cell-free fraction. The cell-enriched fraction can then be pelleted by centrifugation if the recombinant protein is cell-associated.

Cell pellets and conditioned medium should be stored at –70° C if protein purification does not begin immediately.

Materials

1. Bioreactor with infected insect cells
2. Compressed air
3. Tangential flow-concentrating filter (Microgon, 10 l, Cell Flow # C22M-041-01N; Microgon, 100 l, Krosflo # K22M-200-01N)
4. Centrifuge bottles, 1 liter
5. Protease inhibitors
   a. 10 mg/ml PMSF stock. Use at a final concentration of 100 μg/ml.
   Caution: PMSF is very harmful and may be fatal if inhaled, swallowed or by skin contact. Wear protective clothing and a mask when handling PMSF.
   b. 100 μg/ml Pepstatin A stock. Use at a final concentration of 1 μg/ml.
   c. 50 μg/ml Leupeptin stock. Use at a final concentration of 0.5 μg/ml.
6. Bleach

Methods

1. Turn off the controller, stirrer motor, exhaust-condenser water supply, and O_2. Disconnect the O_2 line.
2. Attach compressed air line to sparger inlet and clamp off exhaust vent. Pressurize vessel to 5 psi to 10 psi and blow out medium and cells through silicone tubing into centrifuge bottles or a concentrating reservoir.
3. Concentrate cells by passing culture across a tangential flow filter. Base the size of the filter unit on the volume of culture medium to be processed.
4. Centrifuge culture supernatant or enriched-cell fraction at 1000 × g for 20 minutes to pellet cells.
5. Carefully decant supernatant and pool with cell-free fraction. If recombinant protein is secreted, add protease inhibitors to medium, transfer to sterile bottles, and freeze at –70° C or begin protein purification. If recombinant protein is cell-associated, pool all cell pellets and freeze at –70° C or begin purification.
6. Disinfect vessel with 5% bleach for 30 minutes and rinse thoroughly.

17

SCALE-UP OF PROTEIN PRODUCTION IN AN AIRLIFT FERMENTER

Cheryl Isaac Murphy
Cambridge BioScience, 365 Plantation Street, Biotechnology Research Park
Worcester, MA 01605

The growth of insect cells in large-scale culture, and subsequent infection of these cells by recombinant baculoviruses, presents a number of potential difficulties not encountered on a small scale. Insect cells require relatively high amounts of oxygen for growth (Weiss et al., 1982). While sparging oxygen is necessary to achieve high cell densities, it can lead to decreased viability due to the sparge sensitivity of these cells and their entrapment in the foam that builds up as a result of sparging (Tramper et al., 1986).

The combination of low-protein serum-free medium and the protective copolymer Pluronic F-68 has made large-scale culture of insect cells feasible in airlift fermenters. Several groups have demonstrated high cell density with an accompanying elevated production of recombinant proteins in this system (Maiorella et al., 1988; Murhammer and Goochee, 1988; Godwin et al., 1989). A diagram of an airlift fermenter is shown in Figure 17-1. Cells are suspended in the medium inside the glass vessel and are circulated from the base of the vessel up through the draft tube and down along the sides of the vessel as gas is sparged from the bottom (Onken and Weiland, 1983). The Pluronic F-68 present in the medium has a protective effect on the cells and prevents damage due to sparging (Kilburn and Webb, 1968). Refinements in insect serum-free media formulations now provide for high cell densities and minimum foaming. Most important, the absence of serum proteins greatly facilitates the purification of secreted recombinant proteins (Maiorella et al., 1988; Murhammer and Goochee, 1988; Godwin et al., 1989).

The following description of large-scale cell culture using an airlift fermenter is divided into sections describing assembly of the bioreactor, filling and calibration of the bioreactor, cell growth, virus infection, and harvesting.

AIRLIFT FERMENTER ASSEMBLY

Figure 17-1 shows the assembled fermenter. The temperature control jacket is sealed to the inner vessel with silicone compound, and a silicone "bung" is used to plug the base of the vessel. This is secured to the vessel by an open-ended cap that screws onto the glass base. The silicone bung has two or three ports through which stainless-steel tubing is inserted. The fermenters in our laboratory have two ports, one used for medium addition or harvesting and the other for gas sparging. A third port could be used for supplementary gas sparging or as a separate harvesting or virus-addition port. The airlift fermenter manufacturer supplies a sparge ring for the fermenter that connects through one of the ports in the headplate and extends

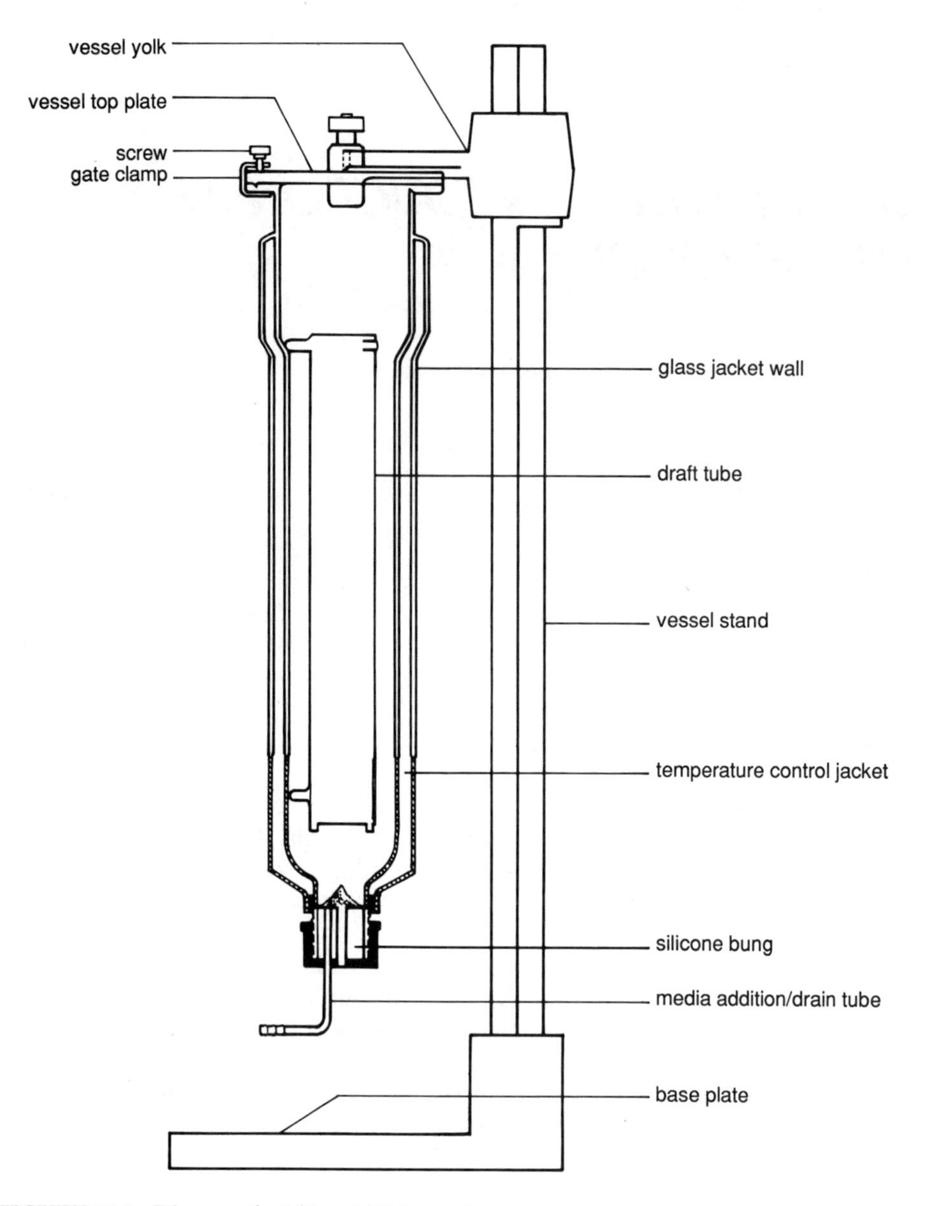

FIGURE 17-1 *Diagram of a 5-liter airlift fermenter.*

down into the vessel through the draft tube. We have found that this ring causes two problems. First, the sparging holes are very small, and the air bubbles generated contribute to premature cell death and low maximum cell densities, presumably due to increased hydrodynamic stress on the cells (see Chapter 16). Second, cells tend to settle on the bottom of the glass vessel because the position of the sparge ring does not allow air to enter the vessel from the bottom. The larger orifice of the steel tubing creates larger gas bubbles, thereby reducing shear stress. As a result, cell density is increased two- to threefold. Also, because air can be sparged from the very bottom of the vessel, cells are prevented from settling.

Materials

The following are available from Porton Instruments/LH Fermentation:

1. 5-liter airlift fermenter system (# V02-01-52)
2. Dissolved-oxygen control module (# 1516)

3. Service module (# 1520CL1)
4. Dissolved-oxygen electrode (# E01-04-16). This is an Ingold, Clark Cell-type polarographic electrode with replaceable membrane. It may be repeatedly sterilized and has an insertion length of 320 mm with a 19 mm stem diameter. It may also be obtained directly from Valley Instrument Company. I recommend getting an electrode such as this with a long insertion length. The shorter ones do not extend very far into the vessel and are only covered when the vessel is filled.

Other materials:

5. Circulating water bath with at least 6 l capacity, which maintains a temperature of 27° C
6. Bioreactor autoclave stand (Bellco, #7800-24027)
7. Popit check valve (Cambridge Valve and Fitting, # 6L-CW454)
8. Tube adapter (Cambridge Valve and Fitting, # 5S-2-HC-A-401)
9. Gelman 0.22 μM filters (Gelman Acro 50, # 4251)
10. Gelman acrodisk 0.45 μM filters (Gelman # 4184)
11. Millex FG50 filters (Millipore, # SLFG05010)
12. Millipak 60 0.22 μM filters (Millipore, # MPGL06SH2)
13. Silicone tubing size 1/4" ID (Cole Palmer, # L-06411-74) and size 3/16" ID (Cole Palmer, # L-06411-03)
14. Y connectors (Cole Palmer, # N-06296-20 and # N-06296-10)
15. Clamps (Cole Palmer, # L-06833-00), cable ties (4 inches; Cole Palmer, # L-06830-52), and tension tool (Cole Palmer, # L-06830-05)
16. 1/4" stainless-steel tubing

Method

1. The fermenter is best assembled when it is strapped into the autoclave stand with the headplate-end tilted up and the base down. Place the draft tube inside the vessel with the pronged feet at the base. It is helpful to cover the feet and side projections on the draft tube with silicone tubing because these glass parts are sensitive to breakage.
2. Insert steel tubing into the bung ports so that one end is flush with the top of the silicone bung and the other end extends about 3 inches outside of the vessel. The length of the steel tubing inserted should be kept to a minimum so the fermenter will fit into its stand and into the autoclave.
3. Attach check valves to the tubes that will be used for gas sparging. This will prevent medium from backing up into the sparge tube if the flow of air is stopped. Attach a tube adapter to the check valve and a piece of silicone tubing to the adapter.
4. Attach the end of the tubing to a Millex FG50 0.22 μm filter. If two tubes are being used for air, connect both pieces of tubing to a Y connector and attach this to the filter with a small piece of tubing.
5. Secure the tubing at all attachment points with cable ties.
6. Attach silicone tubing to the steel tubing extending from the medium addition/harvest ports at the base of the vessel. We use a piece long enough to reach into an adjacent laminar flow hood. Secure the tubing connection with cable ties and attach a small (1 to 2 inch) piece of stainless steel or glass tubing to the free end of the silicone tubing. This will facilitate attachment of the tubing to a medium addition vessel.
7. Cover the open-ended tubing with aluminum foil.
8. Plug the bung assembly firmly into the fermenter base and secure it with the screw cap. The tubing can be rolled up and secured with tape for ease of autoclaving.

9. For each closed port, insert a septum of the proper diameter into the fitting (see Figure 17-2 for a diagram of the headplate of the bioreactor). Make sure the septum is seated properly when the fitting is screwed into the headplate and the O-ring is tight but not crushed.
10. Port #1 is the air-outlet port. If air is not allowed to escape, pressure will build up inside the vessel and eventually shut the system down. We routinely have two air-outlet ports in case one outlet becomes blocked. The fitting for port #1 contains a stainless-steel tube that extends down into the headspace of the fermenter. Attach a 12-inch piece of silicone tubing to this tube with cable ties and a Y-connector to the free end.
11. Attach two more pieces of silicone tubing to the Y-connector and Gelman Acro 50 0.22 μm filters to the other ends. Secure the tubing with cable ties. Tubes should be held up in a vertical position so that any condensation will drain back down the tube and not plug the filters. Alternatively, an air-outlet condenser can be used.
    a. Insert the condenser into the condenser port (#1), tighten the collar, and attach a Gelman filter with silicone tubing to the top of the condenser.
    b. Run silicone tubing from a cold water supply to the bottom connection of the condenser.
    c. Run silicone tubing from the top connection of the condenser to a drain. This will enable the warm water vapor from the vessel to be cooled sufficiently so that it will not block the air filter.
12. Attach a piece of silicone tubing to the steel tubing of port #2 with cable ties, and a Gelman 0.22 μm filter to the free end. This will provide a back-up air outlet, which can be clamped off until needed.

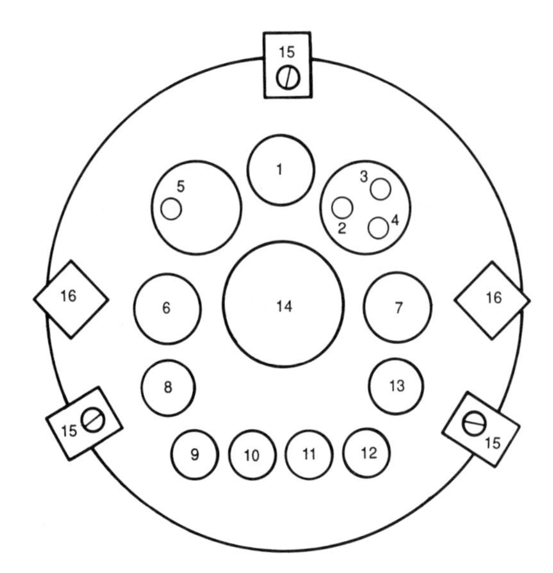

Item	Description	Item	Description
1.	Air out	9.	Closed
2.	Air out	10.	Closed
3.	Closed	11.	Closed
4.	Closed	12.	Closed
5.	Bung closure	13.	Sample
6.	Closed	14.	Closed
7.	O^2 Electrode port	15.	Top plate clamp and screws
8.	Media addition	16.	Vessel retaining screw

FIGURE 17-2 *Headplate of airlift fermenter.* Our preferred uses of the various ports are given in the list below the figure.

13. To close off ports #3 and #4, attach a small piece of silicone tubing to the steel tubing of these ports, one end to #3 and the other end to #4. Secure the tubing with cable ties.
14. Port #6 is for the pH electrode, which may or may not be used. We have not seen any significant pH variation throughout the course of a bioreactor run and therefore feel that monitoring pH continually is unnecessary.
15. The oxygen electrode fits into port #7. Before inserting the electrode, inspect the membrane for signs of age, such as a yellow color or other damage. Membranes can be used repeatedly, but we usually change the electrolyte solution for every run and ensure that there are no air bubbles in it.
16. Port #8 is for adding small volumes of medium or virus. Attach a long piece of silicone tubing (about 4 feet) to the stainless-steel adapter of the fitting. Secure with cable ties. At the other end of the tubing, attach an adapter for a syringe. Cover with aluminum foil before autoclaving.
17. Port #13 (or any of the other ports of the same size) is used for sampling. Attach a short piece of silicone tubing to the side arm of the sample-bottle top holder. Put a small 0.22 μm syringe filter on the free end of the tubing and secure with cable ties. After autoclaving, a 10 ml syringe can be attached to the filter. Because autoclaving may loosen the connection between the sample bottle and its top, cover both with a piece of aluminum foil.
18. Before attaching the headplate to the vessel, put about 50 ml of deionized water into the vessel. Place the O-ring into its groove and screw on the headplate with retaining clamps (Figure 17-2, #15) just tightly enough to hold the headplate on. These clamps tighten further during autoclaving and can crack the vessel.
19. Make sure all tubing ends are covered with aluminum foil and autoclave the assembled fermenter at 123° C for 60 to 70 minutes.
20. After autoclaving, tighten the headplate-retaining clamps and put clamps on the following: the air-inlet tubing, the media-addition tubing, the back-up air-outlet tubing, the sample-bottle side-arm tubing, and the tubing on port #8.
21. After cooling, place the fermenter in the fermenter stand by putting the base of the fermenter in the stand first, then sliding the top into place. Tighten the two side clamps (Figure 17-2, #16).
22. Attach water bath connections so water enters the bottom of the water jacket and flows up.

RUNNING THE BIOREACTOR

The minimum instrumentation required for growing SF cells in the airlift bioreactor consists of the service module and the dissolved oxygen (DO) module. Other modules that can be added to the system include pH, anti-foam, and temperature modules. These modules all connect into the service module. The standard service module has the capacity for blending up to three different gases and has a total gas-flow meter.

Materials

1. Sterile assembled bioreactor
2. Sources of oxygen, nitrogen, and air (if desired)
3. Regulators for gas tanks (two stage)
4. Heavy walled Tygon tubing for regulator to controller connections
5. Serum-free medium. (See Chapter 11 for details of the various media available.)
6. Peristaltic pump (Masterflex, Cole Palmer # L-07523-00)

Methods

1. We connect our nitrogen supply to the N_2-labeled meter immediately to the left of the total flow meter. The oxygen supply is connected next to the nitrogen meter (labeled "air" on the service module). If desired, an air supply could be connected to the last gas-blending port. The gas outlet on the service module is located below the total flow meter. Turn the knobs for each gas flow completely off.
2. Turn on the O_2 and N_2 and set an initial pressure for each tank at about 10 psi.
3. Connect a piece of silicone tubing to the service module outlet and secure with cable ties. Connect the other end of the tubing to the air filter attached to the air inlet tubing of the fermenter.
4. Connect the dissolved oxygen cable from the DO module to the electrode on the fermenter headplate. Turn on the controller.
5. The fermenter is now ready to be filled. Uncoil the media-addition tubing and connect it to a peristaltic pump.
6. Put the free end of the tubing inside the hood adjacent to the fermenter. Remove the aluminum foil and pump out any residual water.
7. Connect the tubing to the bottom of a sterile holding bottle. We use the large size glass bottle that comes in the airlift fermenter accessory pack.
8. Pump prewarmed medium into the fermenter. As an extra sterility precaution, an inline 0.22 μm filter (Millipak 60) can be connected to the tubing if desired. Fill the fermenter with medium until the bottom of the DO probe is covered. This will require approximately 5 l to 5.5 l.
9. Clamp the media-addition tubing and maintain its sterility by rinsing the end in alcohol or covering it with a sterile alcohol pad and sterile aluminum foil.
10. Before adding cells to the fermenter, calibrate the dissolved oxygen probe.
    a. First saturate the medium with nitrogen gas to set the 0 point. Open up the nitrogen flow meter and then the total flow meter. Unclamp the air inlet tubing and adjust the flow to 300 ml/min. Let the nitrogen bubble through for at least four hours. When the dissolved oxygen value stabilizes, adjust the reading to 0.
    b. To set the 100% value, saturate the medium with air. This can be done through the service module if air is connected to it, or air can be introduced by connecting the air-inlet tubing to the peristaltic pump. Saturating the medium with air may take several hours to overnight. When the DO reading stabilizes, adjust it to 100% using the "span" knob.
    c. Switch the gas supply back to the service module. Set the DO value to 50%.
    d. Open the flow meters for both oxygen and nitrogen all the way and adjust the total flow to 300 ml/min. Once the dissolved oxygen concentration approaches 50%, adjustments may have to be made to the oxygen regulator to ensure that pure oxygen is delivered to the vessel when oxygen is demanded. The controller should be set to local control and the output switch set to auto. The flow rate tends to drift downward for the first few hours of operation, and periodic adjustments may have to be made.

CELL GROWTH AND VIRUS INFECTION

Several factors are important for consistent growth of SF cells in the airlift bioreactor. The cells used for seeding the bioreactor should be grown in serum-free medium for several passages prior to seeding. When added to the fermenter, they should be in exponential growth and have a viability greater than 95%. We usually seed our fermenters so that the cell density is $3–5 \times 10^5$ cells/ml. Cells can be grown either in spinner or shake flasks for seeding. We have noticed that cell growth tends to lag more if we use shake flasks for seeding. However, cells grow to higher cell densities in shake flasks and thus can be seeded as a smaller volume. Cells should not be concentrated by centrifugation prior to seeding because they will not grow well in the fermenter.

The quality of the serum-free medium is extremely important. Both JRH Scientific and GIBCO manufacture high-quality serum-free medium for insect cells (see Chapter 11). Different lots of the same medium may vary widely in their ability to support logarithmic cell growth. If possible, obtain a large stock of medium from one lot that has been tested successfully in the fermenter. The same medium can give different cell growth results in a shaker or spinner flask as compared with the bioreactor. Other factors important in airlift fermentation include proper assembly of the fermenter, as covered in the previous sections; maintenance of the fermenter parts; and monitoring of the oxygen supply delivered to the cells. In our own experiments, we maintain the dissolved oxygen at 50% of air saturation throughout the bioreactor run, including the time period of virus infection.

It is difficult to predict *a priori* the optimal multiplicity of infection (MOI) and time of harvest for maximum yield of a particular protein. These parameters will depend on the nature of the recombinant protein expressed and will not necessarily be equivalent to that determined in tissue-culture flasks. Godwin et al. (1989) have shown that the maximum yield of β-galactosidase is obtained with an MOI of 1 and an infection period of five days. This allows for some additional growth of the cells following viral infection and an extended period of higher cell viability. However, the highly glycosylated envelope proteins of human immunodeficiency virus (HIV) are expressed in a soluble, glycosylated form relatively early in viral infection (Murphy et al., 1990). We have found that an MOI of 5 to 10 and a relatively short infection period of 48 hours in the airlift fermenter results in the maximum yield of the HIV envelope proteins. For other recombinant proteins, MOI and time of product harvest will have to be determined experimentally by the individual investigator.

Materials

1. Calibrated bioreactor
2. Exponential-phase SF cell inoculum
3. 15 ml conical centrifuge tubes (Falcon # 2097)
4. Hemocytometer or Coulter counter
5. Recently titered stock of recombinant virus
6. Pellicon cassette-filter acrylic-holder assembly (Millipore # XX42 PS460)
7. Detergent (e.g., Linbro; Flow Laboratories # 76-670-94)
8. Bleach

Method

1. Inoculate the cells into the fermenter. It may be necessary to remove some of the medium from the fermenter before adding the cells if the volume is too large. The medium-addition tube at the base of the fermenter is used for both purposes. Keep the gas sparge rate between 200 and 300 ml/min when adding the cells to the vessel. Rinse enough medium through the tubing after adding the cells to ensure that all the cells have been added to the fermenter.
2. Take an initial sample using the sampling tube (Figure 17-2, # 13). The sampler cap also fits 15 ml Falcon tubes, and these can be substituted for the glass bottle after autoclaving.
3. Count the cells using a hemocytometer or a Coulter counter.
4. On subsequent days, take one or two samples per day for cell density and viability measurements. Cell growth may lag for the first day or two but, by

the third day, cells should be doubling every 20 to 24 hours. We recommend doing initial cell growth studies in the fermenter before trying a viral infection. This permits each investigator to determine growth kinetics and maximum cell density. It is also possible to subculture the cells in the bioreactor as a way of both adapting cells to it and determining cell growth kinetics.

   a. When cell density begins to level off, remove most of the cells from the fermenter and add back fresh medium so that cell density is $3–5 \times 10^5$ cells/ml.
   b. This cycle can be repeated a third time if careful attention is paid to sterility and monitoring the air-outlet filters for potential blockage.

5. The dissolved oxygen in the medium may also become limiting as cell growth approaches maximum cell density. This will become evident if the DO reading drops from 50% to less than 10%. The total flow rate can be turned up without harming the cells, and the controller will deliver pure oxygen as the sparge gas.
6. Once the growth kinetics have been determined, cells can be infected with recombinant baculovirus. We have found that infecting the cells at densities between $1–2.5 \times 10^6$ cells/ml results in maximum recombinant protein expression per cell. Virus can be added to the fermenter either from the medium-addition tube at the bottom of the vessel or from the tubing in the headplate at port #8.
7. Operate the bioreactor for the appropriate period of time following infection before harvesting the cells and/or medium.
8. We have used two methods to harvest recombinant protein: centrifugation and filtration through a Pellicon cell harvester. Centrifugation is more time-consuming and labor-intensive than using the Pellicon, but it may be used to recover a high-titer, sterile virus stock. The Pellicon is a fast and efficient way of concentrating infected cells if the recombinant product is cell-associated. The culture medium can also be recovered free of cells if the protein is secreted. To use the Pellicon, the cell suspension is pumped from the bottom of the vessel through the Pellicon membrane and separated into a cell-enriched fraction and cell-free medium fraction. At the end of the filtration, the cell fraction can be further concentrated by centrifugation. Recovery efficiencies for the Pellicon method are equivalent to those using centrifugation.
9. Before setting up the bioreactor again, all parts should be washed with a detergent solution or dilute acid and base, and rinsed thoroughly with distilled, deionized water. A 10% bleach solution can be used to decontaminate the fermenter but should not be used on stainless-steel parts. Silicone tubing can be reused but should be checked for signs of aging or weakness. Change all O-rings and septums as needed.

Acknowledgments

The author is grateful for the technical support of James McIntire and Sophie Lehar and others at Cambridge Biotech Corporation who made this work possible. She also thanks those from LH Fermentation and IGB Products who made many helpful suggestions on fermenter setup and cell growth.

18

INSECT REARING AND INFECTION

It is often desirable to propagate a virus in insect larvae rather than in cell culture. Larval production can be an extremely useful alternative to growth in cell culture for generating large quantities of recombinant protein. In addition to the very high yields that can be obtained, protein production in insects may facilitate the accurate post-translational modification of the expressed gene product. (See the introduction to Part IV for a more detailed discussion of the relative merits of expression in cell culture or in insects.) Infection of insect larvae is also the most convenient method for the production of large quantities of occlusion bodies (PIBs).

This chapter includes some simple procedures for rearing and infecting the larvae of several lepidopteran species. The species covered are the fall armyworm *Spodoptera frugiperda,* the cabbage looper *Trichoplusia ni,* the tobacco budworm *Heliothis virescens,* and the silkworm *Bombyx mori.* The first three of these species belong to the family Noctuidae, and are permissive hosts for AcMNPV, whereas *B. mori,* the host for BmNPV, is a member of the family Bombycidae. Recent reports suggest that *Manduca sexta* or *Hyalaphora cecropia* may also be useful hosts for AcMNPV-mediated production of proteins (Gretch et al., 1991; Hellers et al., 1991). Neither of these insects are particularly good hosts for the virus, and thus require high doses of virus for infection. However, the poor replication of the virus may be advantageous because the insects survive longer following infection. In addition, these insects are considerably larger than any of the three noctuid hosts. Thus, they may be worth considering, especially for the production of secreted proteins, because relatively large volumes of hemolymph can be obtained from a single insect. *H. cecropia* has the following disadvantages: it has only one generation per year and is not commercially available on a large scale, making it more difficult to ensure that suitable insects are continuously available. We have not provided methods for rearing either *M. sexta* or *H. cecropia* here. However, such methods are widely available in the literature (e.g., Singh and Moore, 1985). The infection and harvesting protocols that follow should be applicable to both these insects.

We have limited our discussion to rearing insects from eggs through larval development and do not deal with the establishment and maintenance of adult colonies. This is because eggs are readily available from a variety of sources for all the species described, and the numbers of insects required for the preparation of PIBs or production of a recombinant protein are generally relatively small. Readers interested in establishing their own colonies should consult the reviews of Singh and Moore (1985) or Smith (1966) and the references therein.

REARING INSECTS

Procuring and Storing Eggs

Eggs of lepidopteran species are available from a wide variety of sources. In the United States, a catalog of colonies of insects reared is available from the Entomology Society of America. It is entitled *Arthropod Species in Culture* (ed. D. Edwards) and may be obtained by sending a check for $7.00 ($4.50 for members), made payable to "ESA," to:

Entomological Society of America
P.O. Box 177
Hyattsville, MD 20781

As of this writing, the most recent edition of the catalog was published in 1987. However, a new edition is currently in preparation and is scheduled for publication in 1992. Readers are advised to request the most recent edition when ordering the catalog.

In the United Kingdom, the Entomology Group of the Association of Applied Biologists maintains a list of laboratories that rear different insects. Inquiries should be directed to:

Convenor of the AAB Entomology Group
c/o AAB office
HRI Wellesbourne
Wellesbourne
Warwick CV35 9EF
United Kingdom

Note that it is necessary to obtain a license to hold any of the noctuids in the United Kingdom. Applications for a license should be made to:

Plant Health Division
MAFF
Room 504, Ergon House
c/o Nobel House
Smith Square
London, SW1P 3HX
United Kingdom

The eggs of all species may be held at reduced temperatures for a period of time. Of the noctuids, *S. frugiperda* and *H. virescens* may be held at 15° C whereas *T. ni* may be held at 10° C. This cold storage slows, but does not arrest, the development of these moths. Thus, the length of time the eggs may be stored before hatching depends on their age at the beginning of cold storage. Generally, one- to two-day-old eggs can be kept for five days to one week.

B. mori undergoes diapause (a period of dormancy, generally for surviving adverse environmental conditions) during the egg stage. Diapausing eggs may be maintained at 2° to 5° C for prolonged periods of time. The length of time they may be stored depends on how they have been treated by the supplier. Thus, we recommend that you consult the supplier for precise details concerning storage conditions and hatching procedures. See Choudary et al. (1992) for details concerning diapause in *B. mori* and the ways it may be manipulated.

Avoiding Contamination and Disease

The problem you will most likely encounter rearing lepidopteran larvae is contamination by molds, bacteria, and/or viruses. Viral contamination is of particular concern, because infectious virus particles will be present in laboratories working with baculovirus expression vectors. This problem is greatest when working with *H. virescens* or *T. ni*, as these species are most susceptible to AcMNPV infection. To avoid these problems, the following precautionary measures should be taken:

- Adhere to good laboratory practices when manipulating viruses. In particular, ensure that all virus-containing solutions are disposed of properly and all labware that has been in contact with the virus is properly sterilized.
- If possible, keep diet components, cups, trays, and utensils for diet preparation physically separated from areas used for virus manipulation. Similarly, prepare uninfected diets and maintain uninfected insects separate from infected insects and away from areas where virus is used.
- Always clean and sterilize utensils used for diet preparation after use. Maintain separate utensils for the preparation of uninfected and infected diets.
- Sterilize eggs before placing them on the diet.
- During larval growth, monitor larvae carefully for signs of contamination or disease. If fungal or bacterial contamination of the food or feces (frass) becomes evident, transfer the insects to fresh containers as soon as possible. Containers with supposedly uninfected insects that show signs of viral infection (see later in this chapter) should be sterilized and disposed of immediately *without opening*.

Diet Preparation

There are many available artificial diets designed for the laboratory growth of all the species discussed here. These diets have been reviewed extensively (Singh, 1977). This manual does not systematically compare different diets for any of these species. Note, however, that artificial diets are commercially available for all the species discussed here. Examples of some of these diets are as follows:

S. frugiperda	Bio-Serv # F9179
T. ni	Bio-Serv # F9283
H. virescens	Bio-Serv # F9786
B. mori	1. Nihon Nosan Kogyo Co. Ltd. (no catalog number)
	2. North Carolina Biologicals # L-905A

Alternatively, diets may be prepared from the individual components in the laboratory. For *B. mori*, however, the commercially available diets are usually superior to diet prepared in the laboratory. In addition, it is necessary to prepare more than one diet to support the larvae at different developmental stages. Thus, recipes for *B. mori* diets are not provided here. Readers who wish to prepare their own *B. mori* diets should consult Choudary et al. (1992).

Recipes for artificial diets for the noctuids are provided here. These recipes have been used successfully in our laboratories. You may wish to try the diet recipe used by the egg suppliers. For all diets, the dry ingredients may be premixed and stored at 4° C. The ingredients will keep for long periods of time, provided they are kept dry. The quantities listed in each table are for the preparation of 1 liter of diet. It is possible to increase the amount of water added to these ingredients (up to twofold) to make the diet easier to pour. Adding water may be useful when the diet is being aliquoted into multiple, small containers.

Method

S. frugiperda (Burton, 1969)

1. Weigh the following dry diet ingredients (all from Bio-Serv):

Ingredients	*g/l*
ground pinto beans (# 1430)	120.0
Brewer's yeast (# 1710)	35.0
ascorbic acid (# 6015)	3.5
wheat germ (# 1660)	55.0
sorbic acid (# 6967)	1.1
methylparaben (# 7685)	2.2

2. Melt 15 g of agar (Bio-Serv # 7060) by autoclaving in 360 ml of H_2O for 20 minutes. Be sure to use a flask that is substantially larger (1–2 l in this case) than the volume being autoclaved.
3. Cool the melted agar to 60° C.
4. While the agar is cooling, mix the dry ingredients with 465 mls of H_2O and 2.5 mls of 37% formaldehyde (Baker # 2106-3). Blend at high speed for two to three minutes.
   Caution: Formaldehyde vapors are toxic. Dispense formaldehyde in a fume hood.
5. Add the melted agar and blend for an additional two to three minutes. Pour 10 ml aliquots into 1 oz. cups [Bio-Serv # 9051 (cups) and # 9053 (lids)]. Store at 4° C.

T. ni (Treat and Halfhill, 1973)

1. Weigh the following dry diet ingredients (Bio-Serv):

Ingredients	*g/l*
alfalfa meal (# 1010)	75.0
ground pinto beans (# 1430)	75.0
Brewer's yeast (# 1710)	36.0
ascorbic acid (# 6015)	3.5
wheat germ (# 1660)	53.5
sorbic acid (# 6967)	1.8
14% aureomycin (# 7135)	0.64
vitamin mix (# F8045)	7.15

2. Melt 15 g of agar (Bio-Serv # 7060) by autoclaving in 340 ml of H_2O for 20 minutes. Be sure to use a flask that is substantially larger (1–2 l in this case) than the volume being autoclaved.
3. Cool the melted agar to 60° C.
4. While the agar is cooling, mix the dry ingredients with 550 mls of H_2O and 2.5 mls of 37% formaldehyde (Baker # 2106-3). Blend at high speed for two to three minutes.
   Caution: Formaldehyde vapors are toxic. Dispense formaldehyde in a fume hood.
5. Add the melted agar and blend for an additional two to three minutes. Pour 200 ml aliquots into 16 oz. cups [Bio-Serv # 9061 (cups) and # 9081 (lids)]. Store at 4° C.

H. virescens (Berger, 1963)

6. Weigh the following dry diet ingredients (Bio-Serv, unless stated):

Ingredients	*g/l*
wheat germ diet (ICN # 901942)	122.0
soy flour (# 1500)	41.0
sorbic acid (# 6967)	1.0
methylparaben (# 7685)	1.0
14% aureomycin (# 7135)	0.64
vitamin mix (ICN # 904654)	10.0

7. Melt 20 g of agar (Bio-Serv # 7060) by autoclaving in 500 ml of H_2O for 20 minutes. Be sure to use a flask that is substantially larger (1–2 l in this case) than the volume being autoclaved.
8. Cool the melted agar to 60° C.
9. While the agar is cooling, mix the dry ingredients with 500 mls of H_2O and 10 mls of mold inhibitor (42% propionic acid and 4.2% phosphoric acid in H_2O). Blend at high speed for two to three minutes.
10. Add the melted agar, blend for an additional two to three minutes and pour 10 ml aliquots into 1 oz. cups [Bio-Serv # 9051 (cups) and # 9053 (lids)]. Store at 4° C.

Hatching Eggs and Raising the Larvae

The number of insects that may be reared per container varies depending on the species. *H. virescens* are highly cannibalistic, so it is necessary to maintain the insects in individual containers. On the other hand, *T. ni* and *B. mori* are not at all cannibalistic and can be easily reared in groups. Remember, however, that *T. ni* are particularly susceptible to virus infection, and contamination problems are exacerbated when there are many insects per container. Thus, extra care is required when rearing these insects. *S. frugiperda* may be reared in groups if desired, especially early in their development. However, they are somewhat cannibalistic, and the number of animals surviving larval development can be variable. For the small number of insects used in expression applications, we prefer to separate the required number of insects into individual 1 oz. containers early in development.

The techniques for rearing *B. mori* are somewhat different from those for the noctuids and are provided separately.

Materials

1. Eggs (see earlier discussion on processing and storing eggs)
2. 0.1% sodium hypochlorite: 2% (v/v) Clorox bleach in water
3. 2% formaldehyde (Baker # 2106-3)
   Caution: Formaldehyde vapors are toxic. Dispense formaldehyde in a fume hood.
4. Diet prepared and aliquoted as described in the previous section.
5. Fine camel-hair brush
6. Light blunt-nosed forceps

Method

S. frugiperda, T. ni, and H. virescens

1. Surface sterilize the eggs as follows:
   a. Place the eggs (usually on pieces of paper towel) in a buchner funnel attached to a vacuum flask.
   b. Add 5 mls 0.1% sodium hypochlorite.

   c. After one minute, draw off the bleach by applying a gentle vacuum. Rinse extensively in water and draw off the water in the same way. Allow the eggs to air-dry briefly.
2. The insects will be hatched and maintained at 27° C (*S. frugiperda* and *T. ni*) or 29° C (*H. virescens*). The synchrony and reproducibility of development will be optimal if the insects are maintained under a controlled light cycle (14:10 hours light:dark is appropriate). However, this is not essential. It is best not to place the eggs in contact with the diet before they hatch, as the mold inhibitor can be harmful. The eggs may be hatched in empty containers, or the paper towels with the eggs may be taped to the underside of the lid of a container with some diet. Ensure that the eggs are hatched in a humidified environment. This is particularly important after sterilization, as the sodium hypochlorite treatment may partially remove the chorion, rendering the eggs prone to desiccation. We advise placing a moistened paper tissue in each container with eggs.
3. The eggs of these three species will hatch five to seven days after being laid if they have not been subjected to cold storage in the meantime. Check the eggs twice daily so that newly hatched insects are not without diet for more than 12 to 15 hours. For *S. frugiperda* and *H. virescens,* transfer one to two neonate larvae into each 1 oz. cup. For *T. ni,* 60–70 neonate larvae may be put in each 16 oz. cup. Use a fine camel-hair brush when transferring small insects.
4. Monitor the condition of the insects closely during growth. Add fresh diet if the diet appears to be too dry. If there is excessive moisture in the container, transfer the insects to a fresh container. Larger insects can be handled with a light blunt-nosed forceps. Guidelines for determining the developmental stage of the insect are provided later.

B. mori

B. mori may be reared either on artificial diet or on fresh mulberry leaves. Some workers find that the insects fare better on leaves. However, this approach is more tedious and does not permit year-round rearing. Both approaches are described here.

Rearing on Artificial Diet

1. Surface-sterilize the eggs as follows:
   a. Place the eggs (usually on pieces of paper towel) in a buchner funnel attached to a vacuum flask.
   b. Add 5 mls 2% formaldehyde.
   c. After one to two minutes draw off the formaldehyde by applying a gentle vacuum. Rinse extensively in water and draw off the water in the same way. Allow the eggs to air-dry briefly.
2. Incubate at 25° C under high humidity for several days (e.g., in a covered plastic box with a ball of water-soaked Kimwipes). The larvae hatch a few days after the eggs become dark blue, ≈8 to 10 days after transferring to 25° C.
3. Transfer newly hatched larvae (within a day) onto thinly sliced diet (≈1 mm thick) with a camel-hair brush. Alternatively, cover neonate larvae with thinly sliced diet. Incubate at 25° to 27° C in a small plastic box. Keep the larvae and diet covered with plastic wrap to retain moisture if necessary.
4. Feed the larvae slices of artificial diet every day or two, making sure moist diet is always available. Remove the frass every three to five days. Ensure that the larvae are kept in a humidified environment, especially during early development. Use a larger container as the larvae increase in size. Thousands of neonate instar larvae may be maintained in a box of ≈600 cm^2 (i.e., about the size of a letter-size sheet of paper). This number should be reduced to about 50 by the time the insects are fully grown.

Rearing on Mulberry Leaves

1. Sterilize and hatch the eggs in a humid environment as described already.
2. Provide young, fresh, clean leaves to the neonate larvae. For young larvae, it is important that the leaves are very tender. Young larvae will not eat older, tougher leaves. Invert the leaves beside or over the larvae. Try to provide only as many leaves as the larvae will consume in a day.
3. Maintain the insects at 25° to 27° C.
4. Older, tougher leaves may be used as the larvae develop. However, it is important to provide fresh leaves and clean the container daily. If desired, leaves can be stored in a plastic box at 5° C for up to a week.
5. With larger larvae, you will need to reduce the number of insects in each container to avoid overcrowding (see the protocol for rearing on diet).

Staging Insects

For maximum yield of heterologous protein or PIBs, it is important to infect the larvae at the appropriate developmental stage. These insects usually molt five (*T. ni, B. mori*) or six (*S. frugiperda, H. virescens*) times as larvae, the last molt being the larval-pupal molt. Each intermolt period is known as a stadium or instar. The optimal time for infection is one to two days after molting into the last larval instar.

The first morphological sign of an impending molt is head-capsule slippage, when the head capsule begins to slide forward on the body of the insect. Head-capsule slippage may be easily seen for all the species discussed, and specific descriptions are given here. Head-capsule slippage begins about 12 to 36 hours before the insect molts. During this period, the insect does not feed and may often be found on the side or under the lid of the cup.

The last instar of these larvae is longer than any of the preceding instars. After molting, the insect feeds extensively for two to eight days (depending on the species). It then stops feeding, empties its gut, and starts to wander in preparation for pupation. This activity is accompanied by a dramatic weight loss and, as a result, the insect shrinks to about half its former size. The prolegs are retracted, and the insect, now known as a pharate pupa, is no longer mobile. It spins a cocoon of one form or another and pupates in the cocoon several days later.

The simplest way for a beginner to learn to identify the developmental stages of an insect is to monitor individual insects closely and count molts. The weight of the insect is also a useful indicator. The following are brief descriptions of specific morphological changes in each species, which may help in identifying the particular developmental stage. Also noted are the expected weights of insects of each species at the time of head-capsule slippage during the molt to the last instar. (This is the stage at which the insects are selected before infection.)

S. frugiperdu

S. frugiperda larvae molt six times as they develop from eggs to pupae. Each of the first five instars lasts approximately two days under the rearing conditions described in the preceding section. Head-capsule slippage is clearly manifested by the appearance of a gap between the head and the first thoracic segment of the old cuticle. The dorsal part of the first thoracic segment, like the head capsule, is black. Thus, separation of this segment from the head capsule creates the appearance of a black collar behind the head. Immediately after the molt, the insect is light gray, with a whitish head capsule. However, within one to two hours, the head capsule and dorsal part of the first thoracic segment become black, and the insect darkens to a dark-gray color. As it progresses through the instar, the color of the cuticle changes to a light brown, while the head capsule and first thoracic segment remain black. At the fifth to sixth instar molt, an insect weighs approximately 120 mg to 170 mg.

The final instar lasts four to five days in *S. frugiperda*. The insects stop feeding and begin wandering 48 to 60 hours after entering the instar. They then burrow into the diet and enclose themselves in a chamber surrounded by a small amount of silk (generally woven into the surrounding food). The insect pupates two to three days later. Thus, larval development lasts about two weeks in this species.

T. ni

T. ni generally molt five times during larval development. In first instar *T. ni* larvae, the head capsule and dorsal part of the first thoracic segment are black, and head- capsule slippage is manifested by the appearance of a black collar, as described for *S. frugiperda*. However, from second instar on, the head capsule and first thoracic segment are green—the same color as the rest of the insect's body. In these instars, larvae with head-capsule slippage are identified as "bubble-heads" due to the presence of air behind the old head capsule, which protrudes from the front of the insect. The new head capsule may be seen as a hard, shiny area visible on the dorsal side of the insect, immediately behind the old head capsule. *T. ni* larvae weigh from 50 mg to 80 mg at the time of head-capsule slippage from fourth to fifth instar.

After feeding ceases in fifth instar, *T. ni* larvae will spin an obvious silk cocoon on the side of the cup or the underside of the lid. The cocoon can be observed from two to three days into the instar. The insects pupate one to two days later. Development from egg to pupa takes 12 to 14 days.

H. virescens

The morphological changes associated with the development of *H. virescens* are very similar to those of *S. frugiperda*, and head-capsule slippage can be identified by the appearance of a collar on the back of the insect. However, the color of *H. virescens* is highly variable, both between instars and between different insects in the same instar. At the time of head-capsule slippage from fifth to sixth instar, the insect weighs 100 mg to 130 mg. Like *S. frugiperda*, these insects burrow into the diet prior to pupation, and any silk that is spun is generally woven into the diet. Development from egg to pupa takes from 12 to 14 days.

B. mori

Under the rearing conditions described in the preceding section for *B. mori*, first instar lasts four days; second instar lasts three days; third instar, four days; fourth instar, six days; and fifth instar, eight to ten days. Thus, development from egg to pupa takes about four weeks in *B. mori*. During head-capsule slippage, the newly synthesized head can be seen as a light-brown triangular patch just behind the old head. The skin of molting insects is also more transparent than usual.

At the molt from fourth to fifth instar, *B. mori* larvae weigh ≈1 g to 1.3 g. They feed for seven to eight days and grow to a maximum size of ≈5 g. At this stage, they cease feeding and shrink to about 2.5 g. Their skin becomes noticeably transparent and yellowish, and they continuously spin a silk cocoon over a two-day period. Pupation takes place one to two days after the cocoon is completed.

INFECTION: PRODUCTION OF RECOMBINANT PROTEIN OR PIBs

In general, infection of insect larvae is accomplished either by feeding insects with PIBs or by injecting budded virus (BV) directly into the hemolymph. Oral (*per os*) infection is less labor-intensive, especially for large numbers of insects. For production of large quantities of PIBs (e.g., for the preparation of DNA), oral infection is the recommended approach. However, most transfer plasmids that are currently available result in the formation of occlusion-negative (occ^-) recombinant baculoviruses. These recombinants do not form PIBs, which are more useful than BV for efficient *per os* infection. Although it is possible to infect larvae by feeding nonoccluded virus, this approach is very inefficient and is not recommended. Instead, a number of different approaches may be used. First, transfer plasmids (pSynXIV VI^+ series plasmids and pAcUW2B) are available for use with AcMNPV, which give occ^+ recombinant viruses (see Chapter 7). Alternatively, one can "co-occlude" an occ^- recombinant virus with wt helper virus (Miller, 1988a). During coinfection of cell cultures with high multiplicities of infection (MOIs) of both wt and recombinant virus, both virus types become occluded in PIBs. These PIBs may be used as the inoculum for *per os* infection of larvae. The disadvantage of this approach is the extra steps it requires: first, the coinfection of cells with both occ^+ and occ^- viruses; and second, the purification of PIBs from these coinfected cells.

Finally, the insects may be infected by injection of BV into the hemolymph. While this approach is somewhat more tedious, it should be considered a valid option for protein production or other applications requiring fewer insects. With some practice, one can inject upwards of 50 to 60 insects per hour. For species that are more resistant to virus infection, such as *S. frugiperda* or *B. mori,* injection of BV is the most efficient method for infection of later instar insects.

Ideally, for production of recombinant protein or PIBs, the insects should be as large as possible when infected. However, as discussed earlier, the larvae cease feeding and undergo radical developmental and behavioral changes during the last larval instar in preparation for pupation. Infection of insects undergoing these changes is somewhat compromised and results in reduced yield of PIBs (and presumably recombinant proteins) (O'Reilly and Miller, 1991). However, wt virus can prevent these processes from taking place, provided the insects are infected early enough in the final instar before any prepupal development has begun (O'Reilly and Miller, 1989). For the insects discussed here, the ideal time for infection is within one to two days of molting into the final instar.

The optimal time to collect the infected insects depends on the particular application. Generally, for protein production, it is best to collect the insects before death. We recommend carrying out an initial trial experiment to evaluate how quickly the infected insects die and to determine the optimal time to harvest the insects. For the production of PIBs, you may wait until the insects die. However, the cadavers should be collected as soon as possible after death, before they disintegrate completely.

The manner in which you process insects infected for protein production will depend on the particular protein being produced. If the protein is secreted, then you may need only to bleed the insects. However, polyphenol oxidases present in the hemolymph catalyze a process known as melanization when the hemolymph is exposed to air. These enzymes catalyze the oxidation of tyrosine derivatives and give rise to highly reactive quinones. The quinones bind irreversibly to themselves and to proteins, causing the formation of large aggregates of hemolymph protein. Once extensive melanization has taken place, purification of any protein from the hemolymph will be virtually impossible. The polyphenol oxidases can be inhibited by addition of phenylthiourea or dithiothreitol to the hemolymph sample. Alternatively, melanization may be avoided by subjecting the hemolymph sample to a chromatography step immediately after collection, thereby separating the components of the melanization cascade from each other. The most appropriate method to use will depend on the particular protein you wish to purify.

For nonsecreted proteins, it may be possible to isolate intracellular proteins by homogenization of whole insects. However, the high levels of proteases and other hydrolases present in the gut can be problematic, and it may help to remove the gut from each insect before homogenization. Another option, which is especially simple with larger insects, is to collect only fat-body tissue from infected insects. The fat body is a major tissue in final instar larvae and is generally very active metabolically. Thus, a significant proportion of a recombinant protein will be produced in the fat body.

The following are techniques for infecting these insects, either by injection or by feeding, and subsequently collecting hemolymph, removing the gut, or collecting fat body from infected larvae. The doses given should ensure 100% mortality of the infected larvae.

CONTROLS: We recommend maintaining a small number of mock-infected insects in addition to the insects infected with the recombinant virus. This will enable you to compare the development of the uninfected and infected larvae. In addition, it allows you to verify that the insects are free from contamination.

Infection of Larvae

Injection of Budded Virus

All the species of insect covered in this manual may be infected by injection.

Materials

1. Late fourth (*T. ni, B. mori*) or fifth (*S. frugiperda, H. virescens*) instar larvae
2. Artificial diet (see recipes earlier in this chapter)
3. Recently titered stock of recombinant virus
4. 10 μl Hamilton syringe with 26-gauge beveled needle (Hamilton # 701N). It helps to clamp the syringe in a stand.

Method

1. Identify insects showing head-capsule slippage prior to ecdysis into the final larval instar. The approximate weights of the different species at this stage are given in the section Staging Insects.
2. The following day examine the insects that were showing head-capsule slippage, and select those insects that completed the molt. *B. mori* larvae should be starved at least 3 to 4 hours prior to injection. Anesthetize each insect by chilling for 10 to 15 minutes (e.g., the insect may be covered with crushed ice or simply placed in a refrigerator). Then, infect as follows:
   a. Grasp the insect gently between the thumb and forefinger.
   b. Insert the needle longitudinally into the third abdominal segment (this segment bears the first pair of prolegs). The needle should be inserted along the midline and directed toward the posterior end. The cuticle may fold a lot before it is actually pierced by the needle. Insert the needle until the point is about three segments posterior of the site of injection. Keep the tip of the needle as close to the cuticle as possible and be careful not to pierce the gut.
   c. Inject the insect with the BV stock. We recommend doses of 5×10^4 plaque-forming units (pfu) for *T. ni* and *H. virescens;* 10^5 pfu for *B. mori;* and 2×10^5 pfu for *S. frugiperda.* The volume used for *T. ni* should not exceed 5 μl. However, the other species may be injected with up to 10 μl. The virus may be in complete or incomplete tissue-culture fluid or in PBS.

d. Hold the insect on the needle for a few seconds, then carefully remove and replace on the diet.

3. Maintain the infected insects as described in the section Hatching Eggs and Raising the Larvae. There will not be many overt signs of infection for the first few days. However, infected insects will not begin to wander or spin a cocoon, in contrast to uninfected insects at the same developmental stage. Instead, infected insects will continue to feed and become somewhat bloated. *T. ni* tend to become pale in color at this stage. Similarly, the cuticle of infected *B. mori* will become transparent. Shortly before death, insects of all these species become lethargic or moribund. Sometimes patches of melanization (blackening) of the cuticle will appear before the insect dies. Often, however, this does not occur until after death. Occasionally, the infected insects will leave the diet before death and hang upside down from the lid of the cup. After death, the cadaver will turn completely black, and the internal tissues will liquefy. The cuticle also becomes very fragile, and the cadavers become difficult to manipulate.
4. If the insects were infected for PIB production, collect the cadavers and store them at –20° C. They may be stored in this form for long periods of time without any loss of infectivity (see Chapter 12). PIBs can be isolated from these cadavers as described in Chapter 12.

   If the insects were infected for production of a recombinant protein, harvest the hemolymph or tissue at the appropriate time as described in the next section.

Infection by Feeding PIBs

Infection by feeding PIBs can be a convenient method of infection, especially when large numbers of insects are required. However, as mentioned already, feeding PIBs is not practical for infection of last instar *S. frugiperda* or *B. mori* larvae because the dose required is excessively high. For the purpose of producing PIBs or recombinant protein, the PIBs may be most easily administered by contamination of the insect diet. This is accomplished by pipeting a solution of PIBs onto a small cube of diet. Formaldehyde or other mold inhibitor should be omitted from diet that comes in contact with the PIBs, to avoid inactivation of the virus.

Materials

1. Late fourth (*T. ni*) or fifth (*H. virescens*) instar larvae
2. Artificial diet prepared without formaldehyde or mold inhibitor
3. Solution of recombinant virus PIBs. (See Chapter 12 for methods for the preparation of PIBs.)

Methods

1. Identify insects showing head-capsule slippage prior to ecdysis to the final larval instar. The approximate weights of the different species at this stage are given in the section Staging Insects. You may remove the insects from the diet at this stage. This period of starvation, once the insects complete the molt, helps ensure rapid consumption of the contaminated diet.
2. The following day, select those insects that completed the molt. Pipet approximately 1×10^7 PIBs of the virus onto each of several small cubes of diet (without mold inhibitors). Let the insects feed on this contaminated diet.
3. When all the contaminated diet has been consumed, transfer the insects back to uncontaminated diet.
4. Proceed as described in Step 3 of the preceding injection protocol.

Harvesting Recombinant Protein

Collecting Hemolymph

Materials

1. Infected insects one to two days before death
2. PTU: phenylthiourea (Sigma # P 5272)
3. 10 mM DTT (dithiothreitol; Sigma # D 0632)
4. Fine scissors and forceps

Method

1. Anesthetize the insect, if preferred, by placing in crushed ice or refrigerating for 10 to 15 minutes.
2. Hold the insect over a chilled plastic tube containing a few crystals of PTU or some 10 mM DTT (the final concentration of DTT required will be 1 mM). Clip one of the hind prolegs with the scissors and allow the hemolymph to flow into the microfuge tube. Alternatively, bend back the larva and hold the head and tail with one hand so that the prolegs are protruding. Clip one of the anterior prolegs. It may help to gently squeeze the insect between your thumb and forefinger. However, be careful not to squeeze too hard as this may cause the extrusion of some gut contents into the tube. Hemolymph collected late in infection from insects infected with an occ^+ virus will have a milky appearance due to the presence of a large number of PIBs. Hemolymph from insects infected with an occ^- virus may be somewhat cloudy but will be clearly less milky in appearance.

Collecting Fat Body or Removing the Gut

Materials

1. Infected insects one to two days before death
2. Fine scissors, forceps, and dissecting pins
3. Paraffin wax
4. Dissecting dishes: Pour melted paraffin wax into a suitable container (e.g., a bacterial Petri dish). Ensure that the container is large enough so that when an insect is pinned to the wax, it can be completely submerged in PBS.
5. PBS, pH 6.2

Methods

1. Anesthetize the insect by placing in crushed ice or refrigerating for 10 to 15 minutes.
2. Pin the head and tail of the animal (dorsal side up) to the wax in the dissecting dish and cover with ice-cold PBS or other buffer of your choice. You may want to include appropriate protease inhibitors in this buffer. (See Chapter 14 for some recommended inhibitors.)
3. Make a longitudinal incision along the insect's back, taking care not to cut the gut. Pull the cuticle aside and pin down. You will see extensive networks of tracheoles branching from the spiracles in the cuticle to all parts of the insect body.
4. The gut should be clearly visible as a brown tube running the length of the insect. It may be covered with fat body, which appears as an amorphous white tissue, particularly in the posterior half of the insect. If you wish to remove the gut, gently tease it away from the remainder of the carcass. Try to avoid removing too much other tissue, especially fat body.

Alternatively, simply collect as much fat body as possible from the body cavity of the insect. There will be substantial amounts of fat body underneath the gut, which may not be immediately visible. It may help to slide the fat body along a piece of weighing paper to separate it from hemolymph.

1. Rinse the carcass or fat body in PBS (or other buffer of your choice) to remove gut contents that may have spilled.
2. Homogenize the dissected tissue in the appropriate buffer.

APPENDIX 1: NUCLEOTIDE SEQUENCE OF AcMNPV POLYHEDRIN REGION

This appendix presents the nucleotide sequence and predicted ORFs within the EcoRI I fragment of AcMNPV C-6, from 0.0 to 5.7 mu (Possee et al., 1991). The recognition sequences of selected restriction endonucleases are shown. Those represented in bold have only one cleavage site in this sequence. Nucleotide numbers in the vicinity of *polh* (shaded) are given relative to the *polh* ATG (+1,+2,+3), as well as relative to the start of the sequence. The start site of *polh* transcription and the *polh* promoter TAAG (double underlined) are indicated. These data are also presented in Figures 7-1 and 13-4.

```
    EcoRI
  1 GAATTCTACTCGTAAAGCGAGTTGAAGGATCATATTTAGTTGCGTTTATGAGATAAGATT 60

                   ORF 504 →
                   M  F  P  A  R  W  H  N  Y  L  Q  C  G  Q  V
 61 GAAAGCACGTGTAAAATGTTTCCCGCGCGTTGGCACAACTATTTACAATGCGGCCAAGTT 120

    I  K  D  S  N  L  I  C  F  K  T  P  L  R  P  E  L  F  A  Y
121 ATAAAAGATTCTAATCTGATATGTTTTAAAACACCTTTGCGGCCCGAGTTGTTTGCGTAC 180

    V  T  S  E  E  D  V  W  T  A  E  Q  I  V  K  Q  N  P  S  I
181 GTGACTAGCGAAGAAGATGTGTGGACCGCAGAACAGATAGTAAAACAAAACCCTAGTATT 240

    G  A  I  I  D  L  T  N  T  S  K  Y  Y  D  G  V  H  F  L  R
241 GGAGCAATAATCGATTTAACCAACACGTCTAAATATTATGATGGTGTGCATTTTTTGCGG 300

    A  G  L  L  Y  K  K  I  Q  V  P  G  Q  T  L  P  P  E  S  I
301 GCGGGCCTGTTATACAAAAAAATTCAAGTACCTGGCCAGACTTTGCCGCCTGAAAGCATA 360

    V  Q  E  F  I  D  T  V  K  E  F  T  E  K  C  P  G  M  L  V
361 GTTCAAGAATTTATTGACACGGTAAAAGAATTTACAGAAAAGTGTCCCGGCATGTTGGTG 420

    G  V  H  C  T  H  G  I  N  R  T  G  Y  M  V  C  R  Y  L  M
421 GGCGTGCACTGCACACACGGTATTAATCGCACCGGTTACATGGTGTGCAGATATTTAATG 480
                                  AgeI

    H  T  L  G  I  A  P  Q  E  A  I  D  R  F  E  K  A  R  G  H
481 CACACCCTGGGTATTGCGCCGCAGGAAGCCATAGATAGATTCGAAAAAGCCAGAGGTCAC 540

    K  I  E  R  Q  N  Y  V  Q  D  L  L  I  *
541 AAAATTGAAAGACAAAATTACGTTCAAGATTTATTAATTTAATTAATATTATTTGCATTC 600
                                           PacI

601 TTTAACAAATACTTTATCCTATTTTCAAATTGTTGCGCTTCTTCCAGCGAACCAAAACTA 660
                   *  G  I  K  L  N  N  R  K  K  W  R  V  L  V
```

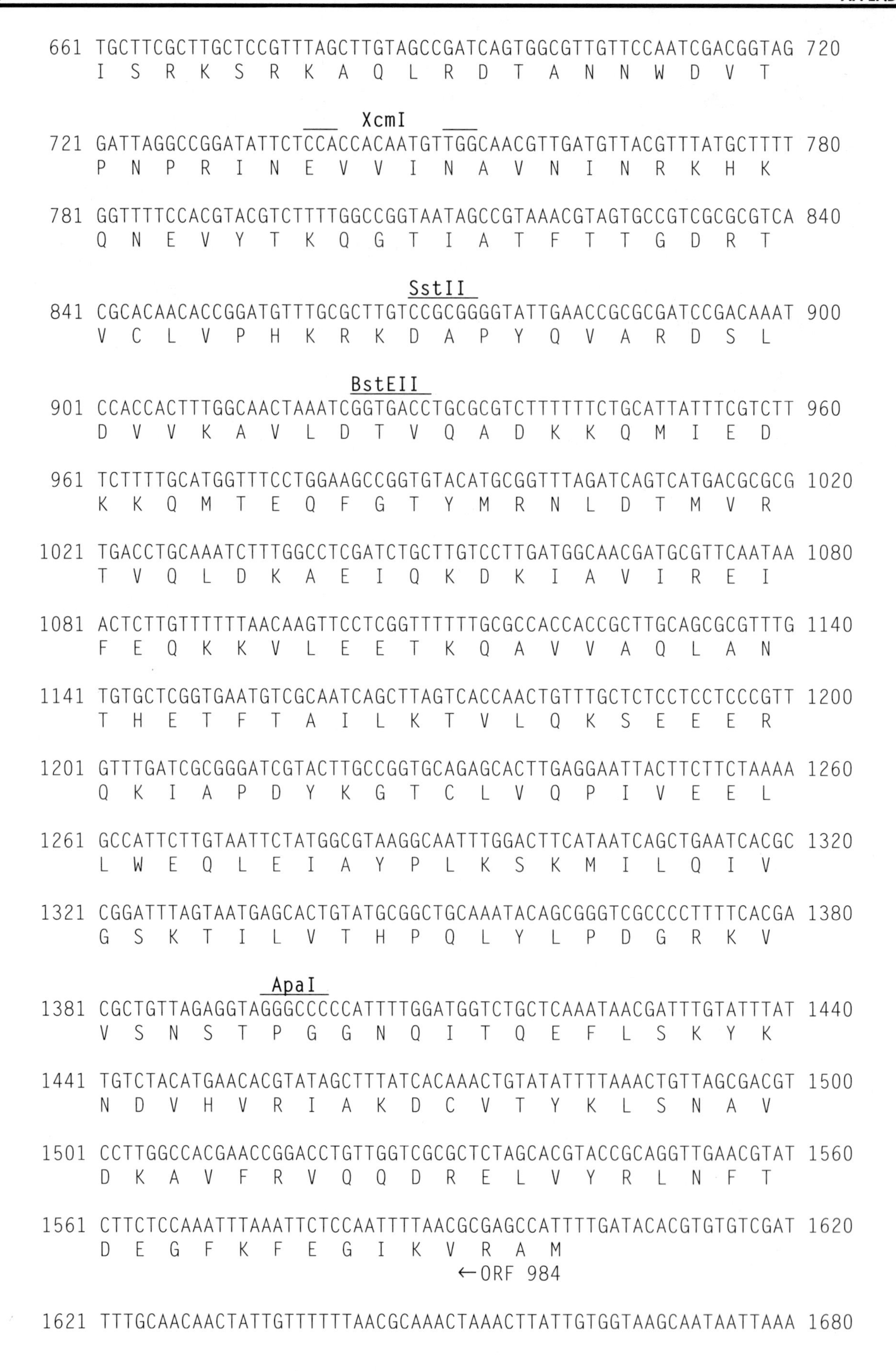

```
 661 TGCTTCGCTTGCTCCGTTTAGCTTGTAGCCGATCAGTGGCGTTGTTCCAATCGACGGTAG 720
     I  S  R  K  S  R  K  A  Q  L  R  D  T  A  N  N  W  D  V  T

                            XcmI
 721 GATTAGGCCGGATATTCTCCACCACAATGTTGGCAACGTTGATGTTACGTTTATGCTTTT 780
     P  N  P  R  I  N  E  V  V  I  N  A  V  N  I  N  R  K  H  K

 781 GGTTTTCCACGTACGTCTTTTGGCCGGTAATAGCCGTAAACGTAGTGCCGTCGCGCGTCA 840
     Q  N  E  V  Y  T  K  Q  G  T  I  A  T  F  T  T  G  D  R  T

                                     SstII
 841 CGCACAACACCGGATGTTTGCGCTTGTCCGCGGGGTATTGAACCGCGCGATCCGACAAAT 900
     V  C  L  V  P  H  K  R  K  D  A  P  Y  Q  V  A  R  D  S  L

                           BstEII
 901 CCACCACTTTGGCAACTAAATCGGTGACCTGCGCGTCTTTTTTCTGCATTATTTCGTCTT 960
     D  V  V  K  A  V  L  D  T  V  Q  A  D  K  K  Q  M  I  E  D

 961 TCTTTTGCATGGTTTCCTGGAAGCCGGTGTACATGCGGTTTAGATCAGTCATGACGCGCG 1020
     K  K  Q  M  T  E  Q  F  G  T  Y  M  R  N  L  D  T  M  V  R

1021 TGACCTGCAAATCTTTGGCCTCGATCTGCTTGTCCTTGATGGCAACGATGCGTTCAATAA 1080
     T  V  Q  L  D  K  A  E  I  Q  K  D  K  I  A  V  I  R  E  I

1081 ACTCTTGTTTTTTAACAAGTTCCTCGGTTTTTTGCGCCACCACCGCTTGCAGCGCGTTTG 1140
     F  E  Q  K  K  V  L  E  E  T  K  Q  A  V  V  A  Q  L  A  N

1141 TGTGCTCGGTGAATGTCGCAATCAGCTTAGTCACCAACTGTTTGCTCTCCTCCTCCCGTT 1200
     T  H  E  T  F  T  A  I  L  K  T  V  L  Q  K  S  E  E  E  R

1201 GTTTGATCGCGGGATCGTACTTGCCGGTGCAGAGCACTTGAGGAATTACTTCTTCTAAAA 1260
     Q  K  I  A  P  D  Y  K  G  T  C  L  V  Q  P  I  V  E  E  L

1261 GCCATTCTTGTAATTCTATGGCGTAAGGCAATTTGGACTTCATAATCAGCTGAATCACGC 1320
     L  W  E  Q  L  E  I  A  Y  P  L  K  S  K  M  I  L  Q  I  V

1321 CGGATTTAGTAATGAGCACTGTATGCGGCTGCAAATACAGCGGGTCGCCCCTTTTCACGA 1380
     G  S  K  T  I  L  V  T  H  P  Q  L  Y  L  P  D  G  R  K  V

                  ApaI
1381 CGCTGTTAGAGGTAGGGCCCCCATTTTGGATGGTCTGCTCAAATAACGATTTGTATTTAT 1440
     V  S  N  S  T  P  G  G  N  Q  I  T  Q  E  F  L  S  K  Y  K

1441 TGTCTACATGAACACGTATAGCTTTATCACAAACTGTATATTTTAAACTGTTAGCGACGT 1500
     N  D  V  H  V  R  I  A  K  D  C  V  T  Y  K  L  S  N  A  V

1501 CCTTGGCCACGAACCGGACCTGTTGGTCGCGCTCTAGCACGTACCGCAGGTTGAACGTAT 1560
     D  K  A  V  F  R  V  Q  Q  D  R  E  L  V  Y  R  L  N  F  T

1561 CTTCTCCAAATTTAAATTCTCCAATTTTAACGCGAGCCATTTTGATACACGTGTGTCGAT 1620
     D  E  G  F  K  F  E  G  I  K  V  R  A  M
                                  ←ORF 984

1621 TTTGCAACAACTATTGTTTTTTAACGCAAACTAAACTTATTGTGGTAAGCAATAATTAAA 1680
```

```
1681 TATGGGGGAACATGCGCCGCTACAACACTCGTCGTTATGAACGCAGACGGCGCCGGTCTC 1740

1741 GGCGCAAGCGGCTAAAACGTGTTGCGCGTTCAACGCGGCAAACATCGCAAAAGCCAATAG 1800

1801 TACAGTTTTGATTTGCATATTAACGGCGATTTTTTAAATTATCTTATTTAATAAATAGTT 1860

                                                    XhoI
1861 ATGACGCCTACAACTCCCCGCCCGCGTTGACTCGCTGCACCTCGAGCAGTTCGTTGACGC 1920
                 *  L  E  G  G  A  N  V  R  Q  V  E  L  L  E  N  V

1921 CTTCCTCCGTGTGGCCGAACACGTCGAGCGGGTGGTCGATGACCAGCGGCGTGCCGCACG 1980
     G  E  E  T  H  G  F  V  D  L  P  H  D  I  V  L  P  T  G  C

1981 CGACGCACAAGTATCTGTACACCGAATGATCGTCGGGCGAAGGCACGTCGGCCTCCAAGT 2040
     A  V  C  L  Y  R  Y  V  S  H  D  D  P  S  P  V  D  A  E  L

2041 GGCAATATTGGCAAATTCGAAAATATATACAGTTGGGTTGTTTGCGCATATCTATCGTGG 2100
     H  C  Y  Q  C  I  R  F  Y  I  C  N  P  Q  K  R  M  D  I  T

                                          SphI
2101 CGTTGGGCATGTACGTCCGAACGTTGATTTGCATGCAAGCCGAAATTAAATCATTGCGAT 2160
     A  N  P  M  Y  T  R  V  N  I  Q  M  C  A  S  I  L  D  N  R

2161 TAGTGCGATTAAAACGTTGTACATCCTCGCTTTTAATCATGCCGTCGATTAAATCGCGCA 2220
     N  T  R  N  F  R  Q  V  D  E  S  K  I  M  G  D  I  L  D  R

                BclI
2221 ATCGAGTCAAGTGATCAAAGTGTGGAATAATGTTTTCTTTGTATTCCCGAGTCAAGCGCA 2280
     L  R  T  L  H  D  F  H  P  I  I  N  E  K  Y  E  R  T  L  R

2281 GCGCGTATTTTAACAAACTAGCCATCTTGTAAGTTAGTTTCATTTAATGCAACTTTATCC 2340
     L  A  Y  K  L  L  S  A  M  K  Y  T  L  K  M
                                             ←ORF 453

               ORF 327 →
               M  Y  R  T  S  R  I  N  N  A  P  V  V  A  S  Q  H
2341 AATAATATATTATGTATCGCACGTCAAGAATTAACAATGCGCCCGTTGTCGCATCTCAAC 2400

       D  Y  D  R  D  Q  I  K  R  E  L  N  S  L  R  R  N  V  H  D
2401 ACGACTATGATAGAGATCAAATAAAGCGCGAATTAAATAGCTTGCGACGCAACGTGCACG 2460

       L  C  T  R  S  G  T  S  F  D  C  N  K  F  L  R  S  D  D  M
2461 ATCTGTGCACGCGTTCCGGCACGAGCTTTGATTGTAATAAGTTTTTACGAAGCGATGACA 2520
             MluI

       T  P  V  V  T  T  I  T  P  K  R  T  A  D  Y  K  I  T  E  Y
2521 TGACCCCCGTAGTGACAACGATCACGCCCAAAAGAACTGCCGACTACAAAATTACCGAGT 2580

       V  G  D  V  K  T  I  K  P  S  N  R  P  L  V  E  S  G  P  L
2581 ATGTCGGTGACGTTAAAACTATTAAGCCATCCAATCGACCGTTAGTCGAATCAGGACCGC 2640
```

```
                          ORF 630 →
                          M  A  N  A  S  Y  N  V  W  S  P  L  I
       V  R  E  A  A  K  Y  G  E  C  I  V  *
2641 TGGTGCGAGAAGCCGCGAAGTATGGCGAATGCATCGTATAACGTGTGGAGTCCGCTCATT 2700

     R  A  S  C  L  D  K  K  A  T  Y  L  I  D  P  D  D  F  I  D
2701 AGAGCGTCATGTTTAGACAAGAAAGCTACATATTTAATTGATCCCGATGATTTTATTGAT 2760

     K  L  T  L  T  P  Y  T  V  F  Y  N  G  G  V  L  V  K  I  S
2761 AAATTGACCCTAACTCCATACACGGTATTCTACAATGGCGGGGTTTTGGTCAAAATTTCC 2820

     G  L  R  L  Y  M  L  L  T  A  P  P  T  I  N  E  I  K  N  S
2821 GGACTGCGATTGTACATGCTGTTAACGGCTCCGCCCACTATTAATGAAATTAAAAATTCC 2880

     N  F  K  K  R  S  K  R  N  I  C  M  K  E  C  V  E  G  K  K
2881 AATTTTAAAAAACGCAGCAAGAGAAACATTTGTATGAAAGAATGCGTAGAAGGAAAGAAA 2940

     N  V  V  D  M  L  N  N  K  I  N  M  P  P  C  I  K  K  I  L
2941 AATGTCGTCGACATGCTGAACAACAAGATTAATATGCCTCCGTGTATAAAAAAAATATTG 3000

     N  D  L  K  E  N  N  V  P  R  G  G  M  Y  R  K  R  F  I  L
3001 AACGATTTGAAAGAAAACAATGTACCGCGCGGCGGTATGTACAGGAAGAGGTTTATACTA 3060

     N  C  Y  I  A  N  V  V  S  C  A  K  C  E  N  R  C  L  I  K
3061 AACTGTTACATTGCAAACGTGGTTTCGTGTGCCAAGTGTGAAAACCGATGTTTAATCAAG 3120

     A  L  T  H  F  Y  N  H  D  S  K  C  V  G  E  V  M  H  L  L
3121 GCTCTGACGCATTTCTACAACCACGACTCCAAGTGTGTGGGTGAAGTCATGCATCTTTTA 3180

     I  K  S  Q  D  V  Y  K  P  P  N  C  Q  K  M  K  T  V  D  K
3181 ATCAAATCCCAAGATGTGTATAAACCACCAAACTGCCAAAAAATGAAAACTGTCGACAAG 3240

     L  C  P  F  A  G  N  C  K  G  L  N  P  I  C  N  Y  *
3241 CTCTGTCCGTTTGCTGGCAACTGCAAGGGTCTCAATCCTATTTGTAATTATTGAATAATA 3300

3301 AAACAATTATAAATGCTAAATTTGTTTTTTATTAACGATACAAACCAAACGCAACAAGAA 3360
                                      *  R  Y  L  G  F  A  V  L

                                 MluI
3361 CATTTGTAGTATTATCTATAATTGAAAACGCGTAGTTATAATCGCTGAGGTAATATTTAA 3420
     V  N  T  T  N  D  I  I  S  F  A  Y  N  Y  D  S  L  Y  Y  K

3421 AATCATTTTCAAATGATTCACAGTTAATTTGCGACAATATAATTTTATTTTCACATAAAC 3480
     F  D  N  E  F  S  E  C  N  I  Q  S  L  I  I  K  N  E  C  L

3481 TAGACGCCTTGTCGTCTTCTTCTTCGTATTCCTTCTCTTTTTCATTTTTCTCCTCATAAA 3540
     S  S  A  K  D  D  E  E  E  Y  E  K  E  K  E  N  K  E  E  Y

3541 AATTAACATAGTTATTATCGTATCCATATATGTATCTATCGTATAGAGTAAATTTTTTGT 3600
     F  N  V  Y  N  N  D  Y  G  Y  I  Y  R  D  Y  L  T  F  K  K

3601 TGTCATAAATATATATGTCTTTTTTAATGGGGTGTATAGTACCGCTGCGCATAGTTTTTC 3660
     N  D  Y  I  Y  I  D  K  K  I  P  H  I  T  G  S  R  M  T  K
```

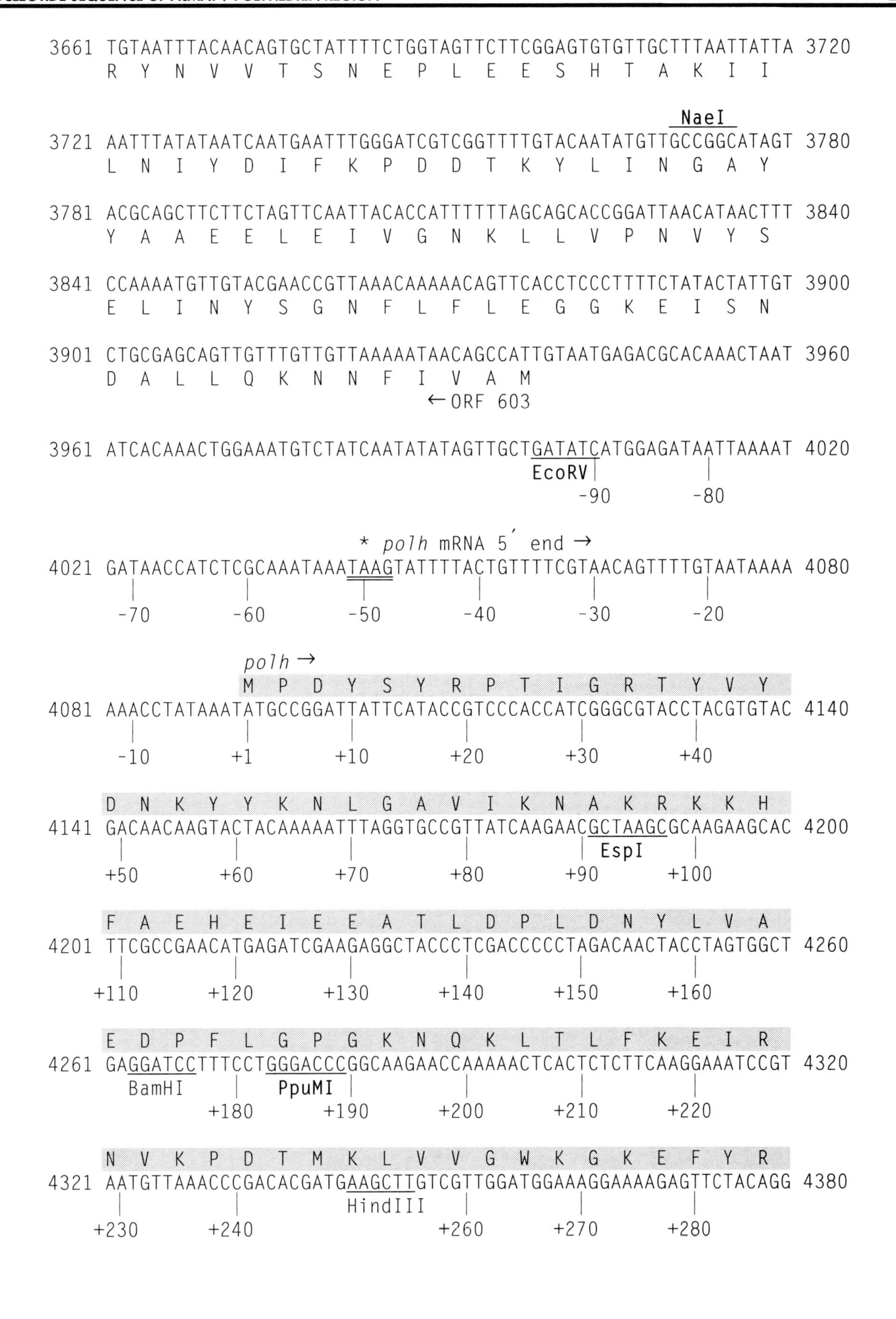
3661 TGTAATTTACAACAGTGCTATTTTCTGGTAGTTCTTCGGAGTGTGTTGCTTTAATTATTA 3720
R Y N V V T S N E P L E E S H T A K I I
NaeI
3721 AATTTATATAATCAATGAATTTGGGATCGTCGGTTTTGTACAATATGTTGCCGGCATAGT 3780
L N I Y D I F K P D D T K Y L I N G A Y
3781 ACGCAGCTTCTTCTAGTTCAATTACACCATTTTTAGCAGCACCGGATTAACATAACTTT 3840
Y A A E E L E I V G N K L L V P N V Y S
3841 CCAAAATGTTGTACGAACCGTTAAACAAAAACAGTTCACCTCCCTTTTCTATACTATTGT 3900
E L I N Y S G N F L F L E G G K E I S N
3901 CTGCGAGCAGTTGTTTGTTGTTAAAAATAACAGCCATTGTAATGAGACGCACAAACTAAT 3960
D A L L Q K N N F I V A M
←ORF 603
3961 ATCACAAACTGGAAATGTCTATCAATATATAGTTGCTGATATCATGGAGATAATTAAAAT 4020
EcoRV
-90 -80
* polh mRNA 5′ end →
4021 GATAACCATCTCGCAAATAAATAAGTATTTTACTGTTTTCGTAACAGTTTTGTAATAAAA 4080
-70 -60 -50 -40 -30 -20
polh →
M P D Y S Y R P T I G R T Y V Y
4081 AAACCTATAAATATGCCGGATTATTCATACCGTCCCACCATCGGGCGTACCTACGTGTAC 4140
-10 +1 +10 +20 +30 +40
D N K Y Y K N L G A V I K N A K R K K H
4141 GACAACAAGTACTACAAAAATTTAGGTGCCGTTATCAAGAACGCTAAGCGCAAGAAGCAC 4200
EspI
+50 +60 +70 +80 +90 +100
F A E H E I E E A T L D P L D N Y L V A
4201 TTCGCCGAACATGAGATCGAAGAGGCTACCCTCGACCCCCTAGACAACTACCTAGTGGCT 4260
+110 +120 +130 +140 +150 +160
E D P F L G P G K N Q K L T L F K E I R
4261 GAGGATCCTTTCCTGGGACCCGGCAAGAACCAAAAACTCACTCTCTTCAAGGAAATCCGT 4320
BamHI PpuMI
+180 +190 +200 +210 +220
N V K P D T M K L V V G W K G K E F Y R
4321 AATGTTAAACCCGACACGATGAAGCTTGTCGTTGGATGGAAAGGAAAAGAGTTCTACAGG 4380
HindIII
+230 +240 +260 +270 +280

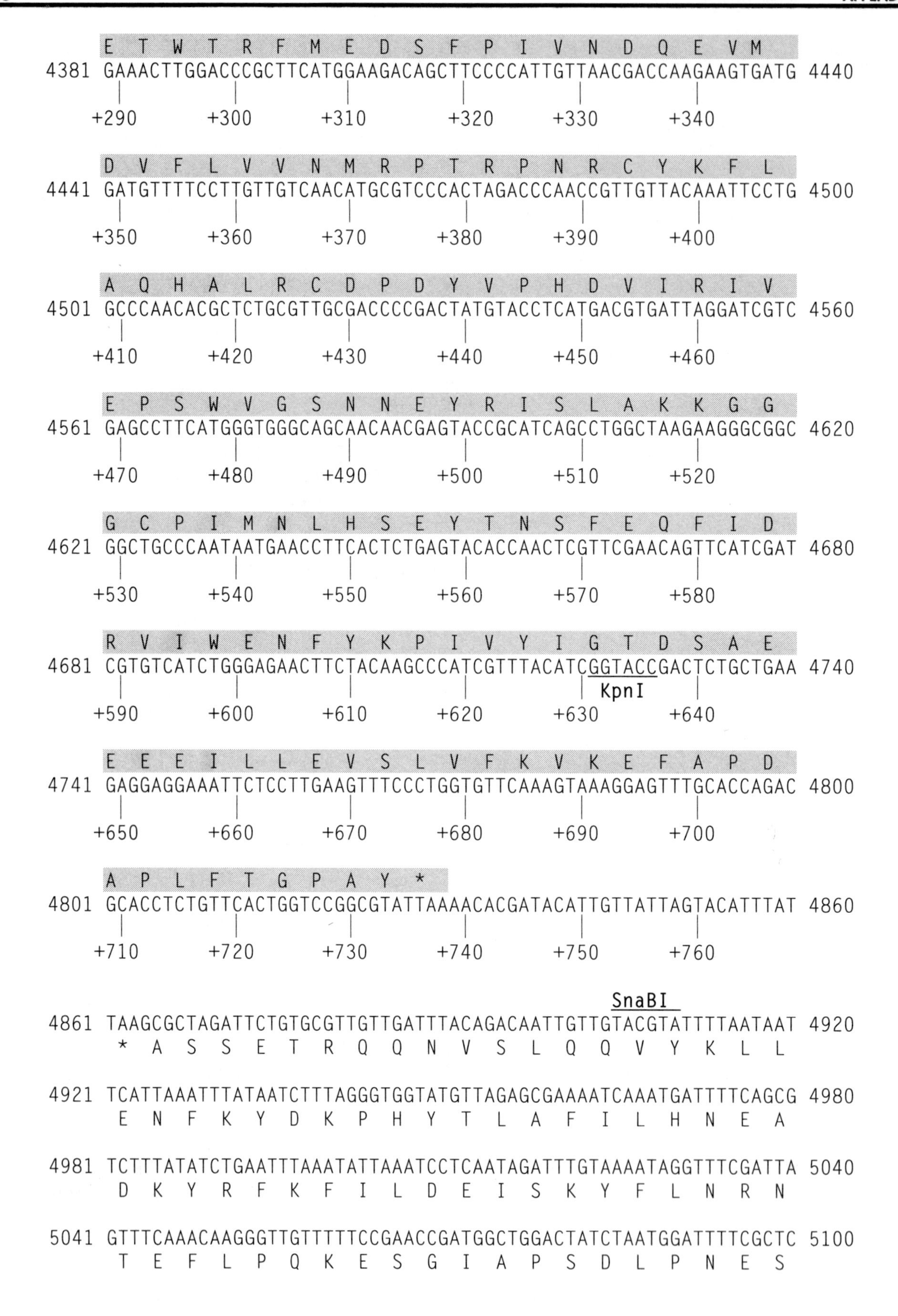
E T W T R F M E D S F P I V N D Q E V M
4381 GAAACTTGGACCCGCTTCATGGAAGACAGCTTCCCCATTGTTAACGACCAAGAAGTGATG 4440
+290 +300 +310 +320 +330 +340
D V F L V V N M R P T R P N R C Y K F L
4441 GATGTTTTCCTTGTTGTCAACATGCGTCCCACTAGACCCAACCGTTGTTACAAATTCCTG 4500
+350 +360 +370 +380 +390 +400
A Q H A L R C D P D Y V P H D V I R I V
4501 GCCCAACACGCTCTGCGTTGCGACCCCGACTATGTACCTCATGACGTGATTAGGATCGTC 4560
+410 +420 +430 +440 +450 +460
E P S W V G S N N E Y R I S L A K K G G
4561 GAGCCTTCATGGGTGGGCAGCAACAACGAGTACCGCATCAGCCTGGCTAAGAAGGGCGGC 4620
+470 +480 +490 +500 +510 +520
G C P I M N L H S E Y T N S F E Q F I D
4621 GGCTGCCCAATAATGAACCTTCACTCTGAGTACACCAACTCGTTCGAACAGTTCATCGAT 4680
+530 +540 +550 +560 +570 +580
R V I W E N F Y K P I V Y I G T D S A E
4681 CGTGTCATCTGGGAGAACTTCTACAAGCCCATCGTTTACATCGGTACCGACTCTGCTGAA 4740
KpnI
+590 +600 +610 +620 +630 +640
E E E I L L E V S L V F K V K E F A P D
4741 GAGGAGGAAATTCTCCTTGAAGTTTCCCTGGTGTTCAAAGTAAAGGAGTTTGCACCAGAC 4800
+650 +660 +670 +680 +690 +700
A P L F T G P A Y *
4801 GCACCTCTGTTCACTGGTCCGGCGTATTAAAACACGATACATTGTTATTAGTACATTTAT 4860
+710 +720 +730 +740 +750 +760
SnaBI
4861 TAAGCGCTAGATTCTGTGCGTTGTTGATTTACAGACAATTGTTGTACGTATTTTAATAAT 4920
* A S S E T R Q Q N V S L Q Q V Y K L L
4921 TCATTAAATTTATAATCTTTAGGGTGGTATGTTAGAGCGAAAATCAAATGATTTTCAGCG 4980
E N F K Y D K P H Y T L A F I L H N E A
4981 TCTTTATATCTGAATTTAAATATTAAATCCTCAATAGATTTGTAAAATAGGTTTCGATTA 5040
D K Y R F K F I L D E I S K Y F L N R N
5041 GTTTCAAACAAGGGTTGTTTTTCCGAACCGATGGCTGGACTATCTAATGGATTTTCGCTC 5100
T E F L P Q K E S G I A P S D L P N E S

```
5101 AACGCCACAAAACTTGCCAAATCTTGTAGCAGCAATCTAGCTTTGTCGATATTCGTTTGT 5160
       L  A  V  F  S  A  L  D  Q  L  L  L  R  A  K  D  I  N  T  Q

5161 GTTTTGTTTTGTAATAAAGGTTCGACGTCGTTCAAAATATTATGCGCTTTTGTATTTCTT 5220
       T  K  N  Q  L  L  P  E  V  D  N  L  I  N  H  A  K  T  N  R

                                                       HindIII
5221 TCATCACTGTCGTTAGTGTACAATTGACTCGACGTAAACACGTTAAATAAAGCTTGGACA 5280
       E  D  S  D  N  T  Y  L  Q  S  S  T  F  V  N  F  L  A  Q  V

5281 TATTTAACATCGGGCGTGTTAGCTTTATTAGGCCGATTATCGTCGTCGTCCCAACCCTCG 5340
       Y  K  V  D  P  T  N  A  K  N  P  R  N  D  D  D  D  W  G  E

5341 TCGTTAGAAGTTGCTTCCGAAGACGATTTTGCCATAGCCACACGACGCCTATTAATTGTG 5400
       D  N  S  T  A  E  S  S  S  K  A  M  A  V  R  R  R  N  I  T

5401 TCGGCTAACACGTCCGCGATCAAATTTGTAGTTGAGCTTTTTGGAATTATTTCTGATTGC 5460
       D  A  L  V  D  A  I  L  N  T  T  S  S  K  P  I  I  E  S  Q

5461 GGGCGTTTTTGGGCGGGTTTCAATCTAACTGTGCCCGATTTTAATTCAGACAACACGTTA 5520
       P  R  K  Q  A  P  K  L  R  V  T  G  S  K  L  E  S  L  V  N

5521 GAAAGCGATGGTGCAGGCGGTGGTAACATTTCAGACGGCAAATCTACTAATGGCGGCGGT 5580
       S  L  S  P  A  P  P  P  L  M  E  S  P  L  D  V  L  P  P  P

5581 GGTGGAGCTGATGATAAATCTACCATCGGTGGAGGCGCAGGCGGGGCTGGCGGCGGAGGC 5640
       P  P  A  S  S  L  D  V  M  P  P  P  A  P  P  A  P  P  P  P

5641 GGAGGCGGAGGTGGTGGCGGTGATGCAGACGGCGGTTTAGGCTCAAATGTCTCTTTAGGC 5700
       P  P  P  P  P  P  P  S  A  S  P  P  K  P  E  F  T  E  K  P

                                                           AgeI
5701 AACACAGTCGGCACCTCAACTATTGTACTGGTTTCGGGCGCCGTTTTTGGTTTGACCGGT 5760
       L  V  T  P  V  E  V  I  T  S  T  E  P  A  T  K  P  K  V  P

5761 CTGAGACGAGTGCGATTTTTTTCGTTTCTAATAGCTTCCAACAATTGTTGTCTGTCGTCT 5820
       R  L  R  T  R  N  K  E  N  R  I  A  E  L  L  Q  Q  R  D  D

5821 AAAGGTGCAGCGGGTTGAGGTTCCGTCGGCATTGGTGGAGCGGGCGGCAATTCAGACATC 5880
       L  P  A  A  P  Q  P  E  T  P  M  P  P  A  P  P  L  E  S  M

5881 GATGGTGGTGGTGGTGGTGGAGGCGCTGGAATGTTAGGCACGGGAGAAGGTGGTGGCGGC 5940
       S  P  P  P  P  P  P  P  A  P  I  N  P  V  P  S  P  P  P  P

5941 GGTGCCGCCGGTATAATTTGTTCTGGTTTAGTTTGTTCGCGCACGATTGTGGGCACCGGC 6000
       P  A  A  P  I  I  Q  E  P  K  T  Q  E  R  V  I  T  P  V  P

6001 GCAGGCGCCGCTGGCTGCACAACGGAAGGTCGTCTGCTTCGAGGCAGCGCTTGGGGTGGT 6060
       A  P  A  A  P  Q  V  V  S  P  R  R  S  R  P  L  A  Q  P  P

6061 GGCAATTCAATATTATAATTGGAATACAAATCGTAAAAATCTGCTATAAGCATTGTAATT 6120
       P  L  E  I  N  Y  N  S  Y  L  D  Y  F  D  A  I  L  M  T  I
```

```
6121 TCGCTATCGTTTACCGTGCCGATATTTAACAACCGCTCAATGTAAGCAATTGTATTGTAA 6180
      E  S  D  N  V  T  G  I  N  L  L  R  E  I  Y  A  I  T  N  Y

                         BamHI
6181 AGAGATTGTCTCAAGCTCGGATCCCGCACGCCGATAACAAGCCTTTTCATTTTTACTACA 6240
      L  S  Q  R  L  S  P  D  R  V  G  I  V  L  R  K  M  K  V  V

6241 GCATTGTAGTGGCGAGACACTTCGCTGTCGTCGACGTACATGTATGCTTTGTTGTCAAAA 6300
      A  N  Y  H  R  S  V  E  S  D  D  V  Y  M  Y  A  K  N  D  F

                 HindIII
6301 ACGTCGTTGGCAAGCTTTAAAATATTTAAAAGAACATCTCTGTTCAGCACCACTGTGTTG 6360
      V  D  N  A  L  K  L  I  N  L  L  V  D  R  N  L  V  V  T  N

6361 TCGTAAATGTTGTTTTTGATAATTTGCGCTTCCGCAGTATCGACACGTTCAAAAAATTGA 6420
      D  Y  I  N  N  K  I  I  Q  A  E  A  T  D  V  R  E  F  F  Q

6421 TGCGCATCAATTTTGTTGTTCCTATTATTGAATAAATAAGATTGTACAGATTCATATCTA 6480
      H  A  D  I  K  N  N  R  N  N  F  L  Y  S  Q  V  S  E  Y  R

               ORF 588 →
               M  A  T  T  N  A  T  L  Q  T  L  V  Q  F  Y  E  N
6481 CGATTCGTCATGGCCACCACAAATGCTACGCTGCAAACGCTGGTACAATTTTACGAAAAC 6540
      R  N  T  M
      ←ORF 1629

     C  K  N  V  K  T  R  Y  K  I  I  N  G  R  F  G  K  I  S  I
6541 TGCAAAAACGTCAAAACTCGGTATAAAATAATCAACGGGCGCTTTGGCAAAATATCTATT 6600

     L  S  H  K  P  T  S  K  L  Y  L  Q  K  T  I  S  A  H  N  F
6601 TTATCGCACAAGCCCACTAGCAAATTGTATTTGCAGAAAACAATTTCGGCGCACAATTTT 6660

     N  A  D  E  I  K  V  H  Q  L  M  S  D  H  P  N  F  I  K  I
6661 AACGCTGACGAAATAAAAGTTCACCAGTTAATGAGCGACCACCCAAATTTTATAAAAATC 6720

     Y  F  N  H  G  S  I  N  N  Q  V  I  V  M  D  Y  I  D  C  P
6721 TATTTTAATCACGGTTCCATCAACAACCAAGTGATCGTGATGGACTACATTGACTGTCCC 6780

     D  L  F  E  T  L  Q  I  K  G  E  L  S  Y  Q  L  V  S  N  I
6781 GATTTATTTGAAACACTACAAATTAAAGGCGAGCTTTCGTACCAACTTGTTAGCAATATT 6840

     I  R  Q  L  C  E  A  L  N  D  L  H  K  H  N  F  I  H  N  D
6841 ATTAGACAGCTGTGTGAAGCGCTCAACGATTTGCACAAGCACAATTTCATACACAACGAC 6900

     I  K  L  E  N  V  L  Y  F  E  A  L  D  R  V  Y  V  C  D  Y
6901 ATAAAACTCGAAAATGTCTTATATTTCGAAGCACTTGATCGCGTGTATGTTTGCGATTAC 6960

     G  L  C  K  H  E  N  S  L  S  V  H  D  G  T  L  E  Y  F  S
6961 GGATTGTGCAAACACGAAAACTCACTTAGCGTGCACGACGGCACGTTGGAGTATTTTAGT 7020

     P  E  K  I  R  H  T  T  M  H  V  S  F  D  W  Y  A  A  C  *
7021 CCGGAAAAAAATTCGACACACAACTATGCACGTTTCGTTTGACTGGTACGCGGCGTGTTAA 7080
```

```
7081 CATACAAGTTGCTAACCGGCGGCCGACACCCATTTGAAAAAAGCGAAGACGAAATGTTGG 7140
                       XmaIII

7141 ACTTGAATAGCATGAAGCGTCGTCAGCAATACAATGACATTGGCGTTTTAAAACACGTTC 7200

7201 GTAACGTTAACGCTCGTGACTTTGTGTACTGCCTAACAAGATACAACATAGATTGTAGAC 7260

7261 TCACAAATTACAAACAAATTATAAAACATGAGTTTTTGTCGTAAAAATGCCACTTGTTTT 7320

7321 ACGAGTAGAATTC 7333
            EcoRI
```

APPENDIX 2: NUCLEOTIDE SEQUENCE OF AcMNPV p10 REGION

This appendix presents the nucleotide sequence and predicted ORFs surrounding the AcMNPV *p10* gene. The sequence presented extends from the HindIII site at 87.3 mu to the EcoRI site at 90.3 mu. The recognition sequences of selected restriction endonucleases are shown. Those represented in bold have only one cleavage site in this sequence. Nucleotide numbers in the vicinity of *p10* (shaded) are given relative to the *p10* ATG (+1,+2,+3), as well as relative to the start of the sequence. The start site of *p10* transcription, and the *p10* promoter TAAG (double underlined) are indicated. These data are also presented in Figures 7-2 and 13-5. The sequence was compiled from data published by Lüebbert and Doerfler (1984), Kuzio et al. (1984), Liu et al. (1986), Friesen and Miller (1987), and Kuzio et al. (1989). Guarino et al. (1986) have also published sequence data for the *hr*5 sequences within this region. Their sequence differs in a number of positions from that of Liu et al. (1986). The significance of these differences is not clear. They do not affect the ORFs or restriction sites shown in this sequence.

```
 ...p35 →
      A  Y  E  K  Y  C  L  P  K  L  V  D  E  R  N  D  Y  Y  V  A
  1 AAGCTTACGAGAAATACTGTTTGCCCAAATTGGTCGACGAACGCAACGACTACTACGTGG 60
    HindIII                          SalI

      V  C  V  L  K  P  G  F  E  N  G  S  N  Q  V  L  S  F  E  Y
 61 CGGTATGCGTGTTGAAGCCGGGATTTGAGAACGGCAGCAACCAAGTGCTATCTTTCGAGT 120

      N  P  I  G  N  K  V  I  V  P  F  A  H  E  I  N  D  T  G  L
121 ACAACCCGATTGGTAACAAAGTTATTGTGCCGTTTGCTCACGAAATTAACGACACGGGAC 180

      Y  E  Y  D  V  V  A  Y  V  D  S  V  Q  F  D  G  E  Q  F  E
181 TTTACGAGTACGACGTCGTAGCTTACGTGGACAGTGTGCAGTTTGATGGCGAACAATTTG 240

      E  F  V  Q  S  L  I  L  P  S  S  F  K  N  S  E  K  V  L  Y
241 AAGAGTTTGTGCAGAGTTTAATATTGCCGTCGTCGTTCAAAAATTCGGAAAAGGTTTTAT 300

      Y  N  E  A  S  K  N  K  S  M  I  Y  K  A  L  E  F  T  T  E
301 ATTACAACGAAGCGTCGAAAAACAAAAGCATGATCTACAAGGCTTTAGAGTTTACTACAG 360

      S  S  W  G  K  S  E  K  Y  N  W  K  I  F  C  N  G  F  I  Y
361 AATCGAGCTGGGGCAAATCCGAAAAGTATAATTGGAAAATTTTTTGTAACGGTTTTATTT 420

      D  K  K  S  K  V  L  Y  V  K  L  H  N  V  T  S  A  L  N  K
421 ATGATAAAAAATCAAAAGTGTTGTATGTTAAATTGCACAATGTAACTAGTGCACTCAACA 480
```

```
       N   V   I   L   N   T   I   K   *
 481 AAAATGTAATATTAAACACAATTAAATAAATGTTAAAATTTATTGCCTAATATTATTTTG 540

 541 TCATTGCTTGTCATTTATTAATTTGGATGATGTCATTTGTTTTTAAAATTGAACTGGCTT 600
               EcoRI
 601 TACGAGTAGAATTCTACGCGTAAAACACAATCAAGTACGAGTCATAAGCTGATGTCATGT 660
                                                EcoRI
 661 TTTGCACACGGCTCATAACCGAACTGGCTTTACGAGTAGAATTCTACTTGTAACGCACGA 720

 721 TCAGTGGATGATGTCATTTGTTTTTCAAATCGAGATGATGTCATGTTTTGCACACGGCTC 780
                               EcoRI
 781 ATAAACTCGCTTTACGAGTAGAATTCTACGTGTAACGCACGATCGATTGCAGAGTCATTT 840

 841 GTTTTGCAATATGATATCATACAATATGACTCATTTGTTTTTCAAAACCGAACTTGATTT 900
              EcoRI
 901 ACGGGTAGAATTCTACTTGTAAAGCACAATCAAAAAGATGATGTCATTTGTTTTTCAAAA 960
                               EcoRI
 961 CTGAACTCGCTTTACGAGTAGAATTCTACGTGTAAAACACAATCAAGAAATGATGTCATT 1020

1021 TGTTATAAAAATAAAAGCTGATGTCATGTTTTGCACATGGCTCATAACTAAACTCGCTTT 1080
              EcoRI
1081 ACGGGTAGAATTCTACGCGTAAAACATGATTGATAATTAAATAATTCATTTGCAAGCTAT 1140

                                  p26 →
                                  M   E   L   Y   N   I   K   Y   A   I   D   P   T
1141 ACGTTAAATCAAACGGACGTTATGGAATTGTATAATATTAAATATGCAATTGATCCAACA 1200

     N   K   I   V   I   E   Q   V   D   N   V   D   A   F   V   H   I   L   E   P
1201 AATAAAATTGTAATAGAGCAAGTCGACAATGTGGACGCGTTTGTGCATATTTTAGAACCG 1260
                           SalI

     G   Q   E   V   F   D   E   T   L   S   Q   Y   H   Q   F   P   G   V   V   S
1261 GGTCAAGAAGTGTTCGACGAAACGCTAAGCCAGTACCACCAATTTCCTGGCGTCGTTAGT 1320
                                                    BstXI

     S   I   I   F   P   Q   L   V   L   N   T   I   I   S   V   L   S   E   D   G
1321 TCGATTATTTTCCCGCAACTCGTGTTAAACACAATAATTAGCGTTTTGAGCGAAGACGGC 1380

     S   L   L   T   L   K   L   E   N   T   C   F   N   F   H   V   C   N   K   R
1381 AGTTTGCTCACGTTGAAACTCGAAAACACTTGTTTTAATTTTCACGTGTGCAATAAACGC 1440
                                                      PmlI

     F   V   F   G   N   L   P   A   A   V   V   N   N   E   T   K   Q   K   L   R
1441 TTTGTGTTTGGCAATTTGCCAGCGGCGGTCGTGAATAATGAAACGAAGCAAAAACTGCGC 1500

     I   G   A   P   I   F   A   G   K   K   L   V   S   V   V   T   A   F   H   R
1501 ATTGGAGCTCCAATTTTTGCCGGCAAAAAGCTGGTTTCGGTCGTGACGGCGTTTCATCGT 1560

     V   G   E   N   E   W   L   L   P   V   T   G   I   R   E   A   S   Q   L   S
1561 GTTGGCGAAAACGAATGGCTGTTACCGGTGACGGGAATTCGAGAGGCGTCCCAGCTGTCG 1620
                                         EcoRI
     G   H   M   K   V   L   N   G   V   R   V   E   K   W   R   P   N   M   S   V
1621 GGACATATGAAGGTGCTGAACGGCGTCCGTGTTGAAAAATGGCGACCCAACATGTCCGTC 1680
```

```
         Y  G  T  V  Q  L  P  Y  D  K  I  K  Q  H  A  L  E  Q  E  N
1681 TACGGGACTGTGCAATTGCCGTACGATAAAATTAAACAGCATGCGCTCGAGCAAGAAAAT 1740
                                               SphI   XhoI

         K  T  P  N  A  L  E  S  C  V  L  F  Y  K  D  S  E  I  R  I
1741 AAAACGCCAAACGCGTTGGAGTCTTGTGTGCTATTTTACAAAGATTCAGAAATACGCATC 1800

         T  Y  N  K  G  D  Y  E  I  M  H  L  R  M  P  G  P  L  I  Q
1801 ACTTACAACAAGGGGGACTATGAAATTATGCATTTGAGGATGCCGGGACCTTTAATTCAA 1860
                                   NsiI

         P  N  T  I  Y  Y  S  *     * p10 mRNA 5' end →        PacI
1861 CCCAACACAATATATTATAGTTAAATAAGAATTATTATCAATCATTTGTATATTAATTA 1920
          |         |         |         |         |         |
         -90       -80       -70       -60       -50       -40

                                              p10 →
                                              M  S  K  P  N  V  L  T
1921 AAATACTATACTGTAAATTACATTTTATTTACAATCATGTCAAAGCCTAACGTTTTGACG 1980
          |         |         |         |         |         |
         -30       -20       -10        +1       +10       +20

      Q  I  L  D  A  V  T  E  T  N  T  K  V  D  S  V  Q  T  Q  L
1981 CAAATTTTAGACGCCGTTACGGAAACTAACACAAAGGTTGACAGTGTTCAAACTCAGTTA 2040
          |         |         |         |         |         |
         +30       +40       +50       +60       +70       +80

      N  G  L  E  E  S  F  Q  L  L  D  G  L  P  A  Q  L  T  D  L
2041 AACGGGCTGGAAGAATCATTCCAGCTTTTGGACGGTTTGCCCGCTCAATTGACCGATCTT 2100
          |         |      PflMI          |         |         |
         +90      +100      +110      +120      +130      +140

      N  T  K  I  S  E  I  Q  S  I  L  T  G  D  I  V  P  D  L  P
2101 AACACTAAGATCTCAGAAATTCAATCCATATTGACCGGCGACATTGTTCCGGATCTTCCA 2160
          | BglII   |         |         |         |         |
         +150      +160      +170      +180      +190      +200

      D  S  L  K  P  K  L  K  S  Q  A  F  E  L  D  S  D  A  R  R
2161 GACTCACTAAAGCCTAAGCTGAAAAGCCAAGCTTTTGAACTCGATTCAGACGCTCGTCGT 2220
          |         |         | HindIII           |         |
         +210      +220      +230      +240      +250      +260

      G  K  R  S  S  K  *
2221 GGTAAACGCAGTTCCAAGTAAATGAATCGTTTTTAAAATAACAAATCAATTGTTTTATAA 2280
          |         |         |          *  F  L  L  D  I  T  K  Y
         +270      +280      +290

                                                    BclI
2281 TATTCGTACGATTCTTTGATTATGTAATAAAATGTGATCATTAGGAAGATTACGAAAAAT 2340
       Y  E  Y  S  E  K  I  I  Y  Y  F  T  I  M  L  F  I  V  F  F

2341 ATAAAAAATATGAGTTCTGTGTGTATAACAAATGCTGTAAACGCCACAATTGTGTTTGTT 2400
       I  F  F  I  L  E  T  H  I  V  F  A  T  F  A  V  I  T  N  T
```

```
2401 GCAAATAAACCCATGATTATTTGATTAAAATTGTTGTTTTCTTTGTTCATAGACAATAGT 2460
       A  F  L  G  M  I  I  Q  N  F  N  N  N  E  K  N  M  S  L  L

2461 GTGTTTTGCCTAAACGTGTACTGCATAAACTCCATGCGAGTGTATAGCGAGCTAGTGGCG 2520
       T  N  Q  R  F  T  Y  Q  M  F  E  M  R  T  Y  L  S  S  T  A

2521 AACGCTTGCCCCACCAAAGTAGATTCGTCAAAATCCTCAATTTCATCACCCTCCTCCAAG 2580
       L  A  Q  G  V  L  T  S  E  D  F  D  E  I  E  D  G  E  E  L

2581 TTTAACATTTGGCCGTCGGAATTAACTTCTAAAGATGCCACATAATCTAATAAATGAAAT 2640
       N  L  M  Q  G  D  S  N  V  E  L  S  A  V  Y  D  L  L  H  F

2641 AGAGATTCAAACGTGGCGTCATCGTCCGTTTCGACCATTTCCGAAAAGAACTCGGGCATA 2700
       L  S  E  F  T  A  D  D  D  T  E  V  M  E  S  F  F  E  P  M

2701 AACTCTATGATTTCTCTGGACGTGGTGTTGTCGAAACTCTCAAAGTACGCAGTCAGGAAC 2760
       F  E  I  I  E  R  S  T  T  N  D  F  S  E  F  Y  A  T  L  F

2761 GTGCGCGACATGTCGTCGGGAAACTCGCGCGGAAACATGTTGTTGTAACCGAACGGGTCC 2820
       T  R  S  M  D  D  P  F  E  R  P  F  M  N  N  Y  G  F  P  D

2821 CATAGCGCCAAAACCAAATCTGCCAGCGTCAATAGAATGAGCACGATGCCGACAATGGAG 2880
       W  L  A  L  V  L  D  A  L  T  L  L  I  L  V  I  G  V  I  S

                                Hpa I
2881 CTGGCTTGGATAGCGATTCGAGTTAACGCTTTGGCAGTCACGGTCAGCGTTTTGATGGCG 2940
       S  A  Q  I  A  I  R  T  L  A  K  A  T  V  T  L  T  K  I  A

2941 ATCACGTTGAGCGAGTGCACTAACGCGGCTTTGTAAGTCTCTCCCAACATGCGCACGGTC 3000
       I  V  N  L  S  H  V  L  A  A  K  Y  T  E  G  L  M  R  V  T

                      EspI
3001 ACGCGCCGAGTCGTGCTAAGCAACATGTGTTTCATGGCCGGAATGAGAGAAGTGTTAATT 3060
       V  R  R  T  T  S  L  L  M  H  K  M  A  P  I  L  S  T  N  I

3061 TTTTTCAACATGCTTTTAAACCCGGACATTAGCATATCAAAGCCAATGTCCGTAGCAATA 3120
       K  K  L  M  S  K  F  G  S  M  L  M  D  F  G  I  D  T  A  I

                                                     PflmI
3121 CCGAAAACGAGCGCGTAATCTTCCAAAAACGATGTTATAATTGACTCCAAGTCTTGGTCG 3180
       G  F  V  L  A  Y  D  E  L  F  S  T  I  I  S  E  L  D  Q  D

                                          PmlI          SstII
3181 CTGATTGAACGGTCGAGCGCCTCGAAATGTTCGACACGTGCACGTTCGTTACCGCGGTAA 3240
       S  I  S  R  D  L  A  E  F  H  E  V  R  A  R  E  N  G  R  Y

                                          XmaIII
3241 TTGTATGCGATCGGAGTTTTAGTAAAGCCGGTTTCGGCCGTGTACGTGATCTGGACGGGC 3300
       N  Y  A  I  P  T  K  T  F  G  T  E  A  T  Y  T  I  Q  V  P

3301 GACCCGTTGACGATCATGCCCAAATCGTTTAGTGTTGGATTTTTGTTAAAAAGTTTTTCA 3360
       S  G  N  V  I  M  G  L  D  N  L  T  P  N  K  N  F  L  K  E
```

```
        ___BstXI ___
3361 AATTCCAAGTCTGTGGCGTTATCGCGCACGCTGCGCCATTGCGCTAGTATTGCGTTGGAG 3420
      F  E  L  D  T  A  N  D  R  V  S  R  W  Q  A  L  I  A  N  S

3421 TCCACGTTGGGTCGTGGCGGTAGTATGCTGGAAGGCGCTTTGTAATCAAAATCGCGCAGT 3480
      D  V  N  P  R  P  P  L  I  S  S  P  A  K  Y  D  F  D  R  L

                       BalI
3481 TCGCTAAAAATGTTGTTGGCCAGCATTTTGAAAGTGACAAAGATCGTGTCGCCCAGCACG 3540
      E  S  F  I  N  N  A  L  M  K  F  T  V  F  I  T  D  G  L  V

3541 AATCCGATGAGCGATTCCCACCATCTAAACGAACAACCGCCGTTGAATAGCTCTCTGCCG 3600
      F  G  I  L  S  E  W  W  R  F  S  C  G  G  N  F  L  E  R  G

        SalI            EcoRI
3601 AAACGTCGACAGTAGGCTTCGTTGAATTC 3629
      F  R  R  C  Y  A  E  N  F  E
                          ← p74...
```

APPENDIX 3: LIST OF GENES EXPRESSED IN THE BACULOVIRUS EXPRESSION SYSTEM

This list was compiled by V.A. Luckow based on literature published before the end of 1991. The entries are ordered according to the organism the gene was derived from, starting with viruses, and then including bacteria, fungi, protozoans, plants, invertebrates, and vertebrates.

Origin	*Gene Product*	*References*
	VIRUSES	
Adenoviruses		
Human Adenovirus 2	DNA polymerase	Watson & Hay, 1990
	E1A protein (13s message)	Patel et al., 1988; Patel & Jones, 1990
	E3 protein (19K)	Lévy & Kvist, 1990
	Preterminal protein (pTP)	Zhao et al., 1991
Arenaviruses		
Lymphocytic choriomeningitis virus	Glycoprotein precursor (GPC)	Matsuura et al., 1986; 1987
	Nucleoprotein (N)	Matsuura et al., 1986; 1987; Emery and Bishop, 1987
Lassa fever virus	Nucleocapsid protein	Barber et al., 1990
Bunyaviruses		
Maguari bunyavirus	Nucleocapsid protein (N)	Elliott and McGregor, 1989
Snowshoe hare bunyavirus	Nonstructural protein (NS_S)	Urakawa et al., 1988
	Nucleoprotein (N)	Urakawa et al., 1988
Punta Toro phlebovirus	Neutralization antigen (N)	Overton et al., 1987
	Nucleoprotein (NS_S)	Overton et al., 1987
Rift Valley fever virus	Envelope glycoprotein (G1)	Schmaljohn et al., 1989; Takehara et al., 1990
	Envelope glycoprotein (G1-G2)	Schmaljohn et al., 1989
Hantaan virus	Envelope glycoproteins G1 and G2	Schmaljohn et al., 1990
	Nucleocapsid protein (N)	Schmaljohn et al., 1988; 1990
Caulimoviruses		
Cauliflower mosaic virus	Gene I protein	Vlak et al., 1990

Origin	Gene Product	References
Comoviruses		
Cowpea mosaic virus	B-RNA-encoded proteins	Van Bokhoven et al., 1990
	RNA polymerase	Van Bokhoven et al., 1991
Coronaviruses		
Bovine coronavirus	Hemagglutinin-esterase (HE)	Parker et al., 1990a; Yoo et al., 1991a
	Spike glycoprotein (S)	Parker et al., 1990b; Yoo et al., 1990; 1991a
	Spike glycoprotein subunit (S1)	Yoo et al., 1991a and b
	Spike glycoprotein subunit (S2)	Yoo et al., 1991a
Murine coronavirus	Spike glycoprotein (S1)	Takase-Yoden et al., 1991
	Spike glycoprotein (E2)	Yoden et al., 1989
Flaviviruses		
Dengue virus type 1	Envelope glycoprotein (E)	Putnak et al., 1991
Dengue virus type 2	Envelope glycoprotein (E)	Deubel et al., 1991
Dengue virus type 4	Capsid protein (C)	Zhang et al., 1988
	Core protein	Makino et al., 1989
	Envelope glycoprotein (E)	Zhang et al., 1988
	Nonstructural protein (NS1)	Zhang et al., 1988; Rothman et al., 1989
	Nonstructural protein (NS2a)	Zhang et al., 1988
	Premembrane protein (PreM)	Zhang et al., 1988
Japanese encephalitis virus	Capsid protein (C)	Matsuura et al., 1989
	Envelope glycoprotein (E)	Aira, 1987; Matsuura et al., 1989
	Nonstructural protein (NS1)	Matsuura et al., 1989
	Premembrane protein (PreM)	Matsuura et al., 1989
Yellow fever virus	Envelope glycoprotein (E)	Despres et al., 1991; Shiu et al., 1991
	Nonstructural protein (NS1)	Despres et al., 1991
Hepadnaviruses		
Hepatitis B virus	Core antigen	Lanford et al., 1988; Takehara et al., 1988; Hilditch et al., 1990; Lanford and Notvall, 1990
	Surface antigen	Cochran et al., 1987; Kang et al., 1987; Lanford et al., 1988; 1989; Price et al., 1988; 1989; Takehara et al., 1988; Lanford and Notvall, 1990
	X protein	Klein et al., 1991
Herpesviruses		
Bovine Herpesvirus-1	Glycoprotein gIV	Van Drunen Littel-Van den Hurk et al., 1991
Epstein-Barr virus	Alkaline DNase	Baylis et al., 1991
	Deoxyribonuclease (BGLF5)	Chen et al., 1991a
	Nuclear antigen (EBNA1)	Frappier and O'Donnell, 1991
	Phosphoprotein pp58 (BMRFI)	Chen et al., 1991a
	Terminal protein 1 (TP-1)	Frech et al., 1990

Origin	Gene Product	References
Herpes simplex virus type 1	DNA polymerase	Hernandez and Lehman, 1990; Marcy et al., 1990
	Glycoprotein D	Krishna et al., 1989
	Glycoprotein-G (gG-1)	Sánchez-Martínez and Pellett, 1991
	UL5 gene product	Dodson et al., 1989; Olivo et al., 1989; Calder and Stow, 1990; Dodson and Lehman, 1991
	UL8 gene product	Calder and Stow, 1990; Dodson et al., 1989; Olivo et al., 1989; Dodson and Lehman, 1991
	UL9 gene product	Olivo et al., 1988; 1989
	UL42 gene product	Gottlieb et al., 1990
	UL52 gene product	Dodson et al., 1989; Olivo et al., 1989; Calder and Stow, 1990; Dodson and Lehman, 1991
	Vmw65 (VP16, αTIF)	Capone, 1989; Kristie et al., 1989; Capone and Werstuck, 1990
Herpes simplex virus type 2	Glycoprotein-G (gG-2)	Sánchez-Martínez and Pellett, 1991
Human cytomegalovirus	DNA polymerase	Ertl et al., 1991
	gp55-116 (gB)	Wells et al., 1990
Marek's disease virus	A antigen	Niikura et al., 1991a
Pseudorabies virus	Glycoprotein gp50	Thomsen et al., 1990
Varicella zoster virus	DNA polymerase	Ertl et al., 1991
Nepoviruses		
Arabis mosaic virus	Coat protein	Bertioli et al., 1991
Orthomyxoviruses		
Influenza virus A		
Ann Arbor strain	Nucleoprotein (NP)	Rota et al., 1990
Fowl plaque virus	Hemagglutinin (HA)	Kuroda et al., 1986; 1987; 1989; 1990; 1991
PR strain	Hemagglutinin (HA)	Possee, 1986
	Polymerase subunits PA, PB1 and PB2	St. Angelo et al., 1987
Shearwater strain	Nucleoprotein (NP)	Harley et al., 1990
Udorn strain	Neuraminidase (NA)	Price et al., 1989
Influenza virus B (Ann Arbor strain)	Nucleoprotein (NP)	Rota et al., 1990
Papovaviruses		
Bovine papilloma virus (BPV1)	E1 protein	Santucci et al., 1990; Blitz and Laimins, 1991
	E2 protein	Tada et al., 1988; McBride et al., 1989; Monini et al., 1991
	E5 oncoprotein	Burkhardt et al., 1989
Human papilloma virus type 6b	E2 protein	Sekine et al., 1988; 1989; Tada et al., 1988
	L2 ORF	Rose et al., 1990
Human papilloma virus type 11	L2 ORF	Rose et al., 1990

Origin	Gene Product	References
Human papilloma virus type 18	E6 protein	Grossman and Laimins, 1989; Grossman et al., 1989
Polyoma virus	Large T antigen	Rice et al., 1987
	Middle T antigen	Piwnica-Worms et al., 1988; 1990; Forstová et al., 1989; Cohen et al., 1990
	VP1 capsid protein	Montross et al., 1991
Simian virus 40	Large T antigen	Lanford, 1988; Murphy et al., 1988; O'Reilly and Miller, 1988; Höss et al., 1990; Shearer et al., 1990; Chen et al., 1991c; Loeber et al., 1991
	Large T antigen, tsA58 temperature-sensitive mutant	Reynisdottir et al., 1990
	Small t antigen	Jeang et al., 1987b; Murphy et al., 1988
Paramyxoviruses		
Human parainfluenza 3	Fusion glycoprotein (F)	Ray et al., 1989; Hall et al., 1991
	Hemagglutinin-neuraminidase	Van Wyke Coelingh et al., 1987
Newcastle disease virus (NDV)	Hemagglutinin-neuraminidase	Nagy et al., 1990; 1991; Niikura et al., 1991b
Measles virus	Hemagglutinin (H)	Vialard et al., 1990
	Membrane fusion protein (F)	Vialard et al., 1990
Respiratory syncytial virus	F glycoprotein	Wathen et al., 1989a
	Chimeric FG glycoprotein	Brideau et al., 1989; Wathen et al., 1989b; 1991a and b; Nicholas et al., 1990; Levely et al., 1991
Parvoviruses		
Adeno-associated virus (AAV)	Rep68 protein	Owens et al., 1991
	Rep78 protein	Owens et al., 1991
Human parvovirus B19	Coat protein VP1	Brown et al., 1990a and b; 1991; Kajigaya et al., 1991
	Coat protein VP2	Brown et al., 1990a; 1991; Kajigaya et al., 1991
Minute virus of mice	Nonstructural protein (NS-1)	Yeung et al., 1991
Picornaviruses		
Foot-and-mouth disease virus	Capsid precursor protein P1-2A	Roosien et al., 1990; Belsham et al., 1991
	RNA polymerase	Van Bokhoven et al., 1991
Hepatitis A virus	Capsid protein (VP1)	Harmon et al., 1988
Poliomyelitis virus	3B protein (VPg)	Neufeld et al., 1991
	3Cpro protein (protease)	Neufeld et al., 1991
	3Dpol protein (RNA polymerase)	Neufeld et al., 1991; Van Bokhoven et al., 1991
	Capsid protein	Urakawa et al., 1989a
	Structural proteins VP0, VP1 and VP2	Urakawa et al., 1989a

Origin	Gene Product	References
Polydnaviruses		
Campoletis sonorensis virus	1.6 kb ORF (segment WHv1)	Blissard et al., 1989
	1.0 kb ORF (segment WHv2)	Blissard et al., 1989
Reoviruses		
African horsesickness disease virus serotype 4	Outer capsid protein VP7	Roy et al., 1991b
Bluetongue virus serotype 1	Outer capsid protein VP2	Loudon et al., 1991
Bluetongue virus serotype 2	Outer capsid protein VP2	Loudon et al., 1991
	Outer capsid protein VP5	Loudon et al., 1991
Bluetongue virus serotype 10	Core protein VP6	Roy et al., 1990a and b; Loudon and Roy, 1991
	E glycoprotein	McCown et al., 1990
	Inner core protein VP3	Inumaru and Roy, 1987; French et al., 1990; Marshall and Roy, 1990; Roy et al., 1990b; Loudon et al., 1991; Loudon and Roy, 1991
	Nonstructural protein NS1	Urakawa and Roy, 1988; Marshall et al., 1990; McCown et al., 1990; Roy et al., 1990b
	Nonstructural proteins NS3, NS3A	French et al., 1989; Roy et al., 1990b
	Outer capsid protein VP2	Inumaru and Roy, 1987; French et al., 1990; Marshall and Roy, 1990; Roy et al., 1990b; Loudon et al., 1991; Loudon and Roy, 1991
	Outer capsid protein VP5	French et al., 1990; Marshall and Roy, 1990; Roy et al., 1990b; Loudon et al., 1991; Loudon and Roy, 1991
	Outer core protein VP7	French et al., 1990; French and Roy, 1990; Oldfield et al., 1990; Roy et al., 1990b; Loudon and Roy, 1991
	Phosphoprotein NS2	Roy et al., 1990b; Thomas et al., 1990
	RNA polymerase VP1	Urakawa et al., 1989b; Roy et al., 1990b; Loudon and Roy, 1991
	VP4	Loudon and Roy, 1991
Bluetongue virus serotype 11	Outer capsid protein VP2	Loudon et al., 1991
Bluetongue virus serotype 13	Outer capsid protein VP2	Loudon et al., 1991
	Outer capsid protein VP5	Loudon et al., 1991
Bluetongue virus serotype 17	Inner core protein VP3	Loudon et al., 1991
	Outer capsid protein VP2	Loudon et al., 1991
Epizootic hemorrhagic disease virus serotype 2	Nonstructural protein NS1	Nel and Huismans, 1991
Bovine rotavirus type 1	Nucleocapsid protein VP6	Caron et al., 1990
	Structural protein VP1	Cohen et al., 1989
Bovine rotavirus strain RF	Structural protein VP2	Labbe et al., 1991

Origin	Gene Product	References
Porcine rotavirus	Outer capsid protein VP4	Nishikawa et al., 1989; Liprandi et al., 1991; Nagesha and Holmes, 1991
Rhesus rotavirus	Hemagglutinin VP8	Fiore et al., 1991
	Outer capsid protein VP4	Mackow et al., 1989; 1990; Nagesha and Holmes, 1991
Simian rotavirus SA11	Capsid antigen VP6	Estes et al., 1987; Au et al., 1989
	Nonstructural phosphoprotein NS26	Au et al., 1989; Wedegaertner and Gill, 1989; Welch et al., 1989
	Nonstructural glycoprotein NS28	Au et al., 1989
	Nonstructural protein NS53	Au et al., 1989
	Outer capsid protein VP4	Au et al., 1989
	Outer capsid protein VP7	Au et al., 1989
	Structural protein VP1	Au et al., 1989
Retroviruses		
Avian leukemia virus	env-gp85	Noteborn et al., 1990
Bovine leukemia virus	$p34^{tax}$	Chen et al., 1989
Bovine immunodeficiencylike virus	Gag protein	Rasmussen et al., 1990
Feline immunodeficiency virus	Gag protein	Morikawa et al., 1991
Feline sarcoma virus	v-*fms*	Morrison et al., 1990
Human immunodeficiency virus type 1	Envelope protein gp120	Cochran et al., 1987; Richardson et al., 1988; McQuade et al., 1989; Morikawa et al., 1990; Murphy et al., 1990a
	Envelope protein gp160	Cochran et al., 1987; Hu et al., 1987; Rusche et al., 1987; Putney et al., 1988; Farmer et al., 1989; Schwaller et al., 1989; Murphy et al., 1990a; Orentas et al., 1990; Redfield et al., 1991; Tacket et al., 1990; Viscidi et al., 1990; Wells and Compans, 1990; Dolin et al., 1991; Keefer et al., 1991; Redfield et al., 1991
	Gag protein	Madisen et al., 1987; Gheysen et al., 1989; Overton et al., 1989; 1990; Royer et al., 1991
	Gag-pol fusion protein	Overton et al., 1990
	Integration protein	Bushman et al., 1990
	Major core p24	Mills and Jones, 1990
	Nef protein	Matsuura et al., 1991
	Pol protein	Hu and Kang, 1991
	Protease	Overton et al., 1989
	Rev protein	Daefler et al., 1990; Kjems et al., 1991

Origin	*Gene Product*	*References*
	Tat protein	Jeang et al., 1988a and b
Human immunodeficiency virus type 2	Gag precursor protein	Luo et al., 1990
Human T-cell lymphotrophic virus type 1 (HTLV-I)	p20E protein	Nyunoya et al., 1990
	gp46 protein	Nyunoya et al., 1990
	$p40^x$ protein	Jeang et al., 1987a; 1988a; Nyunoya et al., 1988
Human T-cell leukemia virus type II (HTLV-II)	Rex protein	Yip et al., 1991
Rous sarcoma virus	v-*src* tyrosine kinase	Abdel-Ghany et al., 1990
Simian immunodeficiency virus	Gag protein	Delchambre et al., 1989
	gp140	Murphey-Corb et al., 1989
Simian sarcoma virus	v-*sis*/platelet-derived growth factor β (PDGF-β) homolog	Giese et al., 1989; Morishita et al., 1991
Rhabdoviruses		
Rabies virus	Glycoprotein (G)	Préhaud et al., 1989
	Nucleoprotein (N)	Préhaud et al., 1990; Reid-Sanden et al., 1990; Fu et al., 1991
Vesicular stomatitis virus	Glycoprotein (G)	Bailey et al., 1989
Infectious hematopoietic necrosis virus	Glycoprotein	Koener and Leong, 1990
Togaviruses		
Rubella virus	Envelope protein (E1)	Oker-Blom et al., 1989; 1990; Seppanen et al., 1991
	Envelope protein (E2)	Oker-Blom et al., 1989; 1990
Sindbis virus	Capsid protein (C)	Oker-Blom and Summers, 1989
	Envelope proteins E1 and E2	Oker-Blom and Summers, 1989
Unclassified viruses		
Hepatitis delta virus	Delta antigen (HDAg)	Kos et al., 1991
Non-A, non-B hepatitis virus (Hepatitis C virus)	Core protein (p22)	Chiba et al., 1991
	BACTERIA	
Bacillus anthracis	Protective antigen (PA)	Iacono-Connors et al., 1990; 1991
Bacillus thuringiensis subspecies *kurstaki*	HD-73 delta endotoxin	Merryweather et al., 1990
Bacillus thuringiensis subspecies *aizawai* 7.21	Crystal protein	Martens et al., 1990
Clostridium tetani	Tetanus toxin fragment-C	Charles et al., 1991
Escherichia coli	Chloramphenicol acetyl transferase	Carbonell et al., 1985; Luckow and Summers, 1988; 1989
	β-Galactosidase	Pennock et al., 1984; Carbonell et al., 1985; Luckow and Summers, 1988; 1989; Vlak et al., 1988; Licari and Bailey, 1991; Ogonah et al., 1991

Origin	Gene Product	References
	β-Glucuronidase	Luckow and Summers, 1989
Pseudomonas diminuta	Phosphotriesterase	Caldwell and Raushel, 1991; Dumas et al., 1990
	FUNGI	
Neurospora crassa	qa-1f activator protein	Baum et al., 1987
Phanerochaete chrysosporium	Manganese peroxidase	Pease et al., 1991
	PLANTS	
Phaseolus vulgaris (French bean)	Phaseolin	Bustos et al., 1988
Ricinus communis (Castor bean)	Ricin and Ricin B-chain	Piatak et al., 1988
Solanum tuberosum (Potato)	Patatin	Andrews et al., 1988
Carica papaya (Papaya)	Papain	Menard et al., 1990; Vernet et al., 1990; Tessier et al., 1991
Zea mays (Corn)	*Ac* transposase	Hauser et al., 1988; Kunze and Starlinger, 1989
	PROTOZOA	
Eimeria acervulina	Merozoite antigen	Jenkins et al., 1991
Plasmodium falciparum (Malaria)	Circumsporozoite protein	Dontfraid et al., 1988; Good et al., 1990; Murphy et al., 1990b; Jacobs et al., 1991
	Major merozoite surface antigen precursor (PMMSA)	Murphy et al., 1990b
	INVERTEBRATES	
Androctonus australis Hector (Scorpion)	Insect neurotoxin (AaIT)	Maeda et al., 1991; Stewart et al., 1991
Bombyx mori (Silkworm)	Chorion proteins	Iatrou and Meidinger, 1990
Buthus eupeus (Scorpion)	BeIT insectotoxin-1	Carbonell et al., 1988
Drosophila melanogaster (Fruitfly)	Krüppel gene product	Ollo and Maniatis, 1987
	Shaker "A-type" K^+ channel	Armstrong and Miller, 1990; Klaiber et al., 1990
Haementaria officianalis (Mexican leach)	Antistasin (ATS)	Han et al., 1989; Nutt et al., 1991
Heliothis virescens (Tobacco budworm)	Juvenile hormone esterase	Hammock et al., 1990
Hyalaphora cecropia (Giant silkmoth)	Basic preproattacin	Gunne et al., 1990
	Preprocecropin-A	Hellers et al., 1991
Hirudo medicinalis (Leach)	Hirudin	Benatti et al., 1991
Manduca sexta (Tobacco hornworm)	Diuretic hormone	Maeda, 1989c
	Eclosion hormone	Eldridge et al., 1991
Photinus luciferin (Firefly)	Luciferase	Hasnain and Nakhai, 1990; Jha et al., 1990
Pyemotes tritici (Mite)	Neurotoxin (TxP-I)	Tomalski and Miller, 1991
Sarcophaga peregrina (Fleshfly)	Sarcotoxin IA	Yamada et al., 1990
Shistosoma mansoni (Shistosomiasis parasite)	Sm32 antigen	Felleisen et al., 1990
Spisula solidissima (Clam)	Cyclins A and B	Roy et al., 1991a

Origin	*Gene Product*	*References*
	VERTEBRATES	
Amphibians		
Xenopus laevis	Peptidylglycine α-hydroxylating monooxygenase (amidating enzyme)	Suzuki et al., 1990
	$pp90^{rsk}$ protein kinase	Vik et al., 1990
	xlcaax-1 protein	Kloc et al., 1991
Birds		
Chicken	Nicotinic acetylcholine receptor (α-subunit)	Atkinson et al., 1990
	$pp60^{c\text{-}src}$	Piwnica-Worms et al., 1988; 1990
	S-cyclophilin	Caroni et al., 1991
Turkey	β-adrenergic receptor	Parker et al., 1991
Fish		
Carp	Gonadotropin α1 and α2	Huang et al., 1991
Mammals		
Baboon	Estradiol-dependent oviduct-specific glycoprotein	Donnelly et al., 1991
Bovine	Calmodulin-sensitive (type I) adenylcyclase	Tang et al., 1991
	Opsin	Janssen et al., 1988; 1991
	Protein kinase C-γ	Patel and Stabel, 1989
Hamster	β-adrenergic receptor	George et al., 1989
	Scrapie prion (PrP)	Scott et al., 1988
Human	Acid β-glucosidase	Grabowski et al., 1989; Grace et al., 1990
	Adenosine deaminase	Medin et al., 1990
	β2-adrenergic receptor	Reilander et al., 1991
	Aldose reductase	Nishimura et al., 1990; 1991
	β-amyloid precursor	Knops et al., 1991a; Ramakrishna et al., 1991
	Antithrombin III (ATIII)	Gillespie et al., 1991
	Apolipoprotein-E	Gretch et al., 1991
	CD2 (T11) erythrocyte receptor	Alcover et al., 1988
	CD4 HIV receptor	Hussey et al., 1988; Webb et al., 1989; Morikawa et al., 1990; Zeira et al., 1991
	CD23 IgE receptor	Janssen et al., 1991
	Choriogonadotropin, α-subunit (hCGα)	Nakhai et al., 1991
	Choriogonadotropin, β-subunit (hCGβ)	Chen and Bahl, 1991a, b, and c; Chen et al., 1991b
	Complement component C1r	Gál et al., 1989; Sárvari et al., 1990
	Corticosteroid binding globulin (hCBG)	Ghose-Dastidar et al., 1991
	CR2 Epstein-Barr virus/ complement-3d receptor (extracellular domain)	Moore et al., 1991
	crk oncogene product (P47gag-crk)	Matsuda et al., 1990
	Cystic fibrosis gene product (CFTR)	Kartner et al., 1991

Origin	Gene Product	References
	Epidermal growth factor receptor	Greenfield et al., 1988; Patel et al., 1988
	Epidermal growth factor receptor (ectodomain)	Greenfield et al., 1989
	Epidermal growth factor receptor (intracellular domain)	Hsu et al., 1991
	Epidermal growth factor receptor (tyrosine kinase domain)	Wedegaertner and Gill, 1989
	Erythropoietin (EPO)	Wojchowski et al., 1987; Quelle et al., 1989
	Estrogen receptor	Brown and Sharp, 1990
	c-*fms* protooncoprotein	Morrison et al., 1990
	c-*fos* protooncoprotein	Allegretto et al., 1990; Tratner et al., 1990
	G25K protein	Munemitsu et al., 1990
	β-galactosidase	Itoh et al., 1990; 1991
	Gastrin-releasing peptide precursor	Lebacq-Verheyden et al., 1988
	Glucocerebrosidase	Martin et al., 1988; Bergh et al., 1990
	Glucocorticoid receptor	Srinivasan and Thompson, 1990; Tsai et al., 1990; Allan et al., 1991
	Granulocyte-macrophage colony stimulating factor (GM-CSF)	Chiou and Wu, 1990
	Haptoglobin	Heinderyckx et al., 1989
	Ha-*ras* transforming protein	Page et al., 1989; Lowe et al., 1990
	hst-1 transforming protein	Miyagawa et al., 1991; Miyagawa et al., 1988
	Immune activation gene Act-2	Lipes et al., 1988
	α-interferon	Maeda et al., 1984; 1985; Horiuchi et al., 1987; Maeda, 1987
	β-interferon	Smith et al., 1983b
	Interleukin-2	Smith et al., 1985
	Interleukin-5	Ingley et al., 1991
	Interleukin-6	Matsuura et al., 1991b; May et al., 1991
	Insulinlike growth factor 2	Marumoto et al., 1987
	Insulin receptor (cytoplasmic domain)	Ellis et al., 1988; Herrera et al., 1988; Cobb et al., 1989; Villalba et al., 1989
	Insulin receptor (extracellular domain)	Sissom and Ellis, 1989; 1991; Paul et al., 1990; Schaefer et al., 1990
	Insulin receptor (tyrosine kinase domain)	Kallen et al., 1990; Ellis et al., 1991; Levine et al., 1991; Tavare et al., 1991
	Kirsten-ras (4B) p21 protein	Lowe et al., 1991
	Ku autoantigen	Allaway et al., 1990
	5-lipoxygenase	Funk et al., 1989; Denis et al., 1991; Percival, 1991
	Lymphocyte activation gene lag-1	Baixeras et al., 1990
	Lymphocyte-specific protein-tyrosine kinase (p56lck)	Carrera et al., 1991; Ramer et al., 1991

Origin	Gene Product	References
	MHC class I HLA-B27 antigen	Lévy and Kvist, 1990
	β_2-micoglobulin	Lévy and Kvist, 1990
	Mineralocorticoid receptor	Alnemri et al., 1991b
	Multidrug transporter	Germann et al., 1990
	Multidrug-resistant gene product (MDRI P-glycoprotein)	Cenciarelli et al., 1991
	Muscarinic cholingeric receptor (MAChR) m1 and m2 forms	Parker et al., 1991
	Myelin-associated glycoprotein (soluble extracellular domain)	Johnson et al., 1989
	c-*myc* gene product	Miyamoto et al., 1985; Naoe et al., 1989
	Myogenic determination factor	Braun et al., 1991
	β-nerve growth factor	Barnett et al., 1990; 1991; Buxser et al., 1991
	Nerve growth-factor receptor (extracellular domain) (NGF-R)	Vissavajjhala and Ross, 1990
	p53 cellular phosphoprotein	Harley et al., 1990; Kraiss et al., 1990; Bargonetti et al., 1991
	P210 BCR-ABL oncogene product	Pendergast et al., 1989
	Platelet-derived growth factor (PDGF) β-receptor	Morrison et al., 1989; 1990
	Plasminogen	Whitefleet-Smith et al., 1989; Davidson et al., 1990; 1991; Davidson and Castellino, 1991a and c
	Protein kinase C	Patel et al., 1988
	Protein kinase C, α, βII and γ	Burns et al., 1990
	Raf-1 protein	Morrison et al., 1989; 1990
	rap1A/K*rev*-1 gene product	Quilliam et al., 1990; Buss et al., 1991
	Retinoblastoma gene product ($pp110^{RB}$)	Lin et al., 1991; Wang et al., 1990
	Na^+/H^+ antiporter protein	Fafournoux et al., 1991
	c-*src* tyrosine kinase	Abdel-Ghany et al., 1990; Morrison et al., 1990; Piwnica-Worms et al., 1990
	Tau microtubule-associated protein	Knops et al., 1991b
	T-cell growth factor	Renauld et al., 1990
	T-cell protein-tyrosine-phosphatase (TC.PTPase)	Zander et al., 1991
	Terminal transferase	Chang et al., 1988
	Thyroid hormone receptor β1	Barkhem et al., 1991; zu Putlitz et al., 1991
	Tissue plasminogen activator	Furlong et al., 1988; Luckow and Summers, 1988; Steiner et al., 1988; Jarvis and Summers, 1989; Jarvis et al., 1990
	Transferrin receptor	Domingo and Trowbridge, 1988; White et al., 1990
	Tumor necrosis factor receptor	Loetscher et al., 1990

Origin	Gene Product	References
	Tyrosine hydroxylase	Ginns et al., 1988
	Urokinase-type plasminogen activator (single-chain, scu-PA)	Devlin et al., 1989
	Vitamin-D receptor	Macdonald et al., 1991; Ross et al., 1991
Mouse	Adipsin	Rosen et al., 1989
	Basement membrane glycoprotein (entactin)	Tsao et al., 1990
	Ca^{2+}/calmodulin dependent kinase II (CaMKII-α)	Brickey et al., 1990
	Egr-1 transcription factor	Ragona et al., 1991
	c-*ets*-1 protooncogene	Chen, 1990
	Immunoglobulins	Hasemann and Capra, 1990; 1991; zu Putlitz et al., 1990; Laroche et al., 1991
	Interferon gamma receptor (extracellular domain)	Fountoulakis et al., 1991
	Interleukin-3 (IL-3)	Miyajima et al., 1987; Svoboda et al., 1991
	Interleukin-4 (IL-4)	Rodewald et al., 1990
	Interleukin-5 (IL-5)	Tavernier et al., 1989; Rodewald et al., 1990; Kunimoto et al., 1991
	p53	O'Reilly and Miller, 1988
	Protein kinase C-e	Schaap and Parker, 1990
	Transcription factor S-II	Hirashima et al., 1989
Rabbit	Cytochrome P450 (P450F1)	Ohta et al., 1991
	Protein phosphatase 1	Berndt and Cohen, 1990
Rat	Atrial natriuretic peptide receptor guanylyl cyclase (GC-A)	Chinkers et al., 1991
	Cytochrome P450 IIA1	Asseffa et al., 1989
	Glucocorticoid receptor	Alnemri et al., 1991a
	Ornithine transcarbamylase	Lithgow et al., 1991
	Muscarinic acetylcholine receptor subtype M3	Vasudevan et al., 1991
	Protein kinase C-βII	Flint et al., 1990
	Protein kinase C-γ	Fiebich et al., 1990
	Tyrosine hydroxylase	Fitzpatrick et al., 1990; Fitzpatrick, 1991
	Yb_1 glutathione-S-transferase	Hsieh et al., 1989

APPENDIX 4: SUPPLIERS

Applikon, Inc.
1165 Chess Drive
Foster City, CA 94404
USA
Tel: (415) 578-1396
Fax: (415) 578-8836

In Europe:
Applikon Dependable Instruments B.V.
P.O. Box 149
NL-3100 AC Schiedam
The Netherlands
Tel: 0 10 462-1855
Fax: 0 10 437-9648

J. T. Baker Chemical Co.
222 Red School Lane
Phillipsburg, NJ 08865
USA
Tel: (800) 582-2537
Fax: (908) 859-9318

In UK:
J. T. Baker
P.O. Box 9, Hayes Gate House
Uxbridge Road
Hayes, Middx., UB4 0JD
United Kingdom
Tel: (081) 569-1191
Fax: (081) 509-1105

Beckman Instruments, Inc.
2500 Harbor Boulevard
Fullerton, CA 92634
USA
Tel: (800) 742-2345
Fax: (714) 773-8898

In UK:
Beckman Instruments (United Kingdom) Ltd.
Progress Road, Sands Industrial Estate
High Wycombe, Bucks., HP12 4JL
United Kingdom
Tel: (0494) 441181
Fax: (0494) 447558

Bellco Glass, Inc.
P.O. Box B, 340 Edrudo Road
Vineland, NJ 08360
USA
Tel: (800) 257-7043
Fax: (609) 691-3247

Bio-Rad Laboratories
3300 Regatta Boulevard
Richmond, CA 94804
USA
Tel: (800) 227-5589
Fax: (415) 232-4257

In UK:
Bio-Rad Laboratories Ltd.
Bio-Rad House, Maylands Avenue
Hemel Hempstead, Hert., HP2 7TD
United Kingdom
Tel: (0442) 232552
Fax: (0442) 259118

Bio-Serv
P.O. Box 450
Frenchtown, NJ 08825
USA
Tel: (800) 473-2155

Boehringer Mannheim Biochemicals
P.O. Box 50414
Indianapolis, IN 46250
USA
Tel: (800) 262-1640
Fax: (317) 576-2754

In Europe:
Boehringer Mannheim GmBH
Biochemica
P.O. Box 320120
D-6800 Mannheim 31
Germany
Tel: 0612 7591

BRL: See GIBCO-BRL

Cambridge Valve and Fitting
50 Manning Road
Billerica, MA 01821
USA
Tel: (617) 272-8270
Fax: (508) 667-5261

Carolina Biological Supply Co.
2700 York Road
Burlington, NC 27215
USA
Tel: (800) 334-5551
Fax: (919) 584-3399

Cole-Parmer Instrument Co.
7425 North Oak Park Avenue
Chicago, IL 60648
USA
Tel: (800) 323-4340
Fax: (708) 647-9660

Corning Science Products
MP-21-5
Corning, NY 14831
USA
Tel: (607) 737-1667

Costar Corp.
One Alewife Center
Cambridge, MA 02140
USA
Tel: (800) 492-1110
Fax: (617) 868-2076

Difco Laboratories
P.O. Box 331058
Detroit, MI 48232
USA
Tel: (313) 961-0800

Fisher Scientific
52 Fadem Road
Springfield, NJ 07081
USA
Tel: (800) 766-7000
Fax: (800) 926-1166

Flow: See ICN-Flow

Fluka Chemical Corp.
980 South 2nd Street
Ronkonkoma, NY 11779
USA
Tel: (800) 358-5287
Fax: (516) 467-0663

In Europe:
Fluka Chemie AG
Industriestrasse 25
CH 9470 Buchs SG
Switzerland
Tel: (085) 60275
Fax: (085) 65449

FMC Bioproducts
5 Maple Street
Rockland, ME 04841
USA
Tel: (800) 431-1754
Fax: (207) 594-3391

Gelman Sciences, Inc.
600 South Wagner Road
Ann Arbor, MI 48106
USA
Tel: (800) 521-1520

In UK:
Gelman Sciences, Inc.
Caswell Road, Brackmills Business Park
Northampton, NN4 OEZ
United Kingdom
Tel: (0604) 765141
Fax: (0604) 761383

GIBCO-BRL
P.O. Box 6009
Gaithersburg, MD 20877
USA
Tel: (800) 874-4266
Fax: (800) 331-2286

In Europe:
GIBCO-BRL
Drongensesteenweg 28
9000 Gent
Belgium
Tel: (091) 277031
Fax: (091) 277525

Gilson Medical Electronics, Inc.
P.O. Box 27
Middleton, WI 53562
USA
Tel: (800) 445-7661
Fax: (608) 831-4451

Gold Biotechnology
7166 Manche
St. Louis, MI 63143
USA
Tel: (800) 248-7609

ICN-Flow
3300 Hyland Avenue
Costa Mesa, CA 92626
USA
Tel: (800) 854-0530
Fax: (800) 334-6999

In UK:
ICN Biomedicals Ltd.
Eagle House, Peregrine Business Park
Gomm Road
High Wycombe, Bucks., HP13 7DL
United Kingdom
Tel: (0494) 443826
Fax: (0494) 473162

Ingold Electrodes, Inc.
261 Ballardvale Street
Wilmington, MA 01887
USA
Tel: (800) 352-8763
Fax: (508) 658-6973

In Europe:
Ingold Messtechnik AG
Industrie Nord
CH-8902 Urdorf
Switzerland
Tel: (01) 736-2211
Fax: (01) 736-2636

Invitrogen
3985B Sorrento Valley Boulevard
San Diego, CA 92121
USA
Tel: (800) 955-6288
Fax: (619) 597-6200

JRH Biosciences
P.O. Box 14848
Leneva, KS 66215
USA
Tel: (800) 255-6032
Fax: (800) 441-1561

LH Fermentation,
Porton Instruments, Inc.
3942 Trust Way
Hayward, CA 94545
USA
Tel: (415) 786-0224
Fax: (415) 786-0229

In UK:
LH Fermentation Ltd.
Porton House, Vanwall Rd, M
Maidenhead, Berks., SL6 4UB
United Kingdom
Tel: (0628) 771471
Fax: (0628) 77008

Miles, Inc.
400 Morgan Lane
West Haven, CT 06516
USA
Tel: (203) 498-6521
Fax: (203) 498-6585

Millipore Corp.
80 Ashby Road
Bedford, MA 01730
USA
Tel: (800) 225-1380
Fax: (617) 275-8200

In UK:
Millipore (UK) Ltd.
The Boulevard,
Blackmoor Lane
Watford, Hert., WD1 2RA
United Kingdom
Tel: (0923) 816375
Fax: (0923) 818297

Nalge Co.
P.O. Box 20365
Rochester, NY 14601
USA
Tel: (716) 586-8800
Fax: (716) 586-8431

New England Biolabs
32 Tozer Road
Beverly, MA 01915
USA
Tel: (800) 632-5277
Fax: (508) 921-1350

In Europe:
New England Biolabs GmBH
Postfach 2750
6231 Schwalbach, Taunus
Germany
Tel: 06196 3031
Fax: 06196 83639

New England Nuclear (NEN)
549 Albany Street
Boston, MA 02118
USA
Tel: (800) 551-2121
Fax: (617) 482-1380

Nihon Nosan Kogyo Co. Ltd.
20-4 Hinode 2-Chome
Funabashi, Chiba 273
Japan
Fax: (81)-473-33-7680

Perkin Elmer-Cetus
761 Main Avenue
Norwalk, CT 06859
USA
Tel: (800) 762-4001
Fax: (203) 762-6000

Pharmacia LKB Biotechnology, Inc.
800 Centennial Avenue
Piscataway, NJ 08854
USA
Tel: (800) 526-3593
Fax: (201) 457-8643

In Europe:
Pharmacia LKB Biotechnology AB
Bjorkgatan 30
Uppsala, Sweden
Tel: 46-18-163000
Fax: 46-18-143820

PharMingen
11555 Sorrento Valley Road
Suite E
San Diego, CA 92121
USA
Tel: (800) 848-6227
(619) 792-5730
Fax: (619) 792-5238

Promega Biotec
2800 Woods Hollow Road
Madison, WI 53711
USA
Tel: (800) 356-9526
Fax: (608) 273-6967

Schleicher and Schuell, Inc.
10 Optical Avenue
Keene, NH 03431
USA
Tel: (800) 245-4024
Fax: (603) 357-3627

Scientific Products
Baxter Healthcare Corp.
1430 Waukegan Road
McGaw Park, IL 60085
USA
Tel: (800) 633-7369

Sigma Chemical Co.
P.O. Box 14509
St. Louis, MO 63718
USA
Tel: (800) 325-3010
Fax: (800) 325-5052

In UK:
Sigma Chemical Co.
Fancy Road
Poole, Dorset, BH17 7NH
United Kingdom
Tel: (0202) 733114

Spectrum Medical Industries
60916 Terminal Annex
Los Angeles, CA 90060
USA
Tel: (800) 634-3300
Fax: (213) 650-8134

United States Biochemical Corp. (USB)
P.O. Box 22400
Cleveland, OH 44122
USA
Tel: (800) 321-9322
Fax: (216) 464-5075

VWR Scientific
P.O. Box 7900
San Francisco, CA 94120
USA
Tel: (415) 468-7150
Fax: (415) 330-4185

In Europe:
VWR Scientific
Stationsstrasse 622
8606 Greifensee
Switzerland
Tel: 1-941-5440
Fax: 1-941-5650

Whatman, Inc.
9 Bridewell Place
Clifton, NJ 07014
USA
Tel: (800) 922-0361
Fax: (201) 472-6949

In UK:
Whatman Biosystems, Inc.
Springfield Mill
Maidstone, Kent, ME14 2LE
United Kingdom
Tel: (0622) 692022
Fax: (0622) 691425

Wheaton
1000 North Tenth Street
Millville, NJ 08332
USA
Tel: (609) 825-1100
Fax: (609) 825-1368

APPENDIX 5: STOCK SOLUTIONS

Chloroform-IAA:	Chloroform:isoamylalcohol (24:1).
50X Denhardt's reagent:	1% (w/v) Ficoll (Type 400, Pharmacia # 170400-01) 1% (w/v) polyvinylpyrrolidone (Sigma # P 5288) 1% (w/v) bovine serum albumin (Sigma # A 7030) in ddH_2O. Filter sterilize and store at –20° C. Alternatively, Denhardt's reagent is available commercially (e.g., Sigma # D 2532).
DEPC-treated water:	Add DEPC (Sigma # D 5758) to ddH_2O at a final concentration of 0.1% (v/v). Mix well and let stand at room temperature overnight. Autoclave for at least two hours to remove the DEPC.
0.5 M EDTA:	Prepare a 0.5 M solution of the disodium salt of EDTA (Sigma # E 5134). Adjust the pH to 8.0 with NaOH. The EDTA will not dissolve until the pH approaches 8.0.
Ethidium bromide stock:	10 mg/ml ethidium bromide (Sigma # E 7637) in ddH_2O. Store at room temperature in a dark bottle. **Caution:** Ethidium bromide is moderately toxic and a powerful mutagen. Wear gloves when handling reagents containing this dye.
Leupeptin stock:	50 µg/ml leupeptin (Fluka # 62070) in ddH_2O. Store at –20° C.
5X loading buffer:	0.25% Bromophenol Blue (Bio-Rad # 161-0404) 0.25% Xylene Cyanole FF (Bio-Rad # 161-0423) 15% Ficoll (Pharmacia # 17-0400-01). Store at room temperature.
NP40 Lysis buffer:	1% NP40 150 mM NaCl 50 mM Tris-HCl, pH 8.0. Store at 4° C.
PBS, pH 6.2:	1 mM $Na_2HPO_4 \cdot 7H_2O$ 10.5 mM KH_2PO_4 140 mM NaCl 40 mM KCl pH 6.2

Pepstatin A stock:	100 µg/ml in pepstatin A (Fluka # 77170) in methanol. Store at –20° C.
Phenol:	Prepare liquefied phenol (Baker # JT2858-1) containing 0.1% (w/v) 8-hydroxyquinoline (Sigma # H 6878). Extract twice with an equal volume of 1 M Tris-HCl, pH 8.0, and once with an equal volume of TE. Store under an equal volume of TE at 4° C in a dark bottle. **Caution:** Phenol is highly corrosive and causes severe burns. Wear protective clothing and safety glasses. In case of contact with skin, wash the affected area with copious amounts of water.
Phenol-chloroform-IAA:	TE-saturated phenol:chloroform:isoamylalcohol (25:24:1). Store under TE at 4° C in a dark bottle.
PMSF stock:	10 mg/ml PMSF (Phenylmethylsulfonyl fluoride, Fluka # 78830) in isopropanol. Store at room temperature. **Caution:** PMSF is very harmful and may be fatal if inhaled, swallowed, or by skin contact. Wear protective clothing and a mask when handling PMSF.
2X SDS gel loading buffer:	4% SDS 125 mM Tris-HCl, pH 6.7 30% (v/v) glycerol 0.002% (w/v) bromophenol blue. Store at 4° C in small aliquots. Before use, add β-mercaptoethanol to a concentration of 2% (v/v). Use each aliquot within two to three 3 weeks of adding the β-mercaptoethanol.
10X SDS gel running buffer:	250 mM Tris base 1.92 M glycine 1% SDS pH 8.3
20X SSC:	3 M NaCl 0.3 M sodium citrate, pH 7.0. Store at room temperature. 20X SSC is also available commercially (e.g., Sigma # S 6630).
10X TAE:	400 mM Tris-acetate, pH 8.0 10 mM EDTA. Store at room temperature.
5X TBE:	450 mM Tris-borate, pH 8.0 10 mM EDTA. Store at room temperature.
TE:	10 mM Tris-HCl, pH 8.0 1 mM EDTA
Virus disruption buffer:	10 mM Tris-HCl, pH 7.6 10 mM EDTA 0.25% SDS
X-gal stock:	20 mg/ml 5-bromo-4-chloro-3-indolyl-β-D-galactopyranoside (X-gal; Gold Biotechnology # X4281C) in dimethylformamide (Sigma # D 8654). Store at –20° C in the dark in a glass container.

APPENDIX 6: EXCEL SPREADSHEET FOR $TCID_{50}$ CALCULATION

Panel A shows a sample Excel spreadsheet for $TCID_{50}$ calculation by the method of Reed and Muench (1938). The data in italics, in boxes B5–B11, will be entered after each experiment. In this case, the data used in the example in Chapter 12 are shown. Panel B shows the results after the spreadsheet routine has executed using the data.

A

	A	B	C	D	E	F
1						
2			Number of	Total	Total	
3		Number of	Uninfected	Number	Number	%
4	Dilutions	Infected Wells	Wells	Infected	Uninfected	Total Infected
5	0.00001	*12*	=B10 - B5	=SUM(B5:B8)	=SUM(C5:C5)	=D5/(D5+E5)
6	0.000001	*8*	=B10 - B6	=SUM(B6:B8)	=SUM(C5:C6)	=D6/(D6+E6)
7	0.0000001	*1*	=B10 - B7	=SUM(B7:B8)	=SUM(C5:C7)	=D7/(D7+E7)
8	0.00000001	*0*	=B10 - B8	=SUM(B8:B8)	=SUM(C5:C8)	=D8/(D8+E8)
9						
10	**Num. Wells**	*12*				
11	**mls/well**	*0.01*				
12						
13	**Prop. Dist.**	=(H10-.05)/(H10-I10)				
14	**Log TCID**	=J10-B13				
15	**TCID50**	=10^B14				
16	**1/TCID50**	=1/B15				
17	**TCID50/ml**	**=B16/B11**				
18						
19	**pfu/ml**	**=B17*0.69**				

A *(continued)*

	G	H	I	J
1				
2		%	%	Log
3	Above	Above	Below	Dilution
4	0.5	0.5	0.5	Above 50%
5	=IF(F5>0.5,"TRUE","FALSE")	=IF(AND(G5="TRUE",G6="FALSE"),F5,0)	=IF(AND(G4="TRUE",G5="FALSE"),F5,0)	=IF(H5>0,LOG(A5),0)
6	=IF(F6>0.5,"TRUE","FALSE")	=IF(AND(G6="TRUE",G7="FALSE"),F6,0)	=IF(AND(G5="TRUE",G6="FALSE"),F6,0)	=IF(H6>0,LOG(A6),0)
7	=IF(F7>0.5,"TRUE","FALSE")	=IF(AND(G7="TRUE",G8="FALSE"),F7,0)	=IF(AND(G6="TRUE",G7="FALSE"),F7,0)	=IF(H7>0,LOG(A7),0)
8	=IF(F8>0.5,"TRUE","FALSE")	=IF(AND(G8="TRUE",G9="FALSE"),F8,0)	=IF(AND(G7="TRUE",G8="FALSE"),F8,0)	=IF(H8>0,LOG(A8),0)
9				
10		=SUM (H5:H8)	=SUM(I5:I8)	=SUM(J5:J8)
11				
12				
13				
14				
15				
16				
17				
18				
19				

B

	A	B	C	D	E	F	G	H	I	J
1										
2			Number of	Total	Total	%		%	%	Log
3		Number of	Uninfected	Number	Number	Total	Above	Above	Below	Dilution
4	Dilutions	Infected Wells	Wells	Infected	Uninfected	Infected	50%	50%	50%	Above 50%
5	1.00E-05	12	0	21	0	100.00	TRUE	0.00	0.00	0
6	1.00E-06	8	4	9	4	69.23	TRUE	69.23	0.00	-6
7	1.00E-07	1	11	1	15	6.30	FALSE	0.00	6.30	0
8	1.00E-08	0	12	0	27	0.00	FALSE	0.00	0.00	0
9										
10	**Num. Wells**	12						69.23	6.30	-6
11	**mls/well**	0.01								
12										
13	**Prop. Dist.**	0.305								
14	**Log TCID**	-6.305								
15	**TCID50**	4.95E-07								
16	**1/TCID50**	2.02E+06								
17	**TCID50/ml**	**2.02E+08**								
18										
19	**pfu/ml**	**1.39E+08**								

REFERENCES

Abdel-Ghany, M., El-Gendy, K., Zhang, S., and Racker, E. 1990. Control of src kinase activity by activators, inhibitors, and substrate chaperones. *Proc. Natl. Acad. Sci. U.S.A.* **87:** 7061–7065.

Adams, J. R., Goodwin, R. H., and Wilcox, T. A. 1977. Electron microscopic investigations on invasion and replication of insect baculoviruses *in vivo* and *in vitro*. *Biologie Cellulaire* **28:** 261–268.

Adang, M. J., and Miller, L. K. 1982. Molecular cloning of DNA complementary to messenger RNA of the baculovirus *Autographa californica* nuclear polyhedrosis virus: location and gene products of RNA transcripts found late in infection. *J. Virol.* **44:** 782–793.

Aira, Y. 1987. Expression of envelope glycoprotein E of Japanese encephalitis virus using *Bombyx mori* nuclear polyhedrosis virus. *Trop. Med.* **29:** 195–210.

Alcover, A., Chang, H. C., Sayre, P. H., Hussey, R. E., and Reinherz, E. L. 1988. The T11 CD2 complementary DNA encodes a transmembrane protein which expresses T111, T112, and T113 epitopes but which does not independently mediate calcium influx. Analysis by gene transfer in a baculovirus system. *Eur. J. Immunol.* **18:** 363–368.

Allan, G. F., Ing, N. H., Tsai, S. Y., Srinivansan, G., Weigel, N. L., Thompson, E. B., Tsai, M. J., and O'Malley, B. W. 1991. Synergism between steroid response and promoter elements during cell-free transcription. *J. Biol. Chem.* **266:** 5905–5910.

Allaway, G. P., Vivino, A. A., Kohn, L. D., Notkins, A. L., and Prabhakar, B. S. 1990. Characterization of the 70kDa component of the human Ku autoantigen expressed in insect cell nuclei using a recombinant baculovirus vector. *Biochem. Biophys. Res. Commun.* **168:** 747–755.

Allegretto, E. A., Smeal, T., Angel, P., Spiegelman, B. M., and Karin, M. 1990. DNA-binding activity of Jun is increased through its interaction with Fos. *J. Cell. Biochem.* **42:** 193–206.

Alnemri, E. S., Maksymowych, A. B., Robertson, N. M., and Litwack, G. 1991a. Characterization and purification of a functional rat glucocorticoid receptor overexpressed in a baculovirus system. *J. Biol. Chem.* **266:** 3925–3936.

Alnemri, E. S., Maksymowych, A. B., Robertson, N. M., and Litwack, G. 1991b. Overexpression and characterization of the human mineralocorticoid receptor. *J. Biol. Chem.* **266:** 18072–18081.

Andrews, D. L., Beames, B., Summers, M. D., and Park, W. D. 1988. Characterization of the lipid acyl hydrolase activity of the major potato *Solanum tuberosum* tuber protein patatin by cloning and abundant expression in a baculovirus vector. *Biochem. J.* **252:** 199–206.

Armstrong, C. M., and Miller, C. 1990. Do voltage-dependent K^+ channels require Ca^{+2}? A critical test employing a heterologous expression system. *Proc. Natl. Acad. Sci. U.S.A.* **87:** 7579–7582.

Asseffa, A., Smith, S. J., Nagata, K., Gillette, J., Gelboin, H. V., and Gonzalez, F. J. 1989. Novel exogenous heme-dependent expression of mammalian cytochrome P450 using baculovirus. *Arch. Biochem. Biophys.* **274:** 481–490.

Atkinson, A. E., Earley, F. G. P., Beadle, D. J., and King, L. A. 1990. Expression and characterization of the chick nicotinic acetylcholine receptor α-subunit in insect cells using a baculovirus vector. *Eur. J. Biochem.* **192:** 451–458.

Au, K.-S., Chan, W.-K., Burns, J. W., and Estes, M. K. 1989. Receptor activity of rotavirus nonstructural glycoprotein NS28. *J. Virol.* **63:** 4553–4562.

Bachmair, A., and Varshavsky, A. 1989. The degradation signal in a short-lived protein. *Cell* **56:** 1019–1032.

Bailey, M. J., McLeod, D. A., Kang, C.-Y., and Bishop, D. H. L. 1989. Glycosylation is not required for the fusion activity of the G protein of vesicular stomatitis virus in insect cells. *Virology* **169:** 322–331.

Baixeras, E., Roman-Roman, S., Jitsukawa, S., Genevee, C., Mechiche, S., Viegas-Pequignot, E., Hercend, T., and Triebel, F. 1990. Cloning and expression of a lymphocyte activation gene lag-1. *Mol. Immunol.* **27:** 1091–1102.

Barber, G. N., Clegg, J. C. S., and Lloyd, G. 1990. Expression of the Lassa virus nucleocapsid protein in insect cells infected with a recombinant baculovirus: application to diagnostic assays for Lassa virus infection. *J. Gen. Virol.* **71:** 19–28.

Bargonetti, J., Fiedman, P. N., Kern, S. E., Vogelstein, B., and Prives, C. 1991. Wild-type but not mutant p53 immunopurified proteins bind to sequences adjacent to the SV-40 origin of replication. *Cell* **65:** 1083–1092.

Barkhem, T., Carlsson, B., Simons, J., Moller, B., Berkenstam, A., Gustafsson, J. A., and Nilsson, S. 1991. High level expression of functional full length human thyroid hormone receptor β1 in insect cells using a recombinant baculovirus. *J. Steroid Biochem. Mol. Biol.* **38:** 667–675.

Barnett, J., Baecker, P., Routledgeward, C., Bursztyn-Pettegrew, H., Chow, J., Nguyen, B., Bach, C., Chan, H., Tuszynski, M. H., Yoshida, K., Rubalcava, R., and Gage, F. H. 1990. Human β nerve growth factor obtained from a baculovirus expression system has potent *in vitro* and *in vivo* neurotrophic activity. *Exp. Neurol.* **110:** 11–24.

Barnett, J., Chow, J., Nguyen, B., Eggers, D., Osen, E., Jarnagin, K., Saldou, N., Straub, K., Gu, L., Erdos, L., Chaing, H. S., Fausnaugh, J., Townsend, R. R., Lile, J., Collins, F., and Chan, H. 1991. Physicochemical characterization of recombinant human nerve growth factor produced in insect cells with a baculovirus vector. *J. Neurochem.* **57:** 1052–1061.

Bassemir, U., Miltenburger, H., and David, P. 1983. Morphogenesis of nuclear polyhedrosis virus from *Autographa californica* in a cell line from *Mamestra brassicae* (cabbage moth). *Cell Tissue Res.* **228:** 587–595.

Baum, J. A., Geever, R., and Giles, N. H. 1987. Expression of qa-1f activator protein: identification of upstream binding sites in the *qa* gene cluster and localization of the DNA-binding domain. *Mol. Cell. Biol.* **7:** 1256–1266.

Baylis, S. A., Purifoy, D. J. M., and Littler, E. 1991. High-level expression of the Epstein-Barr virus alkaline deoxyribonuclease using a recombinant baculovirus: application to the diagnosis of nasopharyngeal carcinoma. *Virology* **181:** 390–394.

Beames, B., Braunagel, S., Summers, M. D., and Lanford, R. E. 1991. Polyhedron initiator codon altered to AUU yields unexpected fusion protein from a baculovirus vector. *Biotechniques* **11:** 378–383.

Beames, B., and Summers, M. D. 1989. Location and nucleotide sequence of the 25k protein missing from baculovirus few polyhedra (FP) mutants. *Virology* **168:** 344–353.

Belsham, G. J., Abrams, C. C., King, A. M. Q., Roosien, J., and Vlak, J. M. 1991. Myristoylation of foot-and-mouth disease virus capsid protein precursors is independent of other viral proteins and occurs in both mammalian and insect cells. *J. Gen. Virol.* **72:** 747–751.

Benatti, L., Scacheri, E., Bishop, D. H. L., and Sarmientos, P. 1991. Secretion of biologically active leech hirudin from baculovirus-infected insect cells. *Gene* **101:** 255–260.

Berger, R. 1963. Laboratory techniques for rearing *Heliothis* species on artificial medium. *U.S.D.A. Pres. PAP ARS* 33–84.

Berger, S., and Kimmel, A. eds. 1987. *Guide to Molecular Cloning Techniques.* Methods in Enzymology, Vol. 152. Academic Press, Inc., San Diego, CA.

Bergh, M. L. E., Naranjo, C., Mentzer, A. F., Barsomian, G. D., Bartlett, C., Hirani, S., and Rasmussen, J. R. 1990. Selection of an expression host for human glucocerebroside: importance of host cell glycosylation. In *Glycobiology,* eds. J. K. Welply, and E. Jaworski, pp. 159–172. New York: Wiley-Liss.

Berndt, N., and Cohen, P. T. W. 1990. Renaturation of protein phosphatase 1 expressed at high levels in insect cells using a baculovirus vector. *Eur. J. Biochem.* **190:** 291–297.

Bertioli, D. J., Harris, R. D., Edwards, M. L., Cooper, J. I., and Hawes, W. S. 1991. Transgenic plants and insect cells expressing the coat protein of arabis mosaic virus produce empty virus-like particles. *J. Gen. Virol.* **72:** 1801–1809.

Bishop, D. H. L., Entwistle, P. F., Cameron, I. R., Allen, C. J., and Possee, R. D. 1988. Field trials of genetically-engineered baculovirus insecticides. In *The release of genetically-engineered micro-organisms*, eds. M. Sussman, C. H. Collins, F. A. Skinner, and D. E. Stewart-Tull, pp. 143–179. London: Academic Press.

Blake, M., Johnson, K., Russell-Jones, G., and Gotschlich, E. 1984. A rapid, sensitive method for detection of alkaline phosphatase-conjugated antibody on western blots. *Anal. Biochem.* **136:** 175–179.

Blissard, G. W., Theilmann, D. A., and Summers, M. D. 1989. Segment W of *Campoletis sonorensis* virus: expression, gene products, and organization. *Virology* **169:** 78–89.

Blitz, I. A., and Laimins, L. A. 1991. The 68-kilodalton E1 protein of bovine papilloma virus is a DNA binding phosphoprotein which associates with the E2 transcriptional activator *in vitro*. *J. Virol.* **65:** 649–656.

Bonner, W., and Laskey, R. 1974. A film detection method for tritium labelled proteins and nucleic acids in polyacrylamide gels. *Eur. J. Biochem.* **46:** 83–88.

Braun, T., Gearing, K., Wright, W. E., and Arnold, H. H. 1991. Baculovirus-expressed myogenic determination factors require E12-complex formation for binding to the myosin-light-chain enhancer. *Eur. J. Biochem.* **198:** 187–193.

Brickey, D. A., Colbran, R. J., Fong, Y. L., and Soderling, T. R. 1990. Expression and characterization of the alpha-subunit of Ca^{2+}/calmodulin-dependent protein kinase II using the baculovirus expression system. *Biochem. Biophys. Res. Commun.* **173:** 578–584.

Brideau, R. J., Walters, R. R., Stier, M. A., and Wathen, M. W. 1989. Protection of cotton rats against human respiratory synctial virus by vaccination with a novel chimeric FG glycoprotein. *J. Gen. Virol.* **70:** 2637–2644.

Brown, C. S., Salimans, M. M. M., Noteborn, M. H. M., and Weiland, H. T. 1990a. Antigenic parvovirus B19 coat proteins VP1 and VP2 produced in large quantities in a baculovirus expression system. *Virus Res.* **15:** 197–211.

Brown, C. S., Van Bussel, M. J. A. W. M., Wassenaar, A. L. M., van Elsacker-Niele, A. M. W., Weiland, H. T., and Salimans, M. M. M. 1990b. An immunofluorescence assay for the detection of parvovirus B19 IgG and IgM antibodies based on recombinant viral antigen. *J. Virol. Methods* **29:** 53–62.

Brown, C. S., Vanlent, J. W. M., Vlak, J. M., and Spaan, W. J. M. 1991. Assembly of empty capsids by using baculovirus recombinants expressing human parvovirus-B19 structural proteins. *J. Virol.* **65:** 2702–2706.

Brown, M., and Sharp, P. A. 1990. Human estrogen receptor forms multiple protein DNA complexes. *J. Biol. Chem.* **265:** 11238–11243.

Burand, J. P., Summers, M. D., and Smith, G. E. 1980. Transfection with baculovirus DNA. *Virology* **101:** 286–290.

Burgess, S. 1977. Molecular weights of Lepidopteran baculovirus DNAs: derivation by electron microscopy. *J. Gen. Virol.* **37:** 501–510.

Burkhardt, A., Willingham, M., Gay, C., Jeang, K. T., and Schlegel, R. 1989. The E5 oncoprotein of bovine papillomavirus is oriented asymmetrically in Golgi and plasma membranes. *Virology* **170:** 334–339.

Burnette, W. 1981. "Western blotting": electrophoretic transfer of proteins from sodium dodecyl sulfate-polyacrylamide gels to unmodified nitrocellulose and radiographic detection with antibody and radioiodinated protein. *Anal. Biochem.* **112:** 195–203.

Burns, D. J., Bloomenthal, J., Lee, M.-H., and Bell, R. M. 1990. Expression of the α, βII, and γ protein kinase C isozymes in the baculovirus-insect cell expression system. Purification and characterization of the individual isoforms. *J. Biol. Chem.* **265:** 12044–12051.

Burton, R. L. 1969. Mass rearing the corn earworm in the laboratory. *USDA ARS (Ser)* 33–34.

Bushman, F.D., Fujiwara, T., and Craigie, R. 1990. Retroviral DNA integration directed by HIV integration protein *in vitro*. *Science* **249:** 1555–1558.

Buss, J. E., Quilliam, L. A., Kato, K., Casey, P. J., Solski, P. A., Wong, G., Clark, R., McCormick, F., Bokoch, G. M., and Der, C. J. 1991. The carboxyl-terminal domain of the rap1a krev-1 protein is isoprenylated and supports transformation by an h-ras rap1a chimeric protein. *Mol. Cell. Biol.* **11:** 1523–1530.

Bustos, M. M., Luckow, V. A., Griffing, L. R., Summers, M. D., and Hall, T. C. 1988. Expression, glycosylation, and secretion of phaseolin in a baculovirus system. *Plant Mol. Biol.* **10:** 475–488.

Butters, T. D., and Hughes, R. C. 1981. Isolation and characterization of mosquito cell membrane glycoproteins. *Biochim. Biophys. Acta* **640:** 655–671.

Butters, T. D., Hughes, R. C., and Vischer, P. 1981. Steps in the biosynthesis of mosquito cell membrane glycoproteins and the effects of tunicamycin. *Biochim. Biophys. Acta* **640:** 672–686.

Buxser, S., Vroegop, S., Decker, D., Hinzmann, J., Poorman, R., Thomsen, D. R., Stier, M., Abraham, I., Greenberg, B. D., Hatzenbuhler, N. T., Shea, M., Curry, K. A., and Tomich, C. S. C. 1991. Single-step purification and biological activity of human nerve growth factor produced from insect cells. *J. Neurochem.* **56:** 1012–1018.

Calder, J. M., and Stow, N. D. 1990. Herpes simplex virus helicase-primase: the UL8 protein is not required for DNA-dependent ATPase and DNA helicase activities. *Nucleic Acids Res.* **18:** 3573–3578.

Caldwell, S. R., and Raushel, F. M. 1991. Detoxification of organophosphate pesticides using an immobilized phosphotriesterase from *Pseudomonas diminuta*. *Biotechnol. Bioeng.* **37:** 103–109.

Capone, J. 1989. Screening recombinant baculovirus plaques *in situ* with antibody probes. *Gene Anal. Techn.* **6:** 62–66.

Capone, J. P., and Werstuck, G. 1990. Synthesis of the herpes simplex virus type 1 *trans*-activator Vmw65 in insect cells using a baculovirus vector. *Mol. Cell. Biochem.* **94:** 45–52.

Carbonell, L. F., Hodge, M. R., Tomalski, M. D., and Miller, L. K. 1988. Synthesis of a gene coding for an insect-specific scorpion neurotoxin and attempts to express it using baculovirus vectors. *Gene* **73:** 409–418.

Carbonell, L. F., Klowden, M. J., and Miller, L. K. 1985. Baculovirus-mediated expression of bacterial genes in dipteran and mammalian cells. *J. Virol.* **56:** 153–160.

Carbonell, L. F., and Miller, L. K. 1987. Baculovirus interaction with nontarget organisms: a virus-borne reporter gene is not expressed in two mammalian cell lines. *Appl. Environ. Microbiol.* **53:** 1412–1417.

Caron, A. W., Archambault, J., and Massie, B. 1990. High-level recombinant protein production in bioreactors using the baculovirus insect cell expression system. *Biotechnol. Bioeng.* **36:** 1133–1140.

Caroni, P., Rothenfluh, A., Mcglynn, E., and Schneider, C. 1991. S cyclophilin. New member of the cyclophilin family associated with the secretory pathway. *J. Biol. Chem.* **266:** 10739–10742.

Carraway, K., and Hull, S. 1989. O-Glycosylation pathway for mucin-type glycoproteins. *Bioessays* **10:** 117–121.

Carrera, A. C., Li, P., and Roberts, T. M. 1991. Characterization of an active, non-myristylated, cytoplasmic form of the lymphoid protein tyrosine kinase pp56lck. *Int. Immunol.* **3:** 673–682.

Carson, D. D., Guarino, L. A., and Summers, M. D. 1988. Functional mapping of an AcNPV immediately-early gene which augments expression of the IE-1 trans-activated 39k gene. *Virology* **162:** 444–451.

Carson, D. D., Summers, M. D., and Guarino, L. A. 1991a. Molecular analysis of a baculovirus regulatory gene. *Virology* **182:** 279–286.

Carson, D. D., Summers, M. D., and Guarino, L. A. 1991b. Transient expression of the *Autographa californica* nuclear polyhedrosis virus immediate-early gene, IE-N, is regulated by three viral elements. *J. Virol.* **65:** 945–951.

Carstens, E. B., Tjia, S. T., and Doerfler, W. 1979. Infection of *Spodoptera frugiperda* cells with *Autographa californica* nuclear polyhedrosis virus. I. Synthesis of intracellular proteins after virus infection. *Virology* **99:** 386–398.

Carstens, E. B., Tjia, S. T., and Doerfler, W. 1980. Infectious DNA from *Autographa californica* nuclear polyhedrosis virus. *Virology* **101:** 311–314.

Cenciarelli, C., Currier, S. J., Willingham, M. C., Thiebault, F., Germann, U. A., Rutherford, A. V., Gottesman, M. M., Barca, S., Tombesi, M., and Et, A. L. 1991. Characterization by somatic cell genetics of a monoclonal antibody to the mdr1 gene product p glycoprotein: determination of p glycoprotein expression in multi-drug-resistant kb and cem cell variants. *Int. J. Cancer* **47:** 533–543.

Chamberlain, J. 1979. Fluorographic detection of radioactivity in polyacrylamide gels with the water-soluble fluor, sodium salicylate. *Anal. Biochem.* **98:** 132–135.

Chang, L. M. S., Rafter, E., Rusquet Valerius, R., Peterson, R. C., White, S. T., and Bollum, F. J. 1988. Expression and processing of recombinant human terminal transferase in the baculovirus system. *J. Biol. Chem.* **263:** 12509–12513.

Charles, I. G., Rodgers, B. C., Makoff, A. J., Chatfield, S. N., Slater, D. E., and Fairweather, N. F. 1991. Synthesis of tetanus toxin fragment-C in insect cells by use of a baculovirus expression system. *Infection. Immun.* **59:** 1627–1632.

Charlton, C. A., and Volkman, L. E. 1991. Sequential rearrangement and nuclear polymerization of actin in baculovirus-infected *Spodoptera frugiperda* cells. *J. Virol.* **65:** 1219–1227.

Chen, G., Willems, L., Portelle, D., Willard-Gallo, K. E., Burny, A., Gheysen, D., and Kettman, R. 1989. Synthesis of functional bovine leukemia virus (BLV) $p34^{tax}$ protein by recombinant baculoviruses. *Virology* **173:** 343–347.

Chen, H. F., Sauter, M., Haiss, P., and Mullerlantzsch, N. 1991a. Immunological characterization of the Epstein-Barr virus phosphoprotein pp58 and deoxyribonuclease expressed in the baculovirus expression system. *Int. J. Cancer* **48:** 879–888.

Chen, J. H. 1990. Cloning, sequencing, and expression of mouse c-*ets*-1 cDNA in baculovirus expression system. *Oncogene Research* **5:** 277–285.

Chen, W. Y., and Bahl, O. P. 1991a. Recombinant carbohydrate variant of human choriogonadotropin β-subunit (hCGβ) descarboxyl terminus (115-145)—expression and characterization of carboxyl-terminal deletion mutant of hCGβ in the baculovirus system. *J. Biol. Chem.* **266:** 6246–6251.

Chen, W. Y., and Bahl, O. P. 1991b. Recombinant carbohydrate and selenomethionyl variants of human choriogonadotropin. *J. Biol. Chem.* **266:** 8192–8197.

Chen, W. Y., and Bahl, O. P. 1991c. Selenomethionyl analog of recombinant human choriogonadotropin. *J. Biol. Chem.* **266:** 9355–9358.

Chen, W. Y., Shen, Q.-X., and Bahl, O. P. 1991b. Carbohydrate variant of the recombinant β-subunit of human choriogonadotropin expressed in baculovirus expression system. *J. Biol. Chem.* **266:** 4081–4087.

Chen, Y. R., Leesmiller, S. P., Tegtmeyer, P., and Anderson, C. W. 1991c. The human DNA-activated protein kinase phosphorylates simian virus-40 T-antigen at amino-terminal and carboxy-terminal sites. *J. Virol.* **65:** 5131–5140.

Chiba, J., Ohba, H., Matsuura, Y., Watanabe, Y., Katayama, T., Kikuchi, S., Saito, I., and Miyamura, T. 1991. Serodiagnosis of hepatitis-C virus (HCV) infection with an HCV core protein molecularly expressed by a recombinant baculovirus. *Proc. Natl. Acad. Sci. U.S.A.* **88:** 4641–4645.

Chinkers, M., Singh, S., and Garbers, D. L. 1991. Adenine nucleotides are required for activation of rat atrial natriuretic peptide receptor guanylyl cyclase expressed in a baculovirus system. *J. Biol. Chem.* **266:** 4088–4093.

Chiou, C.-J., and Wu, M.-C. 1990. Expression of human granulocyte-macrophage colony-stimulating factor gene in insect cells by a baculovirus vector. *FEBS Letters* **259:** 249–253.

Chirgwin, J., Przybyla, A., MacDonald, R., and Rutter, W. 1979. Isolation of biologically active ribonucleic acid from sources enriched in ribonuclease. *Biochemistry* **18:** 5294–5299.

Chisholm, G. E., and Henner, D. J. 1988. Multiple early transcripts and splicing of the *Autographa californica* nuclear polyhedrosis virus IE-1 gene. *J. Virol.* **62:** 3193–3200.

Cho, T., Shuler, M. L., and Granados, R. R. 1989. Current developments in new media and cell culture systems for the large-scale production of insect cells. In *Advances in Cell Culture*, eds. K. Maramorosch, and G. H. Sato, pp. 261–277. San Diego: Academic Press.

Choudary, P., Kamita, S., and Maeda, S. 1992. Expression of foreign genes in insect larvae using baculovirus vectors. In *Methods in Molecular Biology: Protocols for Baculovirus Expression*, ed. C. Richardson. Clifton, NJ: Humana Press. In press.

Chung, K. L., Brown, M., and Faulkner, P. 1980. Studies on the morphogenesis of polyhedral inclusion bodies of a baculovirus *Autographa californica* nuclear polyhedrosis virus. *J. Gen. Virol.* **46:** 335–348.

Clem, R., Fechheimer, M., and Miller, L. 1991. Prevention of apoptosis by a baculovirus gene during infection of insect cells. *Science* **254:** 1388–1390.

Cobb, M. H., Sang, B.-C., Gonzalez, R., Goldsmith, E., and Ellis, L. 1989. Autophosphorylation activates the soluble cytoplasmic domain of the insulin receptor in an intermolecular reaction. *J. Biol. Chem.* **264:** 18701–18706.

Cochran, M. A., Carstens, E. B., Eaton, B. T., and Faulkner, P. 1982. Molecular cloning and physical mapping of restriction endonuclease fragments of *Autographa californica* nuclear polyhedrosis virus DNA. *J. Virol.* **41:** 940–946.

Cochran, M. A., Ericson, B. L., Knell, J. D., and Smith, G. E. 1987. Use of baculovirus recombinants as a general method for the production of subunit vaccines. In *Vaccines 87, Modern Approaches to Vaccines Including Prevention of AIDS*, eds. R. M. Chanock, R. A. Lerner, F. Brown, and H. Ginsberg, pp. 384–388. Cold Spring Harbor, NY: Cold Spring Harbor Laboratory.

Cochran, M. A., and Faulkner, P. 1983. Location of homologous DNA sequences interspersed at 5 regions in the baculovirus *Autographa californica* nuclear polyhedrosis virus genome. *J. Virol.* **45:** 961–970.

Cohen, J., Charpilienne, A., Chilmonczyk, S., and Estes, M. K. 1989. Nucleotide sequence of bovine rotavirus gene 1 and expression of the gene product in baculovirus. *Virology* **171:** 131–140.

Cohen, B., Yoakim, M., Piwnica-Worms, H., Roberts, T. M., and Schaffhausen, B. S. 1990. Tyrosine phosphorylation is a signal for the trafficking of pp85, an 85-kDa phosphorylated polypeptide associated with phosphatidylinositol kinase activity. *Proc. Natl. Acad. Sci. U.S.A.* **87:** 4458–4462.

Cooper, P. 1967. The plaque assay of animal viruses. In *Methods in Virology*, Vol. III, eds. K. Maramorosch, and H. Koprowski, pp. 244–311. New York: Academic Press.

Crawford, A. M., and Miller, L. K. 1988. Characterization of an early gene accelerating expression of late genes of the baculovirus *Autographa californica* nuclear polyhedrosis virus. *J. Virol.* **62:** 2773–2781.

Daefler, S., Klotman, M. E., and Wong-Staal, F. 1990. Trans-activating rev protein of the human immunodeficiency virus 1 interacts directly and specifically with its target RNA. *Proc. Natl. Acad. Sci. U.S.A.* **87:** 4571–4575.

Daugherty, B. L., Zavodny, S. M., Lenny, A. B., Jacobson, M. A., Ellis, R. W., Law, S. W., and Mark, G. E. 1990. The uses of computer-aided signal peptide selection and polymerase chain reaction in gene construction and expression of secreted proteins. *DNA Cell. Biol.* **9:** 453–459.

Davidson, D. J., and Castellino, F. J. 1991a. Asparagine-linked oligosaccharide processing in Lepidopteran insect cells—temporal dependence of the nature of the oligosaccharides assembled on Asparagine-289 of recombinant human plasminogen produced in baculovirus vector infected *Spodoptera frugiperda* (IPLB-SF-21AE) cells. *Biochemistry* **30:** 6167–6174.

Davidson, D. J., and Castellino, F. J. 1991b. Oligosaccharide structures present on Asparagine-289 of recombinant human plasminogen expressed in a Chinese hamster ovary cell line. *Biochemistry* **30:** 625–633.

Davidson, D. J., and Castellino, F. J. 1991c. Structures of the Asparagine-289-linked oligosaccharides assembled on recombinant human plasminogen expressed in a *Mamestra brassicae* cell Line (IZD-MBO503). *Biochemistry* **30:** 6689–6696.

Davidson, D. J., Bretthauer, R. K., and Castellino, F. J. 1991. Alpha-mannosidase-catalyzed trimming of high-mannose glycans in noninfected and baculovirus-infected *Spodoptera frugiperda* cells (IPLB-SF-21AE)—a possible contributing regulatory mechanism for assembly of complex-type oligosaccharides in infected cells. *Biochemistry* **30:** 9811–9815.

Davidson, D. J., Fraser, M. J., and Castellino, F. J. 1990. Oligosaccharide processing in the expression of human plasminogen cDNA by lepidopteran insect (*Spodoptera frugiperda*) cells. *Biochemistry* **29:** 5584–5590.

Davis, B. 1964. Disc electrophoresis. II. Methods and application to human serum proteins. *Ann. N.Y. Acad. Sci.* **121:** 404–427.

Davis, L. G., Dibner, M. D., and Battey, J. F. eds. 1986. *Basic Methods in Molecular Biology*. New York: Elsevier.

Delchambre, M., Gheysen, D., Thines, D., Thiriart, C., Jacobs, E., Verdin, E., Horth, M., Burny, A., and Bex, F. 1989. The GAG precursor of simian immunodeficiency virus assembles into virus-like particles. *EMBO J.* **8:** 2653–2660.

Denis, D., Falgueyret, J. P., Riendeau, D., and Abramovitz, M. 1991. Characterization of the activity of purified recombinant human 5 lipoxygenase in the absence and presence of leukocyte factors. *J. Biol. Chem.* **266:** 5072–5079.

Despres, P., Girard, M., and Bouloy, M. 1991. Characterization of yellow fever virus proteins-E and NS1 expressed in Vero and *Spodoptera frugiperda* cells. *J. Gen. Virol.* **72:** 1331–1342.

Deubel, V., Bordier, M., Megret, F., Gentry, M. K., Schlesinger, J. J., and Girard, M. 1991. Processing, secretion, and immunoreactivity of carboxy terminally truncated dengue-2 virus envelope proteins expressed in insect cells by recombinant baculoviruses. *Virology* **180:** 442–447.

Devlin, J. J., Devlin, P. E., Clark, R., O'Rourke, E. C., Levenson, C., and Mark, D. F. 1989. Novel expression of chimeric plasminogen activators in insect cells. *Bio/Technology* **7:** 286–292.

Diamond, S., Eskin, S., and McIntyre, L. 1989. Fluid flow stimulates tissue plasminogen activator secretion by cultured human endothelial cells. *Science* **243:** 1483–1485.

DiSorbo, D. M., Whitford, W. G., and Weiss, S. A. 1991. Accessory reagents to enhance quantitation of baculovirus expression in serum-free invertebrate cell culture. *Focus* **13:** 16–18.

Dobos, P., and Cochran, M. A. 1980. Protein synthesis in cells infected by *Autographa californica* nuclear polyhedrosis virus (AcNPV): the effect of cytosine arabinoside. *Virology* **103:** 446–464.

Dodson, M. S., Crute, J. J., Bruckner, R. C., and Lehman, I. R. 1989. Overexpression and assembly of the herpes simplex virus type-1 helicase-primase in insect cells. *J. Biol. Chem.* **264:** 20835–20838.

Dodson, M. S., and Lehman, I. R. 1991. Association of DNA helicase and primase activities with a subassembly of the herpes simplex virus 1 helicase-primase composed of the UL5 and UL52 gene products. *Proc. Natl. Acad. Sci. U.S.A.* **88:** 1105–1109.

Dolin, R., Graham, B. S., Greenberg, S. B., Tacket, C. O., Belshe, R. B., Midthun, K., Clements, M. L., Gorse, G. J., Horgan, B. W., Atmar, R. L., Karzon, D. T., Bonnez, W., Fernie, B. F., Montefiori, D. C., Stablein, D. M., Smith, G. E., and Koff, W. C. 1991. The safety and immunogenicity of a human immunodeficiency virus type-1 (HIV-1) recombinant gp160 candidate vaccine in humans. *Ann. Intern. Med.* **114:** 119–127.

Domingo, D. L., and Trowbridge, I. S. 1988. Characterization of the human transferrin receptor produced in a baculovirus expression system. *J. Biol. Chem.* **263:** 13386–13392.

Donnelly, K. M., Tazleabas, A. F., Verhage, H. G., Mavrogianis, P. A., and Jaffe, R. C. 1991. Cloning of a recombinant complementary DNA to a baboon *Papio anubis* estradiol-dependent oviduct-specific glycoprotein. *Mol. Endocrinol.* **5:** 356–364.

Dontfraid, F., Cochran, M. A., Pombo, D., Knell, J. D., Quakyi, I. A., Kumar, S., Houghten, R. A., Berzofsky, J. A., Miller, L. H., and Good, M. F. 1988. Human and murine CD4 T cell epitopes map to the same region of the malaria circumsporozoite protein: limited immunogenicity of sporozoites and circumsporozoite protein. *Mol. Biol. Med.* **5:** 185–196.

Dougherty, R. M. 1964. Animal virus titration techniques. In *Techniques in Experimental Virology,* ed. R. Harris, pp. 169–223. New York: Academic Press.

Dulbecco, R. 1952. Production of plaques in monolayer tissue-cultures caused by single particles of an animal virus. *Proc. Natl. Acad. Sci. U.S.A.* **38:** 747–751.

Dumas, D. P., Wild, J. R., and Raushel, F. M. 1990. Expression of *Pseudomonas* phosphotriesterase activity in the fall armyworm confers resistance to insecticides. *Experientia* **46:** 729–731.

Eldridge, R., Horodyski, F. M., Morton, D. B., O'Reilly, D. R., Truman, J. W., Riddiford, L. M., and Miller, L. K. 1991. Expression of an eclosion hormone gene in insect cells using baculovirus vectors. *Insect Biochem.* **21:** 341–351.

Elliott, R. M., and McGregor, A. 1989. Nucleotide sequence and expression of the small S RNA segment of Magueri bunyavirus. *Virology* **171:** 516–524.

Ellis, L., Levitan, A., Cobb, M. H., and Ramos, P. 1988. Efficient expression in insect cells of a soluble active human insulin receptor protein-tyrosine kinase domain by use of a baculovirus vector. *J. Virol.* **62:** 1634–1639.

Ellis, L., Tavare, J. M., and Levine, B. A. 1991. Insulin receptor tyrosine kinase structure and function. *Biochem. Soc. Trans.* **19:** 426–432.

Emery, V. C., and Bishop, D. H. L. 1987. The development of multiple expression vectors for high level synthesis of eukaryotic proteins expression of LCMV-N and AcNPV polyhedrin protein by a recombinant baculovirus. *Protein Eng.* **1:** 359–366.

Erlich, H. ed. 1989. *PCR Technology—Principles and Applications for DNA Amplification.* New York: Stockton Press.

Ertl, P. F., Thomas, M. S., and Powell, K. L. 1991. High level expression of DNA polymerases from herpes viruses. *J. Gen. Virol.* **72:** 1729–1734.

Estes, M. K., Crawford, S. E., Penaranda, M. E., Petrie, B. L., Burns, J. W., Chan, W. K., Ericson, B., Smith, G. E., and Summers, M. D. 1987. Synthesis and immunogenicity of the rotavirus major capsid antigen using a baculovirus expression system. *J. Virol.* **61:** 1488–1494.

Evans, H. F. 1986. Ecology and epizootiology of baculoviruses. In *The Biology of Baculoviruses,* Vol. 2, eds. R. R. Granados and B. A. Federici, pp. 89–132. Boca Raton, Fla: CRC Press.

Fafournoux, P., Ghysdael, J., Sardet, C., and Pouyssegur, J. 1991. Functional expression of the human growth factor activatable Na^+/H^+ antiporter (NHE-1) in baculovirus-infected cells. *Biochemistry* **30:** 9510–9515.

Farmer, J. L., Hampton, R. G., and Boots, E. 1989. Flow cytometric assays for monitoring production of recombinant HIV-1 gp160 in insect cells infected with a baculovirus expression vector. *J. Virol. Meth.* **26:** 279–290.

Feinberg, A., and Vogelstein, B. 1983. A technique for radiolabeling DNA restriction endonuclease fragments to high specific activity. *Anal. Biochem.* **132:** 6–13.

Felgner, P., Gadek, T., Holm, M., Roman, R., Chan, H., Wenz, M., Northrop, J., Ringold, G., and Danielson, M. 1987. Lipofection: a highly efficient lipid-mediated DNA-transfection procedure. *Proc. Natl. Acad. Sci. U.S.A.* **84:** 7413–7417.

Felleisen, R., Beck, E., Usmany, M., Vlak, J., and Klinkert, M.-Q. 1990. Cloning and expression of *Schistosoma mansoni* protein Sm32 in a baculovirus vector. *Molec. Biochem. Parasitol.* **43:** 289–292.

Fiebich, B., Hug, H., and Marme, D. 1990. High-efficiency expression of rat protein kinase C-γ in baculovirus-infected insect cells. *FEBS Lett.* **277:** 15–18.

Finley, D., and Varshavsky, A. 1984. The ubiquitin system: functions and mechanisms. *TIBS* **10:** 343–346.

Fiore, L., Greenberg, H. B., and Mackow, E. R. 1991. The VP8 fragment of VP4 is the rhesus rotavirus hemagglutinin. *Virology* **181:** 553–563.

Fitzpatrick, P. F. 1991. Steady-state kinetic mechanism of rat tyrosine hydroxylase. *Biochemistry* **30:** 3658–3662.

Fitzpatrick, P. F., Chlumsky, L. J., Daubner, S. C., and O'Malley, K. L. 1990. Expression of rat tyrosine hydroxylase in insect tissue-culture cells and purification and characterization of the cloned enzyme. *J. Biol. Chem.* **265:** 2042–2047.

Flint, A. J., Paladini, R. D., and Koshland, D. E., Jr. 1990. Autophosphorylation of protein kinase C at three separated regions of its primary sequence. *Science* **249:** 408–411.

Forstová, J., Krauzewicz, N., and Griffin, B. E. 1989. Expression of biologically active middle T antigen of polyoma virus from recombinant baculoviruses. *Nucleic Acids Res.* **17:** 1427–1443.

Fountoulakis, M., Schlaeger, E. J., Gentz, R., Juranville, J. F., Manneberg, M., Ozmen, L., and Garotta, G. 1991. Purification and biochemical characterization of a soluble mouse interferon-γ receptor produced in insect cells. *Eur. J. Biochem.* **198:** 441–450.

Frappier, L., and O'Donnell, M. 1991. Overproduction purification and characterization of EBNA1, the origin binding protein of Epstein-Barr virus. *J. Biol. Chem.* **266:** 7819–7826.

Fraser, M. J. 1986. Ultrastructural observations of virion maturation in *Autographa californica* nuclear polyhedrosis virus infected *Spodoptera frugiperda* cell cultures. *J. Ultrastruct. Mol. Struct. Res.* **95:** 189–195.

Fraser, M. J., Smith, G. E., and Summers, M. D. 1983. Acquisition of host cell DNA sequences by baculoviruses relationship between host DNA insertions and FP mutants of *Autographa californica* nuclear polyhedrosis virus and *Galleria mellonella* nuclear polyhedrosis viruses. *J. Virol.* **47:** 287–300.

Frech, B., Zimber-Strobl, U., Suentzenich, K.-O., Pavlish, O., Lenoir, G. M., Bornkamm, G. W., and Mueller-Lantzsch, N. 1990. Identification of Epstein-Barr virus terminal protein 1 (TP1) in extracts of four lymphoid cell lines, expression in insect cells, and detection of antibodies in human sera. *J. Virol.* **64:** 2759–2767.

Freemont, P. S., Hanson, I. M., and Trowsdale, J. 1991. A novel cysteine-rich sequence motif. *Cell* **64:** 483–484.

Freisewinkel, I., Riethmacher, D., and Stabel, S. 1991. Downregulation of protein kinase C-γ is independent of a functional kinase domain. *FEBS Lett.* **280:** 262–266.

French, T. J., Inumara, S., and Roy, P. 1989. Expression of the two related nonstructural proteins of bluetongue virus (BTV) type 10 in insect cells by a recombinant baculovirus: production of polyclonal ascitic fluid and characterization of the gene product in BTV-infected BHK cells. *J. Virol.* **63:** 3270–3278.

French, T. J., Marshall, J. J. A., and Roy, P. 1990. Assembly of double-shelled, virus-like particles of bluetongue virus by the simultaneous expression of 4 structural proteins. *J. Virol.* **64:** 5695–5700.

French, T. J., and Roy, P. 1990. Synthesis of bluetongue virus (BTV) corelike particles by a recombinant baculovirus expressing the two major structural core proteins of BTV. *J. Virol.* **64:** 1530–1536.

Friesen, P. D., and Miller, L. K. 1985. Temporal regulation of baculovirus RNA overlapping early and late transcripts. *J. Virol.* **54:** 392–400.

Friesen, P. D., and Miller, L. K. 1986. The regulation of baculovirus gene expression. In *The Molecular Biology of Baculoviruses*, eds. W. Doerfler, and P. Boehm, pp. 31–50. Berlin: Springer-Verlag.

Friesen, P. D., and Miller, L. K. 1987. Divergent transcription of early 35 and 94-kilodalton protein genes encoded by the Hind-III K genome fragment of the baculovirus *Autographa californica* nuclear polyhedrosis virus. *J. Virol.* **61:** 2264–2272.

Fu, Z. F., Dietzschold, B., Schumacher, C. L., Wunner, W. H., Ertl, H. C. J., and Koprowski, H. 1991. Rabies virus nucleoprotein expressed in and purified from insect cells is efficacious as a vaccine. *Proc. Natl. Acad. Sci. U.S.A.* **88:** 2001–2005.

Fuchs, L. Y., Woods, M. S., and Weaver, R. F. 1983. Viral transcription during *Autographa californica* nuclear polyhedrosis virus infection: a novel RNA polymerase induced in infected *Spodoptera frugiperda* cells. *J. Virol.* **43:** 641–646.

Funk, C. D., Gunne, H., Steiner, H., Izumi, T., and Samuelsson, B. 1989. Native and mutant 5-lipoxygenase expression in a baculovirus/insect cell system. *Proc. Natl. Acad. Sci. U.S.A.* **86:** 2592–2596.

Furlong, A. M., Thomsen, D. R., Marotti, K. R., Post, L. E., and Sharma, S. K. 1988. Active human tissue plasminogen activator secreted from insect cells using a baculovirus vector. *Biotechnol. Appl. Biochem.* **10:** 454–464.

Gál, P., Sárvári, M., Szilágyi, K., Závodszky, P., and Schumaker, V. N. 1989. Expression of hemolytically active human complement component C1r proenzyme in insect cells using a baculovirus vector. *Complement Inflamm.* **6:** 433–441.

Gardiner, G. R., and Stockdale, H. 1975. Two tissue-culture media for production of Lepidopteran cells and nuclear polyhedrosis viruses. *J. Invertebr. Path.* **25:** 363–370.

Gearing, K. L., and Possee, R. D. 1990. Functional analysis of a 603 nucleotide open reading frame upstream of the polyhedrin gene of *Autographa californica* nuclear polyhedrosis virus. *J. Gen. Virol.* **71:** 251–262.

George, S. T., Arbabian, M. A., Ruoho, A. E., Kiely, J., and Malborn, C. C. 1989. High-efficiency expression of mammalian β-adrenergic receptors in baculovirus-infected cells. *Biochem. Biophys. Res. Commun.* **163:** 1265–1269.

Germann, U. A., Willingham, M. C., Pastan, I., and Gottesman, M. M. 1990. Expression of the human multidrug transporter in insect cells by a recombinant baculovirus. *Biochemistry* **29:** 2295–2303.

Gheysen, D., Jacobs, E., de Foresta, F., Thiriart, C., Francotte, M., Thines, D., and De Wilde, M. 1989. Assembly and release of HIV-1 precursor $Pr55^{gag}$ virus-like particles from baculovirus-infected insect cells. *Cell* **59:** 103–112.

Ghose-Dastidar, J., Ross, J. B. A., and Green, R. 1991. Expression of biologically active human corticosteroid binding globulin by insect cells—acquisition of function requires glycosylation and transport. *Proc. Natl. Acad. Sci. U.S.A.* **88:** 6408–6412.

Giese, N., May-Siroff, M., Larochelle, W. J., Van Wyke Coelingh, K., and Aaronson, S. A. 1989. Expression and purification of biologically active v-*sis*/platelet-derived growth factor β protein by using a baculovirus vector system. *J. Virol.* **63:** 3080–3086.

Gillespie, L. S., Hillesland, K. K., and Knauer, D. J. 1991. Expression of biologically active human antithrombin-III by recombinant baculovirus in *Spodoptera frugiperda* cells. *J. Biol. Chem.* **266:** 3995–4001.

Goldblum, S. D., Bae, Y.-K., Hink, W. F., and Chalmers, J. 1990. The protective effect of methylcellulose on suspended insect cells subjected to laminar shear stress. *Biotechnol. Prog.* **6:** 383–390.

Goldstein, N. I., and Mcintosh, A. H. 1980. Glycoproteins of nuclear polyhedrosis viruses. *Arch. Virol.* **64:** 119–126.

Gombart, A. F., Pearson, M. N., Rohrmann, G. F., and Beaudreau, G. S. 1989. A baculovirus polyhedral envelope-associated protein: genetic location nucleotide sequence and immunocytochemical characterization. *Virology* **169:** 182–193.

Goochee, C. F., and Monica, T. 1990. Environmental effects on protein glycosylation. *Bio/Technology* **8:** 421–427.

Good, M. F., Kumar, S., Degroot, A. S., Weiss, W. R., Quakyi, I. A., Dontfraid, F., Smith, G. E., Cochran, M., Berzofsky, J. A., and Miller, L. H. 1990. Evidence implicating MHC genes in the immunological nonresponsiveness to the *Plasmodium falciparum* CS protein. *Bull. WHO* **68:** 80–84.

Gordon, J. D., and Carstens, E. B. 1984. Phenotypic characterization and physical mapping of a temperature sensitive mutant of *Autographa californica* nuclear polyhedrosis virus defective in DNA synthesis. *Virology* **138:** 69–81.

Goswami, B. B., and Glazer, R. I. 1991. A simplified method for the production of recombinant baculovirus. *Biotechniques* **10:** 626–630.

Gottlieb, J., Marcy, A. I., Coen, D. M., and Challberg, M. D. 1990. The herpes simplex virus type 1 UL42 gene product: a subunit of DNA polymerase that functions to increase productivity. *J. Virol.* **64:** 5976–5987.

Grabowski, G. A., White, W. R., and Grace, M. E. 1989. Expression of functional human acid β-glucosidase in COS-1 and *Spodoptera frugiperda* cells. *Enzyme* **41:** 131–142.

Grace, M. E., Graves, P. N., Smith, F. I., and Grabowski, G. A. 1990. Analyses of catalytic activity and inhibitor binding of human acid β-glucosidase by site-directed mutagenesis. Identification of residues critical to catalysis and evidence for causality of two Ashkenazi Jewish Gaucher disease type 1 mutations. *J. Biol. Chem.* **265:** 6827–6835.

Grace, T. D. C. 1962. Establishment of four strains of cells from insect tissues grown *in vitro*. *Nature* **195:** 788–789.

Graham, F. L., and van der Eb, A. J. 1973. A new technique for the assay of infectivity of human adenovirus 5 DNA. *Virology* **52:** 456–467.

Granados, R., and Williams, K. 1986. *In vivo* infection and replication of baculoviruses. In *The Biology of Baculoviruses*, Vol. 1, eds. R. Granados, and B. Federici, pp. 89–108. Boca Raton, Fla: CRC Press.

Greenfield, C., Patel, G., Clark, S., Jones, N., and Waterfield, M. D. 1988. Expression of the human EGF receptor with ligand-stimulatable kinase activity in insect cells using a baculovirus vector. *EMBO J.* **7:** 139–146.

Greenfield, C., Hiles, I., Waterfield, M. D., Federwisch, M., Wollmer, A., Blundell, T. L., and McDonald, N. 1989. Epidermal growth factor binding induces a conformational change in the external binding of its receptor. *EMBO J.* **8:** 4115–4124.

Gretch, D. G., Sturley, S. L., Friesen, P. D., Beckage, N. E., and Attie, A. D. 1991. Baculovirus-mediated expression of human apolipoprotein-E in *Manduca sexta* larvae generates particles that bind to the low density lipoprotein receptor. *Proc. Natl. Acad. Sci. U.S.A.* **88:** 8530–8533.

Gröner, A. 1986. Specificity and safety of baculoviruses. In *The Biology of Baculoviruses*. Vol. 1, eds. R. R. Granados, and B. A. Federici, pp. 177–202. Boca Raton, Fla: CRC Press.

Grossman, S. R., Mora, R., and Laimins, L. A. 1989. Intracellular localization and DNA-binding properties of human papillomavirus type 18 E6 protein expressed with a baculovirus vector. *J. Virol.* **63:** 366–374.

Grossman, S. R., and Laimins, L. A. 1989. E6 protein of human papillomavirus binds zinc. *Oncogene* **4:** 1089–1094.

Grula, M. A., Buller, P. L., and Weaver, R. F. 1981. α-Amanitin-resistant viral RNA synthesis in nuclei isolated from nuclear polyhedrosis virus-infected *Heliothis zea* larvae and *Spodoptera frugiperda* cells. *J. Virol.* **38:** 916–921.

Guarino, L. A. 1990. Identification of a viral gene encoding a ubiquitin-like protein. *Proc. Natl. Acad. Sci. U.S.A.* **87:** 409–413.

Guarino, L. A., and Dong, W. 1991. Expression of an enhancer-binding protein in insect cells transfected with the *Autographa californica* nuclear polyhedrosis virus-IE1 gene. *J. Virol.* **65:** 3676–3680.

Guarino, L. A., Gonzalez, M. A., and Summers, M. D. 1986. Complete sequence and enhancer function of the homologous DNA regions of *Autographa californica* nuclear polyhedrosis virus. *J. Virol.* **60:** 224–229.

Guarino, L. A., and Smith, M. W. 1990. Nucleotide sequence and characterization of the 39K gene region of *Autographa californica* nuclear polyhedrosis virus. *Virology* **179:** 1–8.

Guarino, L. A., and Summers, M. D. 1986a. Functional mapping of a trans-activating gene required for expression of a baculovirus delayed-early gene. *J. Virol.* **57:** 563–571.

Guarino, L. A., and Summers, M. D. 1986b. Interspersed homologous DNA of *Autographa californica* nuclear polyhedrosis virus enhances delayed-early gene expression. *J. Virol.* **60:** 215–223.

Guarino, L. A., and Summers, M. D. 1987. Nucleotide sequence and temporal expression of a baculovirus regulatory gene. *J. Virol.* **61:** 2091–2099.

Guarino, L. A., and Summers, M. D. 1988. Functional mapping of *Autographa californica* nuclear polyhedrosis virus genes required for late gene expression. *J. Virol.* **62:** 463–471.

Gunne, H., Hellers, M., and Steiner, H. 1990. Structure of preproattacin and its processing in insect cells infected with a recombinant baculovirus. *Eur. J. Biochem.* **187:** 699–703.

Hall, R., and Moyer, R. 1991. Identification, cloning, and sequencing of a fragment of *Amsacta moorei* entomopoxvirus DNA containing the spheroidin gene and three vaccinia virus-related open reading frames. *J. Virol.* **65:** 6516–6527.

Hall, S. L., Murphy, B. R., and Coelingh, K. L. V. 1991. Protection of cotton rats by immunization with the human parainfluenza virus type-3 fusion (F) glycoprotein expressed on the surface of insect cells infected with a recombinant baculovirus. *Vaccine* **9:** 659–667.

Hammock, B. D., Bonning, B. C., Possee, R. D., Hanzlik, T. N., and Maeda, S. 1990. Expression and effects of the juvenile hormone esterase in a baculovirus vector. *Nature* **344:** 458–461.

Han, J. H., Law, S. W., Keller, P. M., Kniskern, P. J., Silberklang, M., Tung, J. S., Gasic, T. B., Gasic, G. J., Friedman, P. A., and Ellis, R. W. 1989. Cloning and expression of complementary DNA encoding antistasin, a leech-derived protein having anti-coagulant and anti-metastatic properties. *Gene* **75:** 47–58.

Handa-Corrigan, A., Emery, A., and Spier, R. 1989. Effects of gas-liquid interfaces on the growth of suspended mammalian cells: mechanisms of cell damage by bubbles. *Enzyme Microb. Technol.* **11:** 230–235.

Harley, V. R., Mather, K. A., Power, B. E., Mckimm-Breschkin, J. L., and Hudson, P. J. 1990. Characterization of an avian influenza virus nucleoprotein expressed in *E. coli* and in insect cells. *Arch. Virol.* **113:** 267–277.

Harlow, E., and Lane, D. 1988. *Antibodies: A Laboratory Manual.* Cold Spring Harbor, NY: Cold Spring Harbor Laboratory.

Harmon, S. A., Johnston, J. M., Ziegelhoffer, T., Richards, O. C., Summers, D. F., and Ehrenfeld, E. 1988. Expression of hepatitis A virus capsid sequences in insect cells. *Virus Res.* **10:** 273–280.

Harrap, K. 1972a. The structure of nuclear polyhedrosis viruses. I. The inclusion body. *Virology* **50:** 114–123.

Harrap, K. 1972b. The structure of nuclear polyhedrosis viruses. II. The virus particle. *Virology* **50:** 124–132.

Harrap, K. 1972c. The stucture of nuclear polyhedrosis viruses. III. Virus assembly. *Virology* **50:** 133–139.

Harrap, K., and Longworth, J. 1974. An evaluation of purification methods for baculoviruses. *J. Invertbr. Pathol.* **24:** 55–62.

Hart, G., Haltiwanger, R., Holt, G., and Kelly, W. 1989. Glycosylation in the nucleus and the cytoplasm. *Ann. Rev. Biochem.* **58:** 841–874.

Hartig, P. C., Cardon, M. C., and Kawanishi, C. Y. 1991. Insect virus—assays for viral replication and persistence in mammalian cells. *J. Virol. Methods* **31:** 335–344.

Hartig, P. C., Chapman, M. A., Hatch, G. G., and Kawanishi, C. Y. 1989. Insect virus—assays for toxic effects and transformation potential in mammalian cells. *Appl. Environ. Microbiol.* **55:** 1916–1920.

Hasemann, C. A., and Capra, J. D. 1990. High-level production of a functional immunoglobulin heterodimer in a baculovirus expression system. *Proc. Natl. Acad. Sci. U.S.A.* **87:** 3942–3946.

Hasemann, C. A., and Capra, J. D. 1991. Mutational analysis of arsonate binding by a $CRIA^+$ antibody. VH and VL junctional diversity are essential for binding activity. *J. Biol. Chem.* **266:** 7626–7632.

Hasnain, S. E., and Nakhai, B. 1990. Expression of the gene encoding firefly luciferase in insect cells using a baculovirus vector. *Gene* **91:** 135–138.

Hauser, C., Fusswinkel, H., Li, J., Oellig, C., Kunze, R., Mueller-Neumann, M., Heinlein, M., Starlinger, P., and Doerfler, W. 1988. Overproduction of the protein encoded by the maize transposable element *Ac* in insect cells by a baculovirus vector. *Mol. Gen. Genet.* **214:** 373–378.

Heinderyckx, M., Jacobs, P., and Bollen, A. 1989. Secretion of glycosylated human recombinant haptoglobin in baculovirus-infected insect cells. *Mol. Biol. Rep.* **13:** 225–232.

Hellers, M., Gunne, H., and Steiner, H. 1991. Expression and post-translational processing of preprocecropin-A using a baculovirus vector. *Eur. J. Biochem.* **199:** 435–439.

Hernandez, T. R., and Lehman, I. R. 1990. Functional interaction between the herpes simplex-1 DNA polymerase and UL42 protein. *J. Biol. Chem.* **265:** 11227–11232.

Herrera, R., Lebwohl, D., de Herreros, A. G., Kallen, R. G., and Rosen, O. M. 1988. Synthesis, purification, and characterization of the cytoplasmic domain of the human insulin receptor using a baculovirus expression system. *J. Biol. Chem.* **263:** 5560–5568.

Hilditch, C. M., Rogers, L. J., and Bishop, D. H. L. 1990. Physicochemical analysis of the hepatitis B virus core antigen produced by a baculovirus expression vector. *J. Gen. Virol.* **71:** 2755–2759.

Hill-Perkins, M. S., and Possee, R. D. 1990. A baculovirus expression vector derived from the basic protein promoter of *Autographa californica* nuclear polyhedrosis virus. *J. Gen. Virol.* **71:** 971–976.

Hink, W. F. 1970. Established insect cell line from the cabbage looper, *Trichoplusia ni. Nature* **226:** 466–467.

Hink, W. F. 1976. A compilation of invertebrate cell lines and culture media. In *Invertebrate Tissue-culture: Research Applications,* ed. K. Maramorosch, pp. 319–369. New York: Academic Press.

Hink, W. F. 1980. The 1979 compilation of invertebrate cell lines and culture media. In *Invertebrate Systems in Vitro,* eds. E. Kurstak, K. Maramorosch, and A. Dubendorfer, pp. 553–578. New York: Elsevier.

Hink, W. F. 1989. Recently established invertebrate cell lines. In *Invertebrate Cell System Applications,* ed. J. Mitsuhashi, pp. 269–293. Boca Raton, Fla: CRC Press.

Hink, W. F. 1991. A serum-free medium for the culture of insect cells and production of recombinant proteins. *In Vitro Cell. Dev. Biol.* **27:** 397–401.

Hink, W. F., Thomsen, D. R., Davidson, D. J., Meyer, A. L., and Castellino, F. J. 1991. Expression of three recombinant proteins using baculovirus vectors in 23 insect cell lines. *Biotechnol. Prog.* **7:** 9–14.

Hink, W. F., and Vail, P. V. 1973. A plaque assay for titration of alfalfa looper nuclear polyhedrosis virus in a cabbage looper TN-368 cell line. *J. Invertebr. Pathol.* **22:** 168–174.

Hirashima, S., Horikoshi, N., Sekimizu, K., and Natori, S. 1989. Production of functional S-II in *Bombyx mori* cells. *Biochem. Biophys. Res. Commun.* **160:** 1093–1099.

Hohmann, A. W., and Faulkner, P. 1983. Monoclonal antibodies to baculovirus structural proteins: determination of specificities by western blot analysis. *Virology* **125:** 432–444.

Hoopes, R. R., and Rohrmann, G. F. 1991. *In vitro* transcription of baculovirus immediate early genes—accurate messenger RNA initiation by nuclear extracts from both insect and human cells. *Proc. Natl. Acad. Sci. U.S.A.* **88:** 4513–4517.

Horiuchi, T., Marumoto, Y., Saeki, Y., Sato, Y., Furusawa, M., Kondo, A., and Maeda, S. 1987. High-level expression of the human α-interferon gene through the use of an improved baculovirus vector in the silkworm *Bombyx mori. Agric. Biol. Chem.* **51:** 1573–1580.

Höss, A., Moarefi, I., Scheidtmann, K. H., Cisek, L. J., Corden, J. L., Dornreiter, I., Arthur, A. K., and Fanning, E. 1990. Altered phosphorylation pattern of simian virus 40 T antigen expressed in insect cells by using a baculovirus vector. *J. Virol.* **64:** 4799–4807.

Hsieh, J. C., Liu, L. F., Chen, W. L., and Tam, M. F. 1989. Expression of Yb_1 glutathione-S-transferase using a baculovirus expression system. *Biochem. Biophys. Res. Comm.* **162:** 1147–1154.

Hsieh, P., and Robbins, P. W. 1984. Regulation of asparagine-linked oligosaccharide processing. Oligosaccharide processing in *Aedes albopictus* mosquito cells. *J. Biol. Chem.* **259:** 2375–2382.

Hsu, C. Y. J., Hurwitz, D. R., Mervic, M., and Zilberstein, A. 1991. Autophosphorylation of the intracellular domain of the epidermal growth factor receptor results in different effects on its tyrosine kinase activity with various peptide substrates. Phosphorylation of peptides representing tyrosine phosphorylation sites of phospholipase C-γ. *J. Biol. Chem.* **266:** 603–608.

Hu, S. L., Kosowski, S. G., and Schaaf, K. F. 1987. Expression of envelope glycoproteins of human immunodeficiency virus by an insect virus vector. *J. Virol.* **61:** 3617–3620.

Hu, Y. W., and Kang, C. Y. 1991. Enzyme activities in 4 different forms of human immunodeficiency virus-1 pol gene products. *Proc. Natl. Acad. Sci. U.S.A.* **88:** 4596–4600.

Huang, C. J., Huang, F. L., Chang, G. D., Chang, Y. S., Lo, C. F., Fraser, M. J., and Lo, T. B. 1991. Expression of two forms of carp gonadotropin α subunit in insect cells by recombinant baculovirus. *Proc. Natl. Acad. Sci. U.S.A.* **88:** 7486–7490.

Hughes, P. R., and Wood, H. A. 1986. *In vivo* and *in vitro* bioassay methods for baculoviruses. In *The Biology of Baculoviruses,* Vol. 2, eds. R. R. Granados, and B. A. Federici, pp. 1–30. Boca Raton, Fla: CRC Press.

Huh, N. E., and Weaver, R. F. 1990a. Categorizing some early and late transcripts directed by the *Autographa californica* nuclear polyhedrosis virus. *J. Gen. Virol.* **71:** 2195–2200.

Huh, N. E., and Weaver, R. F. 1990b. Identifying the RNA polymerases that synthesize specific transcripts of the *Autographa californica* nuclear polyhedrosis virus. *J. Gen. Virol.* **71:** 195–202.

Hussey, R. E., Richardson, N. E., Kowalski, M., Brown, N. R., Chang, H. C., Siliciano, R. F., Dorfman, T., Walker, B., Sodroski, J., and Reinherz, E. L. 1988. A soluble CD4 protein selectively inhibits HIV replication and synctium formation. *Nature* **331:** 78–81.

Iacono-Connors, L. C., Schmaljohn, C. S., and Dalrymple, J. M. 1990. Expression of the *Bacillus anthracis* protective antigen gene by baculovirus and vaccinia virus recombinants. *Infect. Immun.* **58:** 366–372.

Iacono-Connors, L. C., Welkos, S. L., Ivins, B. E., and Dalrymple, J. M. 1991. Protection against anthrax with recombinant virus-expressed protective antigen in experimental animals. *Infect. Immun.* **59:** 1961–1965.

Iatrou, K., and Meidinger, R. G. 1990. Tissue-specific expression of silkmoth chorion genes *in vivo* using *Bombyx mori* nuclear polyhedrosis virus as a transducing vector. *Proc. Natl. Acad. Sci. U.S.A.* **87:** 3650–3654.

Iatrou, K., Meidinger, R. G., and Goldsmith, M. R. 1989. Recombinant baculoviruses as vectors for identifying proteins encoded by intron-containing members of complex multigene families. *Proc. Natl. Acad. Sci. U.S.A.* **86:** 9129–9133.

Ingley, E., Cutler, R. L., Fung, M. C., Sanderson, C. J., and Young, I. G. 1991. Production and purification of recombinant human interleukin-5 from yeast and baculovirus expression systems. *Eur. J. Biochem.* **196:** 623–629.

Inlow, D., Shauger, A., and Maiorella, B. 1989. Insect cell culture and baculovirus propagation in protein-free medium. *J. Tissue-culture Meth.* **12:** 13–16.

Innis, M., Gelfand, D., Sninsky, J., and White, T. eds. 1990. *PCR Protocols—A Guide to Methods and Applications.* New York: Academic Press.

Inumaru, S., Ghiasi, H., and Roy, P. 1987. Expression of bluetongue virus group-specific antigen VP3 in insect cells by a baculovirus vector. Its use for the detection of bluetongue virus antibodies. *J. Gen. Virol.* **68:** 1627–1636.

Inumaru, S., and Roy, P. 1987. Production and characterization of the neutralization antigen VP2 of bluetongue virus serotype 10 using a baculovirus expression vector. *Virology* **157:** 472–479.

Itoh, K., Oshima, A., Sakuraba, H., and Suzuki, Y. 1990. Expression, glycosylation, and intracellular distribution of human β-galactosidase in recombinant baculovirus-infected *Spodoptera frugiperda* cells. *Biochem. Biophys. Res. Commun.* **167:** 746–753.

Itoh, K., Oshima, A., Sakuraba, H., and Suzuki, Y. 1991. Characterization and purification of human β-galactosidase overexpressed in recombinant baculovirus-infected *Spodoptera frugiperda* cells. *J. Inherited Metab. Dis.* **14:** 813–818.

Jacobs, P., Massaer, M., Heinderyckx, M., Milican, F., Gilles, P., Vanopstal, O., Voet, P., Gheysen, D., and Bollen, A. 1991. *Plasmodium falciparum*—recombinant baculoviruses direct the expression of circumsporozoite proteins in *Spodoptera frugiperda* cell cultures. *Mol. Biol. Rep.* **15:** 73–79.

Jansen, K. U., Shields, J., Gordon, J., Cairns, J., Graber, P., and Bonnefoy, J. Y. 1991. Expression of human recombinant CD23 in insect cells. *J. Recept. Res.* **11:** 507–520.

Janssen, J. J. M., Mulder, W. R., Decaluwe, G. L. J., Vlak, J. M., and Degrip, W. J. 1991. *In vitro* expression of bovine opsin using recombinant baculovirus—the role of glutamic acid (134) in opsin biosynthesis and glycosylation. *Biochim. Biophys. Acta* **1089:** 68–76.

Janssen, J. J. M., Van de Ven, W. J. M., Van groningen Luyben, W. A. H. M., Roosien, J., Vlak, J. M., and de Grip, W. J. 1988. Synthesis of functional bovine opsin in insect cells under control of the baculovirus polyhedrin promoter. *Mol. Biol. Rep.* **13:** 65–72.

Jarvis, D., Bohlmeyer, D., and Garcia, A. 1991. Requirements for nuclear localization and supramolecular assembly of a baculovirus polyhedrin protein. *Virology* **185:** 795–810.

Jarvis, D. L., Fleming, J. A. G. W., Kovacs, G. R., Summers, M. D., and Guarino, L. A. 1990a. Use of early baculovirus promoters for continuous expression and efficient processing of foreign gene products in stably transformed lepidopteran cells. *Bio/Technology* **8:** 950–955.

Jarvis, D. L., Oker-Blom, C., and Summers, M. D. 1990b. Role of glycosylation in the transport of recombinant glycoproteins through the secretory pathway of Lepidopteran insect cells. *J. Cell. Biochem.* **42:** 181–191.

Jarvis, D. L., and Summers, M. D. 1989. Glycosylation and secretion of human tissue plasminogen activator in recombinant baculovirus-infected insect cells. *Mol. Cell. Biol.* **9:** 214–223.

Jeang, K. T., Giam, C. Z., Nerenberg, M., and Khoury, G. 1987a. Abundant synthesis of functional human T-cell leukemia virus type I p40^{x} protein in eukaryotic cells by using a baculovirus expression vector. *J. Virol.* **61:** 708–713.

Jeang, K. T., Holmgren, K. M., and Khoury, G. 1987b. A baculovirus vector can express intron-containing genes. *J. Virol.* **61:** 1761–1764.

Jeang, K. T., Shank, P. R., and Kumar, A. 1988a. Transcriptional activation of homologous long terminal repeats by the human immunodeficiency virus type 1 or the human T-cell leukemia virus type I *tat* proteins occurs in the absence of *de novo* protein synthesis. *Proc. Natl. Acad. Sci. U.S.A.* **85:** 8291–8295.

Jeang, K. T., Shank, P. R., Rabson, A. B., and Kumar, A. 1988b. Synthesis of functional human immunodeficiency virus *tat* protein in baculovirus as determined by a cell-cell fusion assay. *J. Virol.* **62:** 3874–3878.

Jenkins, M. C., Dougherty, E. M., and Braun, S. K. 1991. Protection against coccidiosis with recombinant *Eimeria acervulina* merozoite antigen expressed in baculovirus. In *Viruses of Invertebrates,* ed. E. Kurstak, pp. 127–140. New York: Marcel Dekker.

Jha, P. K., Nakhai, B., Sridhar, P., Talwar, G. P., and Hasnain, S. E. 1990. Firefly luciferase synthesized to very high levels in caterpillars infected with a recombinant baculovirus can also be used as an efficient reporter enzyme *in vivo. FEBS Lett.* **274:** 23–26.

Jobses, I., Martens, D., and Tramper, J. 1991. Lethal events during gas sparging in animal cell culture. *Biotechnol. Bioeng.* **37:** 484–490.

Johansen, J., Van Der Straten, A., Sweet, R., Otto, E., Maroni, G., and Rosenberg, M. 1989. Regulated expression at high copy number allows production of a growth-inhibiting oncogene product in *Drosophila* Schneider cells. *Genes and Development* **3:** 882–889.

Johnson, P. W., Attia, J., Richardson, C. D., Roder, J. C., and Dunn, R. J. 1989. Synthesis of soluble myelin-associated glycoprotein in insect and mammalian cells. *Gene* **77:** 287–296.

Jun-Chuan, Q., and Weaver, R. 1982. Capping of viral RNA in cultured *Spodoptera frugiperda* cells infected with *Autographa californica* nuclear polyhedrosis virus. *J. Virol.* **43:** 234–240.

Kajigaya, S., Fujii, H., Field, A., Anderson, S., Rosenfeld, S., Anderson, L. J., Shimada, T., and Young, N. S. 1991. Self-assembled B19 parvovirus capsids, produced in a baculovirus system are antigenically and immunogenically similar to native virions. *Proc. Natl. Acad. Sci. U.S.A.* **88:** 4646–4650.

Kallen, R. G., Smith, J. E., Sheng, Z., and Tung, L. 1990. Expression, purification and characterization of a 41 kDa insulin receptor tyrosine kinase domain. *Biochem. Biophys. Res. Commun.* **168:** 616–624.

Kang, C. Y., Bishop, D. H. L., Seo, J. S., Matsuura, Y., and Choe, M. 1987. Secretion of particles of hepatitis B surface antigen from insect cells using a baculovirus vector. *J. Gen. Virol.* **68:** 2607–2614.

Kartner, N., Hanrahan, J. W., Jensen, T. J., Naismith, A. L., Sun, S. Z., Ackerley, C. A., Reyes, E. F., Tsui, L. C., Rommens, J. M., Bear, C. E., and Riordan, J. R. 1991. Expression of the cystic fibrosis gene in non-epithelial invertebrate cells produces a regulated anion conductance. *Cell* **64:** 681–691.

Keddie, B. A., Aponte, G. W., and Volkman, L. E. 1989. The pathway of infection of *Autographa californica* nuclear polyhedrosis virus in an insect host. *Science* **243:** 1728–1730.

Keddie, B. A., and Volkman, L. E. 1985. Infectivity difference between the two phenotypes of *Autographa californica* nuclear polyhedrosis virus: importance of the 64K envelope glycoprotein. *J. Gen. Virol.* **66:** 1195–1200.

Keefer, M. C., Bonnez, W., Roberts, N. J. J., Dolin, R., and Reichman, R. C. 1991. Human immunodeficiency virus HIV-1 gp160-specific lymphocyte proliferative responses of mononuclear leukocytes from HIV-1 recombinant gp160 vaccine recipients. *J. Infect. Dis.* **163:** 448–453.

Kelly, D. 1981. Baculovirus replication: electron microscopy of the sequence of infection of *Trichoplusia ni* nuclear polyhedrosis virus in *Spodoptera frugiperda* cells. *J. Gen. Virol.* **52:** 209–219.

Kelly, D. C., Brown, D. A., Ayres, M. D., Allen, C. J., and Walker, I. O. 1983. Properties of the major nucleocapsid protein of *Heliothis zea* singly enveloped nuclear polyhedrosis virus. *J. Gen. Virol.* **64:** 399–408.

Kelly, D. C., and Lescott, T. 1981. Baculovirus replication: protein synthesis in *Spodoptera frugiperda* cells infected with *Trichoplusia ni* nuclear polyhedrosis virus. *Microbiologica* **4:** 35–37.

Kelly, D. C., and Lescott, T. 1983. Baculovirus replication; glycosylation of polypeptides synthesized in *Trichoplusia ni* nuclear polyhedrosis virus-infected cells and the effect of tunicamycin. *J. Gen. Virol.* **64:** 1915–1926.

Kessler, S. 1975. Rapid isolation of antigens from cells with a staphylococcal protein A-antibody adsorbent: parameters of the interaction of antibody-antigen complexes with protein A. *J. Immunol.* **115:** 1617–1623.

Kilburn, D., and Webb, F. 1968. The cultivation of animal cells at controlled dissolved oxygen partial pressure. *Biotechnol. Bioeng.* **10:** 801–814.

King, L. A., Mann, S. G., Lawrie, A. M., and Mulshaw, S. H. 1991. Replication of wild-type and recombinant *Autographa californica* nuclear polyhedrosis virus in a cell line derived from *Mamestra brassicae*. *Virus Res.* **19:** 93–104.

Kitts, P. A., Ayres, M. D., and Possee, R. D. 1990. Linearization of baculovirus DNA enhances the recovery of recombinant virus expression vectors. *Nucleic Acids Res.* **18:** 5667–5672.

Kjems, J., Brown, M., Chang, D. D., and Sharp, P. A. 1991. Structural analysis of the interaction between the human immunodeficiency virus rev protein and the rev response element. *Proc. Natl. Acad. Sci. U.S.A.* **88:** 683–687.

Klaiber, K., Williams, N., Roberts, T. M., Papazian, D. M., Jan, L. Y., and Miller, C. 1990. Functional expression of *Shaker* K^+ channels in a baculovirus-infected insect cell line. *Neuron* **5:** 221–226.

Kleczkowski, A. 1968. Experimental design and statistical methods of assay. In *Methods in Virology,* Vol. IV, eds. K. Maramorosch, and H. Koprowski, pp. 615–730. New York: Academic Press.

Klein, R., Schroder, C. H., and Zentgraf, H. 1991. Expression of the X-protein of hepatitis-B virus in insect cells using recombinant baculoviruses. *Virus Genes* **5:** 157–174.

Kloc, M., Reddy, B., Crawford, S., and Etkin, L. D. 1991. A novel 110-kDa maternal CAAX box-containing protein from *Xenopus* is palmitoylated and isoprenylated when expressed in baculovirus. *J. Biol. Chem.* **266:** 8206–8212.

Knops, J., Johnsonwood, K., Schenk, D. B., Sinha, S., Lieberburg, I., and McConlogue, L. 1991a. Isolation of baculovirus-derived secreted and full-length β-amyloid precursor protein. *J. Biol. Chem.* **266:** 7285–7290.

Knops, J., Kosik, K. S., Lee, G., Pardee, J. D., Cohengould, L., and McConlogue, L. 1991b. Overexpression of tau in a nonneuronal cell induces long cellular processes. *J. Cell. Biol.* **114:** 725–733.

Knudson, D., and Harrap, K. 1976. Replication of a nuclear polyhedrosis virus in a continuous cell culture of *Spodoptera frugiperda:* microscopy study of the sequence of events of the virus infection. *J. Virol.* **17:** 254–268.

Koener, J. F., and Leong, J. A. C. 1990. Expression of the glycoprotein gene from a fish rhabdovirus by using baculovirus vectors. *J. Virol.* **64:** 428–430.

Kool, M., Voncken, J. W., Vanlier, F. L. J., Tramper, J., and Vlak, J. M. 1991. Detection and analysis of *Autographa californica* nuclear polyhedrosis virus mutants with defective interfering properties. *Virology* **183:** 739–746.

Kornfeld, R., and Kornfeld, S. 1985. Assembly of asparagine-linked oligosaccharides. *Ann. Rev. Biochem.* **54:** 631–664.

Kos, T., Molijn, A., Blauw, B., and Schellekens, H. 1991. Baculovirus-directed high level expression of the hepatitis-delta antigen in *Spodoptera frugiperda* cells. *J. Gen. Virol.* **72:** 833–842.

Kovacs, G. R., Guarino, L. A., and Summers, M. D. 1991. Novel regulatory properties of the IE1 and IE0 transactivators encoded by the baculovirus *Autographa californica* multicapsid nuclear polyhedrosis virus. *J. Virol.* **65:** 5281–5288.

Kozak, M. 1987. At least six nucleotides preceding the AUG initiator codon enhance translation in mammalian cells. *J. Mol. Biol.* **196:** 947–950.

Kraiss, S., Barnekow, A., and Montenarh, M. 1990. Protein kinase activity associated with immunopurified p53 protein. *Oncogene* **5:** 845–856.

Krappa, R., and Knebel-Mörsdorf, D. 1991. Identification of the very early transcribed baculovirus gene PE-38. *J. Virol.* **65:** 805–812.

Krishna, S., Blacklaws, B. A., Overton, H. A., Bishop, D. H. L., and Nash, A. A. 1989. Expression of glycoprotein D of herpes simplex virus type 1 in a recombinant baculovirus—protective responses and T cell recognition of the recombinant-infected cell extracts. *J. Gen. Virol.* **70:** 1805–1814.

Kristie, T. M., LeBowitz, J. H., and Sharp, P. A. 1989. The octamer-binding proteins form multi-protein-DNA complexes with the HSV αTIF regulatory protein. *EMBO J.* **8:** 4429–4238.

Kumar, S., and Miller, L. K. 1987. Effects of serial passage of *Autographa californica* nuclear polyhedrosis virus in cell culture. *Virus Res.* **7:** 335–350.

Kunimoto, D. Y., Allison, K. C., Watson, C., Fuerst, T., Armstrong, G. D., Paul, W., and Strober, W. 1991. High-level production of murine interleukin-5 (IL-5) utilizing recombinant baculovirus expression. Purification of the rIL-5 and its use in assessing the biologic role of IL-5 glycosylation. *Cytokine* **3:** 224–230.

Kunze, R., and Starlinger, P. 1989. The putative transposase of transposable element *Ac* from *Zea mays* L. interacts with subterminal sequences of *Ac. EMBO J.* **8:** 3177–3186.

Kuroda, K., Geyer, H., Geyer, R., Doerfler, W., and Klenk, H.-D. 1990. The oligosaccharides of influenza virus hemagglutinin expressed in insect cells by a baculovirus vector. *Virology* **174:** 418–429.

Kuroda, K., Gröner, A., Frese, K., Drenckhahn, D., Hauser, C., Rott, R., Doerfler, W., and Klenk, H.-D. 1989. Synthesis of biologically active influenza virus hemagglutinin in insect larvae. *J. Virol.* **63:** 1677–1685.

Kuroda, K., Hauser, C., Rott, R., Klenk, H. D., and Doerfler, W. 1986. Expression of the influenza virus hemagglutinin in insect cells by a baculovirus vector. *EMBO J.* **5:** 1359–1366.

Kuroda, K., Hauser, C., Rott, R., Klenk, H. D., and Doerfler, W. 1987. Biologically active influenza virus hemagglutinin expressed in insect cells by a baculovirus vector. In *Biotechnology In Invertebrate Pathology And Cell Culture*, ed. K. Maramorosch, pp. 235–250. San Diego and London: Academic Press.

Kuroda, K., Veit, M., and Klenk, H. D. 1991. Retarded processing of influenza virus hemagglutinin in insect cells. *Virology* **180:** 159–165.

Kuzio, J., Jaques, R., and Faulkner, P. 1989. Identification of P74, a gene essential for virulence of baculovirus occlusion bodies. *Virology* **173:** 759–763.

Kuzio, J., Rohel, D. Z., Curry, C. J., Krebs, A., Carstens, E. B., and Faulkner, P. 1984. Nucleotide sequence of the p10 polypeptide gene of *Autographa californica* nuclear polyhedrosis virus. *Virology* **139:** 414–418.

Labbe, M., Charpilienne, A., Crawford, S. E., Estes, M. K., and Cohen, J. 1991. Expression of rotavirus-VP2 produces empty corelike particles. *J. Virol.* **65:** 2946–2952.

Laemmli, U. K. 1970. Cleavage of structural proteins during the assembly of the head of bacteriophage T4. *Nature* **227:** 680–685.

Lanford, R. E. 1988. Expression of SV-40 T antigen in insect cells using a baculovirus expression vector. *Virology* **167:** 72–81.

Lanford, R. E., Luckow, V., Kennedy, R. C., Dreesman, G. R., Notvall, L., and Summers, M. D. 1989. Expression and characterization of hepatitis B virus surface antigen polypeptides in insect cells with a baculovirus expression system. *J. Virol.* **63:** 1549–1557.

Lanford, R. E., and Notvall, L. 1990. Expression of hepatitis B virus core and precore antigens in insect cells and characterization of a core-associated kinase activity. *Virology* **176:** 222–233.

Lanford, R. E., White, R. G., Dunham, R. G., and Kanda, P. 1988. Effect of basic and nonbasic amino acid substitutions on transport induced by simian virus 40 T-antigen synthetic peptide nuclear transport signals. *Mol. Cell. Biol.* **8:** 2722–2729.

Laroche, Y., Demaeyer, M., Stassen, J. M., Gansemans, Y., Demarsin, E., Matthyssens, G., Collen, D., and Holvoet, P. 1991. Characterization of a recombinant single-chain molecule comprising the variable domains of a monoclonal antibody specific for human fibrin fragment D-dimer. *J. Biol. Chem.* **266:** 16343–16349.

Lebacq-Verheyden, A. M., Kasprzyk, P. G., Raum, M. G., Van Wyke Coelingh, K., Lebacq, J. A., and Battey, J. F. 1988. Posttranslational processing of endogenous and of baculovirus-expressed human gastrin-releasing peptide precursor. *Mol. Cell. Biol.* **8:** 3129–3135.

Lee, H. H., and Miller, L. K. 1978. Isolation of genotypic variants of *Autographa californica* nuclear polyhedrosis virus. *J. Virol.* **27:** 754–767.

Lee, H. H., and Miller, L. K. 1979. Isolation, complementation, and initial characterization of temperature-sensitive mutants of the baculovirus *Autographa californica* nuclear polyhedrosis virus. *J. Virol.* **31:** 240–252.

Lery, X., and Fediere, G. 1990. A new serum-free medium for lepidopteran cell culture. *J. Invertebr. Pathol.* **55:** 342–349.

Levely, M. E., Bannow, C. A., Smith, C. W., and Nicholas, J. A. 1991. Immunodominant T-cell epitope on the F-protein of respiratory syncytial virus recognized by human lymphocytes. *J. Virol.* **65:** 3789–3796.

Levine, B. A., Clack, B., and Ellis, L. 1991. A soluble insulin receptor kinase catalyzes ordered phosphorylation at multiple tyrosines of dodecapeptide substrates. *J. Biol. Chem.* **266:** 3565–3570.

Lévy, F., and Kvist, S. 1990. Co-expression of the human HLA-B27 class-I antigen and the E3/19K protein of adenovirus-2 in insect cells using a baculovirus vector. *Int. Immunol.* **2:** 995–1002.

Licari, P., and Bailey, J. E. 1991. Factors influencing recombinant protein yields in an insect cell-baculovirus expression system: multiplicity of infection and intracellular protein degradation. *Biotechnol. Bioeng.* **37:** 238–246.

Lin, B. T. Y., Gruenwald, S., Morla, A. O., Lee, W. H., and Wang, J. Y. J. 1991. Retinoblastoma cancer suppressor gene product is a substrate of the cell cycle regulator cdc2 kinase. *EMBO J.* **10:** 857–864.

Lipes, M. A., Napolitano, M., Jeang, K. T., Chang, N. T., and Leonard, W. J. 1988. Identification cloning and characterization of an immune activation gene. *Proc. Natl. Acad. Sci. U.S.A.* **85:** 9704–9708.

Liprandi, F., Rodriguez, I., Pina, C., Larralde, G., and Gorziglia, M. 1991. VP4 monotype specificities among porcine rotavirus strains of the same VP4 serotype. *J. Virol.* **65:** 1658–1661.

Lithgow, T., Ristevski, S., Hoj, P., and Hoogenraad, N. 1991. High-level expression of a mitochondrial enzyme, ornithine transcarbamylase from rat liver, in a baculovirus expression system. *DNA Cell. Biol.* **10:** 443–449.

Liu, A., Qin, J., Rankin, C., Hardin, S. E., and Weaver, R. F. 1986. Nucleotide sequence of a portion of the *Autographa californica* nuclear polyhedrosis virus genome containing the EcoRI site-rich region hr-5 and an open reading frame just 5′ of the p10 gene. *J. Gen. Virol.* **67:** 2565–2570.

Livingstone, C., and Jones, I. 1989. Baculovirus expression vector with single-strand capability. *Nucleic Acids Res.* **17:** 2366.

Loeber, G., Stenger, J. E., Ray, S., Parsons, R. E., Anderson, M. E., and Tegtmeyer, P. 1991. The zinc finger region of simian virus-40 large T-antigen is needed for hexamer assembly and origin melting. *J. Virol.* **65:** 3167–3174.

Loetscher, H., Pan, Y.-C. E., Lahm, H.-W., Gentz, R., Brockhaus, M., Tabuchi, H., and Lesslauer, W. 1990. Molecular cloning and expression of the human 55 kd tumor necrosis factor receptor. *Cell* **61:** 351–360.

Loudon, P. T., and Roy, P. 1991. Assembly of five bluetongue virus proteins expressed by recombinant baculoviruses: inclusion of the largest protein VP1 in the core and virus-like particles. *Virology* **180:** 798–802.

Loudon, P. T., Hirasawa, T., Oldfield, S., Murphy, M., and Roy, P. 1991. Expression of the outer capsid protein VP5 of two bluetongue viruses, and synthesis of chimeric double-shelled virus-like particles using combinations of recombinant baculoviruses. *Virology* **182:** 793–801.

Lowe, P. N., Page, M. J., Bradley, S., Rhodes, S., Sydenham, M., Paterson, H., and Skinner, R. H. 1991. Characterization of recombinant human Kirsten-*ras*(4B) p21 produced at high levels in *Escherichia coli* and insect baculovirus expression systems. *J. Biol. Chem.* **266:** 1672–1678.

Lowe, P. N., Sydenham, M., and Page, M. J. 1990. The Ha-*ras* protein, p21, is modified by a derivative of mevalonate and methyl-esterified when expressed in the insect/baculovirus system. *Oncogene* **5:** 1045–1048.

Lu, A., and Carstens, E. B. 1991. Nucleotide sequence of a gene essential for viral DNA replication in the baculovirus *Autographa californica* nuclear polyhedrosis virus. *Virology* **181:** 336–347.

Lübbert, H., and Doerfler, W. 1984. Transcription of overlapping sets of RNAs from the genome of *Autographa californica* nucler polyhedrosis virus: a novel method for mapping RNAs. *J. Virol.* **52:** 255–265.

Luckow, V. A. 1991. Cloning and expression of heterologous genes in insect cells with baculovirus vectors. In *Recombinant DNA Technology and Applications*, eds. A. Prokop, R. K. Bajpai, and C. Ho, pp. 97–152. New York: McGraw-Hill.

Luckow, V. A., and Summers, M. D. 1988a. Signals important for high-level expression of foreign genes in *Autographa californica* nuclear polyhedrosis virus expression vectors. *Virology* **167:** 56–71.

Luckow, V. A., and Summers, M. D. 1988b. Trends in the development of baculovirus expression vectors. *Bio/Technology* **6:** 47–55.

Luckow, V. A., and Summers, M. D. 1989. High level expression of nonfused foreign genes with *Autographa californica* nuclear polyhedrosis virus expression vector. *Virology* **170:** 31–39.

Luo, L., Li, Y., and Kang, C. Y. 1990. Expression of gag precursor protein and secretion of virus-like gag particles of HIV-2 from recombinant baculovirus-infected insect cells. *Virology* **179:** 874–880.

Lynn, D., and Hink, W. 1977. Infection of synchronized TN-368 cell cultures with alfalfa looper nuclear polyhedrosis virus. *J. Invertebr. Pathol.* **32:** 1–5.

MacDonald, P. N., Haussler, C. A., Terpening, C. M., Galligan, M. A., Reeder, M. C., Whitfield, G. K., and Haussler, M. R. 1991. Baculovirus-mediated expression of the human vitamin-D receptor—functional characterization, vitamin-D response element interactions, and evidence for a receptor auxiliary factor. *J. Biol. Chem.* **266:** 18808–18813.

Mackow, E. R., Barnett, J. W., Chan, H., and Greenberg, H. B. 1989. The rhesus rotavirus outer capsid protein VP4 functions as a hemagglutinin and is antigenically conserved when expressed by a baculovirus recombinant. *J. Virol.* **63:** 1661–1668.

Mackow, E. R., Vo, P. T., Broome, R., Bass, D., and Greenberg, H. B. 1990. Immunization with baculovirus-expressed VP4 protein passively protects against simian and murine rotavirus challenge. *J. Virol.* **64:** 1698–1703.

Madisen, L., Travis, B., Hu, S. L., and Purchio, A. F. 1987. Expression of the human immunodeficiency virus *gag* gene in insect cells. *Virology* **158:** 248–250.

Maeda, S. 1987. Expression of human interferon α in silkworms with a baculovirus vector. In *Biotechnology In Invertebrate Pathology And Cell Culture*, ed. K. Maramorosch, pp. 221–234. San Diego and London: Academic Press.

Maeda, S. 1989a. Expression of foreign genes in insects using baculovirus vectors. *Ann. Rev. Entomol.* **34:** 351–372.

Maeda, S. 1989b. Gene transfer vectors of a baculovirus, *Bombyx mori* nuclear polyhedrosis virus, and their use for expression of foreign genes in insect cells. In *Invertebrate Cell System Applications*, ed. J. Mitsuhashi, pp. 167–182. Boca Raton, Fla: CRC Press.

Maeda, S. 1989c. Increased insecticidal effect by a recombinant baculovirus carrying a synthetic diuretic hormone gene. *Biochem. Biophys. Res. Commun.* **165:** 1177–1183.

Maeda, S., Kawai, T., Obinata, M., Chika, T., Horiuchi, T., Maekawa, K., Nakasuji, K., Saeki, Y., Sato, Y., Yamada, K., and Furusawa, M. 1984. Characteristics of human interferon-α produced by a gene transferred by a baculovirus in the silkworm, *Bombyx mori. Proc. Japan. Acad.* **60:** 423–426.

Maeda, S., Kawai, T., Obinata, M., Fujiwara, H., Horiuchi, T., Saeki, Y., Sato, Y., and Furusawa, M. 1985. Production of human α-interferon in silkworm using a baculovirus vector. *Nature* **315:** 592–594.

Maeda, S., and Majima, K. 1990. Molecular cloning and physical mapping of the genome of *Bombyx mori* nuclear polyhedrosis virus. *J. Gen. Virol.* **71:** 1851–1855.

Maeda, S., Volrath, S. L., Hanzlik, T. N., Harper, S. A., Kei, M. M., Maddox, D. W., Hammock, B. D., and Fowler, E. 1991. Insecticidal effects of an insect-specific neurotoxin expressed by a recombinant baculovirus. *Virology* **184:** 777–780.

Maiorella, B., Inlow, D., Shauger, A., and Harano, D. 1988. Large-scale insect cell-culture for recombinant protein production. *Bio/Technology* **6:** 1406–1410.

Makino, Y., Tadano, M., Anzai, T., Ma, S. P., Yasuda, S., and Fukunaga, T. 1989. Detection of Dengue 4 virus core protein in the nucleus II. Antibody against Dengue 4 core protein produced by a recombinant baculovirus reacts with the antigen in the nucleus. *J. Gen. Virol.* **70:** 1417–1426.

Malitschek, B., and Schartl, M. 1991. Rapid identification of recombinant baculoviruses using PCR. *Biotechniques* **11:** 177–178.

Marcy, A. I., Olivo, P. D., Challberg, M. D., and Coen, D. M. 1990. Enzymatic activities of overexpressed herpes simplex virus DNA polymerase purified from recombinant baculovirus-infected insect cells. *Nucleic Acids Res.* **18:** 1207–1215.

Marshall, J. J. A., Fayard, B., and Roy, P. 1990. Biophysical studies on the morphology of baculovirus-expressed bluetongue virus tubules. *J. Gen. Virol.* **71:** 1839–1844.

Marshall, J. J. A., and Roy, P. 1990. High level expression of the two outer capsid proteins of bluetongue virus serotype 10: their relationship with the neutralization of virus infection. *Virus Res.* **15:** 189–196.

Martens, J. W. M., Honee, G., Zuidema, D., van Lent, J. W. M., Visser, B., and Vlak, J. M. 1990. Insecticidal activity of a bacterial crystal protein expressed by a recombinant baculovirus in insect cells. *Appl. Environ. Microbiol.* **56:** 2764–2770.

Martin, B. M., Tsuji, S., Lamarca, M. E., Maysak, K., Eliason, W., and Ginns, E. I. 1988. Glycosylation and processing of high levels of active human glucocerebroside in invertebrate cells using a baculovirus expression vector. *DNA* **7:** 99–106.

Marumoto, Y., Sato, Y., Fujiwara, H., Sakano, K., Saeki, Y., Agata, M., Furusawa, M., and Maeda, S. 1987. Hyperproduction of polyhedrin-IGF II fusion protein in silkworm larvae infected with recombinant *Bombyx mori* nuclear polyhedrosis virus. *J. Gen. Virol.* **68:** 2599–2606.

Matsuda, M., Marshall, C. P., and Hanafusa, H. 1990. Purification and characterization of $P47^{gag\text{-}crk}$ expressed in insect cells. *J. Biol. Chem.* **265:** 12000–12004.

Matsuura, Y., Maekawa, M., Hattori, S., Ikegami, N., Hayashi, A., Yamazaki, S., Morita, C., and Takebe, Y. 1991a. Purification and characterization of human immunodeficiency virus type-1 nef gene product expressed by a recombinant baculovirus. *Virology* **184:** 580–586.

Matsuura, Y., Miyamoto, M., Sato, T., Morita, C., and Yasui, K. 1989. Characterization of Japanese encephalitis virus envelope protein expressed by recombinant baculoviruses. *Virology* **173:** 674–682.

Matsuura, Y., Possee, R. D., and Bishop, D. H. L. 1986. Expression of the S-coded genes of lymphocytic choriomeningitis arenavirus using a baculovirus vector. *J. Gen. Virol.* **67:** 1515–1530.

Matsuura, Y., Possee, R. D., Overton, H. A., and Bishop, D. H. L. 1987. Baculovirus expression vectors: the requirements for high level expression of proteins including glycoproteins. *J. Gen. Virol.* **68:** 1233–1250.

Matsuura, Y., Tatsumi, M., Enami, K., Morikawa, S., Yamazaki, S., and Kohase, M. 1991b. Biological function of recombinant IL-6 expressed in a baculovirus system. *Lymphokine Cytokine Res.* **10:** 201–205.

May, L. T., Shaw, J. E., Khanna, A. K., Zabriskie, J. B., and Sehgal, P. B. 1991. Marked cell-type-specific differences in glycosylation of human interleukin-6. *Cytokine* **3:** 204–211.

McBride, A. A., Bolen, J. B., and Howley, P. M. 1989. Phosphorylation sites of the E2 transcriptional regulatory proteins of bovine papillomavirus type 1. *J. Virol.* **63:** 5076–5085.

McCown, J., Cochran, M., Putnak, R., Feighny, R., Burrous, J., Henchal, E., and Hoke, C. 1990. Protection of mice against lethal Japanese encephalitis with a recombinant baculovirus vaccine. *Amer. J. Tropic. Med. Hygiene* **42:** 491–499.

McCutchan, J., and Pagano, J. 1968. Enhancement of the infectivity of simian virus 40 deoxyribonucleic acid with diethyl-aminoethyl-dextran. *J. Natl. Cancer Inst.* **41:** 351–357.

McMaster, G., and Carmichael, G. 1977. Analysis of single- and double-stranded nucleic acids on polyacrylamide and agarose gels by using glyoxal and acridine orange. *Proc. Natl. Acad. Sci. U.S.A.* **74:** 4835–4839.

McQuade, T. J., Pitts, T. W., and Tarpley, W. G. 1989. A rapid solution immunoassay to quantify binding of the human immunodeficiency virus envelope glycoprotein to soluble CD4. *Biochem. Biophys. Res. Commun.* **163:** 172–176.

Medin, J. A., Hunt, L., Gathy, K., Evans, R. K., and Coleman, M. S. 1990. Efficient, low-cost protein factories: expression of human adenosine deaminase in baculovirus-infected insect larvae. *Proc. Natl. Acad. Sci. U.S.A.* **87:** 2760–2764.

Menard, R., Kouri, H. E., Plouffe, C., Dupras, R., Ripoll, D., Vernet, T., Tessier, D. C., Laliberte, F., Thomas, D. Y., and Storer, A. C. 1990. A protein engineering study of the role of aspartate 158 in the catalytic mechanism of papain. *Biochemistry* **29:** 6706–6713.

Merryweather, A. T., Weyer, U., Harris, M. P. G., Hirst, M., Booth, T., and Possee, R. D. 1990. Construction of genetically engineered baculovirus insecticides containing the *Bacillus thuringiensis* subsp. *kurstaki* HD-73 delta endotoxin. *J. Gen. Virol.* **71:** 1535–1544.

Michaels, J., Peterson, J., McIntyre, L., and Papoutsakis, E. 1991. Protection mechanisms of freely suspended animal cells (CRL 8018) from fluid mechanical injury. Viscometric and bioreactor studies using serum, Pluronic F68 and polyethylene glycol. *Biotechnol. Bioeng.* **38:** 169–180.

Mikhailov, V., Ataeva, J., Marlyev, K., and Kullyev, P. 1986. Changes in DNA polymerase activities in pupae of the silkworm *Bombyx mori* after infection with nuclear polyhedrosis virus. *J. Gen. Virol.* **67:** 175–179.

Miller, D. W. 1988a. Genetically engineered viral insecticides. In *Biotechnology For Crop Protection,* eds. P. A. Hedin, J. J. Menn, and R. M. Hollingworth, pp. 405–421. Washington, D.C.: American Chemical Society.

Miller, D. W., and Miller, L. K. 1982. A virus mutant with an insertion of a copia-like transposable element. *Nature* **299:** 562–564.

Miller, D. W., Safer, P., and Miller, L. K. 1986. An insect baculovirus host-vector for high-level expression of foreign genes. In *Genetic Engineering,* eds. J. K. Setlow, and A. Hollaender, pp. 277–298. New York: Plenum.

Miller, L. K. 1981. A virus vector for genetic engineering in invertebrates. In *Genetic Engineering in the Plant Sciences,* ed. N. Panopoulos, pp. 203–224. New York: Praeger.

Miller, L. K. 1986. The genetics of baculoviruses. In *The Biology of Baculoviruses,* Vol. 1, eds. R.R. Granados, and B.A. Federici, pp. 217–238. Boca Raton, Fla: CRC Press.

Miller, L. K. 1988b. Baculoviruses as gene expression vectors. *Ann. Rev. Microbiol.* **42:** 177–199.

Miller, L. K. 1989. Insect baculoviruses—powerful gene expression vectors. *Bioessays* **11:** 91–95.

Miller, L. K., Adang, M. J., and Browne, D. 1983a. Protein kinase activity associated with the extracellular and occluded forms of the baculovirus *Autographa californica* nuclear polyhedrosis virus. *J. Virol.* **46:** 275–278.

Miller, L. K., and Dawes, K. P. 1978. Restriction endonuclease analysis for the identification of baculovirus pesticides. *Appl. Environ. Microbiol.* **35:** 411–421.

Miller, L. K., Jewell, J. E., and Browne, D. 1981. Baculovirus induction of a DNA polymerase. *J. Virol.* **40:** 305–308.

Miller, L. K., Miller, D., and Adang, M. 1983b. An insect virus for genetic engineering: developing baculovirus polyhedrin substitution vectors. In *Genetic Engineering In Eukaryotes,* eds. P. Lurquin, and A. Kleinhofs, pp. 89–97. New York: Plenum.

Miller, L. K., Trimarchi, R. E., Browne, D., and Pennock, G. D. 1983c. A temperature-sensitive mutant of the baculovirus *Autographa californica* nuclear polyhedrosis virus defective in an early function required for further gene expression. *Virology* **126:** 376–380.

Mills, H. R., and Jones, I. M. 1990. Expression and purification of p24, the core protein of HIV, using a baculovirus-insect cell expression system. *AIDS* **4:** 1125–1131.

Mitsuhashi, J. 1989. Nutritional requirements of insect cells *in vitro.* In *Invertebrate Cell System Applications,* ed. J. Mitsuhashi, pp. 3–20. Boca Raton, Fla: CRC Press.

Mitsuhashi, J., and Goodwin, R. H. 1989. The serum-free culture of insect cells *in vitro.* In *Invertebrate Cell System Applications,* ed. J. Mitsuhashi, pp. 167–182. Boca Raton, Fla: CRC Press.

Miyagawa, K., Sakamoto, H., Yoshida, T., Yamashita, Y., Mitsui, Y., Furusawa, M., Maeda, S., Takaku, F., Sugimura, T., and Terada, M. 1988. Hst-1 transforming protein expression in silkworm cells and characterization as a novel heparin-binding growth factor. *Oncogene* **3:** 383–390.

Miyagawa, K., Kimura, S., Yoshida, T., Sakamoto, H., Takaku, F., Sugimura, T., and Terada, M. 1991. Structural analysis of a mature hst-1 protein with transforming growth factor activity. *Biochem. Biophys. Res. Commun.* **174:** 404–410.

Miyajima, A., Schreurs, J., Otsu, A., Kondo, A., Arai, K.I., and Maeda, S. 1987. Use of the silkworm *Bombyx mori* and an insect baculovirus vector for high-level expression and secretion of biologically active mouse interleukin-3. *Gene* **58:** 273–282.

Miyamoto, C., Smith, G. E., Farrell-Towt, J., Chizzonite, R., Summers, M. D., and Ju, G. 1985. Production of human c-*myc* protein in insect cells infected with a baculovirus expression vector. *Mol. Cell. Biol.* **5:** 2860–2865.

Monini, P., Grossman, S. R., Pepinsky, B., Androphy, E. J., and Laimins, L. A. 1991. Cooperative binding of the E2 protein of bovine papillomavirus to adjacent E2-responsive sequences. *J. Virol.* **65:** 2124–2130.

Montross, L., Watkins, S., Moreland, R. B., Mamon, H., Caspar, D. L. D., and Garcea, R. L. 1991. Nuclear assembly of polyomavirus capsids in insect cells expressing the major capsid protein VP1. *J. Virol.* **65:** 4991–4998.

Moore, M. D., Cannon, M. J., Sewall, A., Finlayson, M., Okimoto, M., and Nemerow, G. R. 1991. Inhibition of Epstein-Barr virus infection *in vitro* and *in vivo* by soluble CR2 CD21 containing two short consensus repeats. *J. Virol.* **65:** 3559–3565.

Morikawa, S., Booth, T. F., and Bishop, D. H. L. 1991. Analyses of the requirements for the synthesis of virus-like particles by feline immunodeficiency virus gag using baculovirus vectors. *Virology* **183:** 288–297.

Morikawa, Y., Overton, H. A., Moore, J. P., Wilkinson, A. J., Brady, R. L., Lewis, S. J., and Jones, I. M. 1990. Expression of HIV-1 gp120 and human soluble CD4 by recombinant baculoviruses and their interaction in vitro. *AIDS Res. Hum. Retroviruses* **6:** 765–773.

Morishita, K., Sakano, K., Takeda, K., and Maeda, S. 1991. Characterization of v-sis protein expressed in silkworm larvae using the *Bombyx mori* nuclear polyhedrosis virus vector. *J. Biochem. Tokyo* **109:** 36–44.

Morrison, D., Kaplan, D. R., Escobedo, J. A., Rapp, U. R., Roberts, T. M., and Williams, L. T. 1989. Direct activation of the serine/threonine kinase activity of Raf-1 through tyrosine phosphorylation by the PDGF β-receptor. *Cell* **58:** 649–657.

Morrison, D., Kaplan, D. R., Rhee, S. G., and Williams, L. T. 1990. Platelet-derived growth factor (PDGF)-dependent association of phospholipase C-γ with the PDGF receptor signalling complex. *Mol. Cell. Biol.* **10:** 2359–2366.

Munemitsu, S., Innis, M. A., Clark, R., McCormick, F., Ullrich, A., and Polakis, P. 1990. Molecular cloning and expression of a g25k complementary DNA; the human homolog of the yeast cell cycle gene cdc42. *Mol. Cell. Biol.* **10:** 5977–5982.

Murhammer, D. 1989. Scale-up of insect cell cultures: protective effects of Pluronic F-68. Ph.D. diss., University of Houston.

Murhammer, D. W., and Goochee, C. F. 1988. Scale-up of insect cell cultures: Protective effects of Pluronic F-68. *Bio/Technology* **6:** 1411–1415.

Murhammer, D. W., and Goochee, C. F. 1990a. Sparged animal cell bioreactors: mechanism of cell damage and Pluronic F-68 protection. *Biotechnol. Prog.* **6:** 391–397.

Murhammer, D. W., and Goochee, C. F. 1990b. Structural features of nonionic polyglycol polymers responsible for the protective effect in sparged animal cell bioreactors. *Biotechnol. Prog.* **6:** 142–148.

Murphey-Corb, M., Martin, L. N., Davison-Fairburn, B., Montelaro, R. C., Miller, M., West, M., Ohkawa, S., Baskin, G. B., Zhang, J.-Y., Putney, S. D., Allison, A. C., and Eppstein, D. A. 1989. A formalin-inactivated whole SIV vaccine confers protection in macaques. *Science* **246:** 1293–1297.

Murphy, C. I., Lennick, M., Lehar, S. M., Beltz, G. A., and Young, E. 1990a. Temporal expression of HIV-1 envelope proteins in baculovirus-infected insect cells: implications for glycosylation and CD4 Binding. *Genet. Anal-Tech. Appl.* **7:** 160–171.

Murphy, C. I., Weiner, B., Bikel, I., Piwnica-Worms, H., Bradley, M. K., and Livingston, D. M. 1988. Purification and functional properties of SV-40 large and small T antigens overproduced in insect cells. *J. Virol.* **62:** 2951–2959.

Murphy, V. F., Rowan, W. C., Page, M. J., and Holder, A. A. 1990b. Expression of hybrid malaria antigens in insect cells and their engineering for correct folding and secretion. *Parasitology* **100:** 177–183.

Nagesha, H. S., and Holmes, I. H. 1991. VP4 relationships between porcine and other rotavirus serotypes. *Arch. Virol.* **116:** 107–118.

Nagy, E., Derbyshire, J. B., Dobos, P., and Krell, P. J. 1990. Cloning and expression of NDV hemagglutinin-neuraminidase cDNA in a baculovirus expression vector system. *Virology* **176:** 426–438.

Nagy, E., Huber, P., Krell, P. J., and Derbyshire, J. B. 1991. Synthesis of Newcastle disease virus (NDV)-like envelopes in insect cells infected with a recombinant baculovirus expressing the hemagglutinin-neuraminidase of NDV. *J. Gen. Virol.* **72:** 753–756.

Nakhai, B., Pal, R., Sridhar, P., Talwar, G. P., and Hasnain, S. E. 1991. The α-subunit of human chorionic gonadotropin hormone synthesized in insect cells using a baculovirus vector is biologically active. *FEBS Lett.* **283:** 104–108.

Naoe, T., Nozaki, N., Yamada, K., Okazaki, T., Nakayama, E., Kurosawa, Y., and Shiku, H. 1989. Diversity of cellular molecules in human cells detected by monoclonal antibodies reactive with c-*myc* proteins produced in *Escherichia coli*. *Jpn. J. Cancer Res.* **80:** 747–753.

Nel, L. H., and Huismans, H. 1991. Synthesis of virus-specified tubules of epizootic haemorrhagic disease virus using a baculovirus expression system. *Virus Res.* **19:** 139–152.

Neufeld, K. L., Richards, O. C., and Ehrenfeld, E. 1991. Expression and characterization of poliovirus proteins 3Bvpg, 3Cpro, and 3Dpol in recombinant baculovirus-infected *Spodoptera frugiperda* cells. *Virus Res.* **19:** 173–188.

Nicholas, J. A., Rubino, K. L., Levely, M. E., Adams, E. G., and Collins, P. L. 1990. Cytolytic T-lymphocyte responses to respiratory syncytial virus: effector cell phenotype and target proteins. *J. Virol.* **64:** 4232–4241.

Niikura, M., Matsuura, Y., Hattori, M., Onuma, M., and Mikami, T. 1991. Expression of the A-antigen (gp57-65) of Marek's disease virus by a recombinant baculovirus. *J. Gen. Virol.* **72:** 1099–1104.

Niikura, M., Matsuura, Y., Hattori, M., Onuma, M., and Mikami, T. 1991. Characterization of haemagglutinin-neuraminidase glycoprotein of Newcastle disease virus expressed by a recombinant baculovirus. *Virus Res.* **20:** 31–43.

Nishikawa, K., Fukuhara, N., Liprandi, F., Green, K., Kapikian, A. Z., Chanock, R. M., and Gorziglia, M. 1989. VP4 protein of porcine rotavirus strain OSU expressed by a baculovirus recombinant induces neutralizing antibodies. *Virology* **173:** 631–637.

Nishimura, C., Matsuura, Y., Kokai, Y., Akera, T., Carper, D., Morjana, N., Lyons, C., and Flynn, T. G. 1990. Cloning and expression of human aldose reductase. *J. Biol. Chem.* **265:** 9788–9792.

Nishimura, C., Yamaoka, T., Mizutani, M., Yamashita, K., Akera, T., and Tanimoto, T. 1991. Purification and characterization of the recombinant human aldose reductase expressed in baculovirus system. *Biochim. Biophys. Acta* **1078:** 171–178.

Nissen, M. S., and Friesen, P. D. 1989. Molecular analysis of the transcriptional regulatory region of an early baculovirus gene. *J. Virol.* **63:** 493–503.

Noteborn, M. H. M., de Boer, G. F., Kant, A., Koch, G., Bos, J. L., Zantema, A., and van der Eb, A. J. 1990. Expression of avian leukemia virus *env*-gp85 in *Spodoptera frugiperda* cells by use of a baculovirus expression vector. *J. Gen. Virol.* **71:** 2641–2648.

Nutt, E. M., Jain, D., Lenny, A. B., Schaffer, L., Siegl, P. K., and Dunwiddie, C. T. 1991. Purification and characterization of recombinant antistasin: a leech-derived inhibitor of coagulation factor Xa. *Arch. Biochem. Biophys.* **285:** 37–44.

Nyunoya, H., Akagi, T., Ogura, T., Maeda, S., and Shimotohno, K. 1988. Evidence for phosphorylation of trans-activator $p40^X$ of human T-cell leukemia virus type I produced in insect cells with baculovirus expression vector. *Virology* **167:** 538–544.

Nyunoya, H., Ogura, T., Kikuchi, M., Iwamoto, H., Yamashita, K., Maekawa, M., Takebe, Y., Miyamura, K., Yamazaki, S., and Shimotohno, K. 1990. Expression of HTLV-I envelope protein fused to hydrophobic amino-terminal peptide of baculovirus polyhedron in insect cells and its application for serological assays. *AIDS Res. Hum. Retroviruses* **6:** 1311–1321.

Oellig, C., Happ, B., Mueller, T., and Doerfler, W. 1987. Overlapping sets of viral RNAs reflect the array of polypeptides in the EcoRI J and N fragments (map positions 81.2 to 85.0) of the *Autographa californica* nuclear polyhedrosis virus genome. *J. Virol.* **61:** 3048–3057.

Ogonah, O., Shuler, M. L., and Granados, R. R. 1991. Protein production (β-galactosidase) from a baculovirus vector in *Spodoptera frugiperda* and *Trichpolusia ni* cells in suspension culture. *Biotechnol. Lett.* **13:** 265–270.

Ohta, D., Matsuura, Y., and Sato, R. 1991. Expression and characterization of a rabbit liver cytochrome-p450 belonging to p450IIB-subfamily with the aid of the baculovirus expression vector system. *Biochem. Biophys. Res. Comm.* **175:** 394–399.

Oker-Blom, C., Jarvis, D. L., and Summers, M. D. 1990. Translocation and cleavage of rubella virus envelope glycoproteins: identification and role of the E2 signal sequence. *J. Gen. Virol.* **71:** 3047–3053.

Oker-Blom, C., Pettersson, R. F., and Summers, M. D. 1989. Baculovirus polyhedrin promoter-directed expression of rubella virus envelope glycoproteins E1 and E2 in *Spodoptera frugiperda* cells. *Virology* **172:** 82–91.

Oker-Blom, C., and Summers, M. D. 1989. Expression of sindbis virus 26S complementary DNA in *Spodoptera frugiperda* Sf9 cells using a baculovirus expression vector. *J. Virol.* **63:** 1256–1264.

Oldfield, S., Adachi, A., Urakawa, T., Hirasawa, T., and Roy, P. 1990. Purification and characterization of the major group-specific core antigen VP7 of bluetongue virus synthesized by a recombinant baculovirus. *J. Gen. Virol.* **71:** 2649–2656.

Olivo, P. D., Nelson, N. L., and Challberg, M. D. 1988. Herpes simplex virus DNA replication: the UL9 gene encodes an origin-binding protein. *Proc. Natl. Acad. Sci. U.S.A.* **85:** 5414–5418.

Olivo, P. D., Nelson, N. L., and Challberg, M. D. 1989. Herpes simplex virus type 1 gene products required for DNA replication: Identification and overexpression. *J. Virol.* **63:** 196–204.

Ollo, R., and Maniatis, T. 1987. *Drosophila* Krüeppel gene product produced in a baculovirus expression system is a nuclear phosphoprotein that binds to DNA. *Proc. Natl. Acad. Sci. U.S.A.* **84:** 5700–5704.

Onken, U., and Weiland, P. 1983. Airlift fermentors: construction, behavior and uses. *Adv. Biotechnol. Proc.* **1:** 67–95.

Ooi, B. G., and Miller, L. K. 1990. Transcription of the baculovirus polyhedrin gene reduces the levels of an antisense transcript initiated downstream. *J. Virol.* **64:** 3126–3129.

Ooi, B. G., and Miller, L. K. 1988. Regulation of host RNA levels during baculovirus infection. *Virology* **166:** 515–523.

Ooi, B. G., and Miller, L. K. 1991. The influence of antisense RNA on transcriptional mapping of the 5′ terminus of a baculovirus RNA. *J. Gen. Virol.* **72:** 527–534.

Ooi, B. G., Rankin, C., and Miller, L. K. 1989. Downstream sequences augment transcription from the essential initiation site of a baculovirus polyhedrin gene. *J. Mol. Biol.* **210:** 721–736.

O'Reilly, D. R., Brown, M. R., and Miller, L. K. 1992. Alteration of ecdysteroid metabolism due to baculovirus infection of the fall armyworm *Spodoptera frugiperda:* host ecdysteroids are conjugated with galactose. *Insect Biochem. Mol. Biol.* In press.

O'Reilly, D. R., Crawford, A. M., and Miller, L. K. 1989. Viral proliferating cell nuclear antigen. *Nature* **337:** 606.

O'Reilly, D. R., and Miller, L. K. 1988. Expression and complex formation of simian virus 40 large T antigen and mouse p53 in insect cells. *J. Virol.* **62:** 3109–3119.

O'Reilly, D. R., and Miller, L. K. 1989. A baculovirus blocks insect molting by producing ecdysteroid UDP-glucosyl transferase. *Science* **245:** 1110–1112.

O'Reilly, D., and Miller, L. 1991. Improvement of a baculovirus pesticide by deletion of the *egt* gene. *Bio/Technology* **9:** 1086–1089.

O'Reilly, D. R., Passarelli, A. L., Goldman, I. F., and Miller, L. K. 1990. Characterization of the DA26 gene in a hypervariable region of the *Autographa californica* nuclear polyhedrosis virus genome. *J. Gen. Virol.* **71:** 1029–1037.

Orentas, R. J., Hildreth, J. E. K., Obah, E., Polydefkis, M., Smith, G. E., Clements, M. L., and Siliciano, R. F. 1990. Induction of $CD4^+$ human cytolytic T cells specific for HIV-infected cells by a gp160 subunit vaccine. *Science* **248:** 1234–1237.

Ornstein, L. 1964. Disc electrophoresis. I. Background and theory. *Ann. N.Y. Acad. Sci.* **121:** 321–349.

Overton, H. A., Fujii, Y., Price, I. R., and Jones, I. M. 1989. The protease and *gag* gene products of the human immunodeficiency virus: authentic cleavage and post-translational modification in an insect cell expression system. *Virology* **170:** 107–116.

Overton, H. A., Ihara, T., and Bishop, D. H. L. 1987. Identification of the N and NS-S proteins coded by the ambisense S RNA of Punta Toro phlebovirus using monospecific antisera raised to baculovirus expressed N and NS-S proteins. *Virology* **157:** 338–350.

Overton, H. A., McMillan, D. J., Gridley, S. J., Brenner, J., Redshaw, S., and Mills, J. S. 1990. Effect of two novel inhibitors of the human immunodeficiency virus protease on the maturation of the HIV gag and gag-pol polyproteins. *Virology* **179:** 508–511.

Owens, R. A., Trempe, J. P., Chejanovsky, N., and Carter, B. J. 1991. Adeno-associated virus rep proteins produced in insect and mammalian expression systems—wild-type and dominant-negative mutant proteins bind to the viral replication origin. *Virology* **184:** 14–22.

Page, M. J. 1989. p36c, an improved baculovirus expression vector for producing high levels of mature recombinant proteins. *Nucleic Acids Res.* **17:** 454.

Page, M. J., Hall, A., Rhodes, S., Skinner, R. H., Murphy, V., Sydenham, M., and Lowe, P. N. 1989. Expression and characterization of the Ha-*ras* p21 protein produced at high levels in the insect/baculovirus system. *J. Biol. Chem.* **264:** 19147–19154.

Parker, E. M., Kameyama, K., Higashijima, T., and Ross, E. M. 1991. Reconstitutively active G-protein-coupled receptors purified from baculovirus-infected insect cells. *J. Biol. Chem.* **266:** 519–527.

Parker, L. L., Atherton Fessler, S., Lee, M. S., Ogg, S., Falk, J. L., Swenson, K. I., and Piwnica-Worms, H. 1991. Cyclin promotes the tyrosine phosphorylation of p34cdc2 in a wee1-positive dependent manner. *EMBO J.* **10:** 1255–1264.

Parker, M. D., Yoo, D., and Babiuk, L. A. 1990a. Expression and secretion of the bovine coronavirus hemagglutinin-esterase glycoprotein by insect cells infected with recombinant baculoviruses. *J. Virol.* **64:** 1625–1629.

Parker, M. D., Yoo, D., Cox, G. J., and Babiuk, L. A. 1990b. Primary structure of the S peplomer gene of bovine coronavirus and surface expression in insect cells. *J. Gen. Virol.* **71:** 263–270 (and correction p. 1885).

Partington, S., Yu, H., Lu, A., and Carstens, E. B. 1990. Isolation of temperature-sensitive mutants of *Autographa californica* nuclear polyhedrosis virus: phenotype characterization of baculovirus mutants defective in very late gene expression. *Virology* **175:** 91–102.

Patel, G., Greenfield, C., Stabel, S., Waterfield, M. D., Parker, P. J., and Jones, N. C. 1988. The overproduction of biologically active E1a, EGF receptor, and protein kinase C proteins in the baculovirus expression system. In *Current Communications In Molecular Biology: Viral Vectors,* eds. Y. Gluzman, and S. H. Hughes, pp. 98–103. Cold Spring Harbor, NY: Cold Spring Harbor Laboratory.

Patel, G., and Jones, N. C. 1990. Activation *in vitro* of RNA polymerase II and III directed transcription by baculovirus produced E1A protein. *Nucleic Acids Res.* **18:** 2909–2915.

Patel, G., Nasmyth, K., and Jones N. C. 1992. A new method for the isolation of recombinant baculovirus. *Nucleic Acids Res.* **20:** 97–104.

Patel, G., and Stabel, S. 1989. Expression of a functional protein kinase C-γ using a baculovirus vector: purification and characterization of a single protein kinase-C iso-enzyme. *Cellular Signalling* **1:** 227.

Paul, J. I., Tavaré, J., Denton, R. M., and Steiner, D. F. 1990. Baculovirus-directed expression of the human insulin receptor and an insulin-binding ectodomain. *J. Biol. Chem.* **265:** 13074–13083.

Pease, E. A., Aust, S. D., and Tien, M. 1991. Heterologous expression of active manganese peroxidase from *Phanerochaete chrysosporium* using the baculovirus expression system. *Biochem. Biophys. Res. Comm.* **179:** 897–903.

Pendergast, A. M., Clark, R., Kawasaki, E. S., McCormick, F. P., and Witte, O. N. 1989. Baculovirus expression of functional P210 BCR-ABL oncogene product. *Oncogene* **4:** 759–766.

Pennock, G. D., Shoemaker, C., and Miller, L. K. 1984. Strong and regulated expression of *Escherichia coli* β-galactosidase in insect cells with a baculovirus vector. *Mol. Cell. Biol.* **4:** 399–406.

Percival, M. D. 1991. Human 5 lipoxygenase contains an essential iron. *J. Biol. Chem.* **266:** 10058–10061.

Piatak, M., Lane, J. A., O'Rourke, E., Clark, R., Houston, L. L., and Apell, R. 1988. Expression of ricin and ricin B-chain in insect cells. *ICSU Short Reports* **8:** 62.

Piwnica-Worms, H., Williams, N. G., Cheng, S. H., and Roberts, T. M. 1990. Regulation of $pp60^{c\text{-}src}$ and its interaction with polyomavirus middle-T-antigen in insect cells. *J. Virol.* **64:** 61–68.

Piwnica-Worms, H., Williams, N. G., and Roberts, T. M. 1988. Use of a baculovirus expression system to study the regulation of $pp60^{c\text{-}src}$ and its interactions with the middle T antigen of polyomavirus. In *Current Communications In Molecular Biology: Viral Vectors,* eds. Y. Gluzman, and S. H. Hughes, pp. 104–110. Cold Spring Harbor, NY: Cold Spring Harbor Laboratory.

Possee, R. D. 1986. Cell-surface expression of influenza virus hemagglutinin in insect cells using a baculovirus vector. *Virus Res.* **5:** 43–60.

Possee, R. D., and Howard, S. C. 1987. Analysis of the polyhedrin gene promoter of the *Autographa californica* nuclear polyhedrosis virus. *Nucleic Acids Res.* **15:** 10233–10248.

Possee, R., Sun, T.-P., Howard, S., Ayres, M., Hill-Perkins, M., and Gearing, K. 1991. Nucleotide sequence of the *Autographa californica* nuclear polyhedrosis 9.4 kbp EcoRI-I and -R (polyhedrin gene) region. *Virology* **185:** 229–241.

Potter, K., Faulkner, P., and MacKinnon, E. 1976. Strain selection during serial passage of *Trichoplusia ni* nuclear polyhedrosis virus. *J. Virol.* **18:** 1040–1050.

Potter, K. N., Jaques, R. P., and Faulkner, P. 1978. Modification of *Trichoplusia ni* nuclear polyhedrosis virus passaged *in vivo. Intervirology* **9:** 76–85.

Potter, K. N., and Miller, L. K. 1980. Transfection of 2 invertebrate cell lines with DNA of *Autographa californica* nuclear polyhedrosis virus. *J. Invertebr. Pathol.* **36:** 431–432.

Préhaud, C., Harris, R. D., Fulop, V., Koh, C. L., Wong, J., Flamand, A., and Bishop, D. H. L. 1990. Expression, characterization, and purification of a phosphorylated rabies nucleoprotein synthesized in insect cells by baculovirus vectors. *Virology* **178:** 486–497.

Préhaud, C., Takehara, K., Flamand, A., and Bishop, D. H. L. 1989. Immunogenic and protective properties of rabies virus glycoprotein expressed by baculovirus vectors. *Virology* **173:** 390–399.

Price, P. M., Mohamad, A., Zelent, A., Neurath, A. R., and Acs, G. 1988. Translational selection in the expression of the hepatitis B virus envelope proteins. *DNA* **7:** 417–422.

Price, P. M., Reichelderfer, C. F., Johansson, B. E., Kilbourne, E. D., and Acs, G. 1989. Complementation of recombinant baculoviruses by coinfection with wild-type virus facilitates production in insect larvae of antigenic proteins of hepatitis B virus and influenza virus. *Proc. Natl. Acad. Sci. U.S.A.* **86:** 1453–1456.

Putnak, R., Feighny, R., Burrous, J., Cochran, M., Hackett, C., Smith, G., and Hoke, C. 1991. Dengue-1 virus envelope glycoprotein gene expressed in recombinant baculovirus elicits virus-neutralizing antibody in mice and protects them from virus challenge. *Am. J. Trop. Med. Hyg.* **45:** 159–167.

Putney, S., Rusche, J., Matthews, T., Krohn, K., Carson, H., Lynn, D., Jackson, J., Robey, W. G., Ranki, A., Robert-Guroff, M., Gallo, R., and Bolegnesi, D. 1988. HIV neutralizing antibodies elicited by recombinant envelope proteins. In *Human Retroviruses, Cancer, and AIDS: Approaches to Prevention and Therapy,* ed. D. Bolognesi, pp. 149–160. New York: Alan R. Liss.

Qin, J., Liu, A., and Weaver, R. F. 1989. Studies on the control region of the p10 gene of the *Autographa californica* nuclear polyhedrosis virus. *J. Gen. Virol.* **70:** 1273–1279.

Quelle, F. W., Caslake, L. F., Burkert, R. E., and Wojchowski, D. M. 1989. High-level expression and purification of a recombinant human erythropoietin produced using a baculovirus vector. *Blood* **74:** 652–657.

Quilliam, L. A., Der, C. J., Clark, R., O'Rourke, E. C., Zhang, K., McCormick, F., and Bokoch, G. M. 1990. Biochemical characterization of baculovirus-expressed *rap*1A/*Krev*-1 and its regulation by GTPase-activating proteins. *Mol. Cell. Biol.* **10:** 2901–2908.

Raghow, R., and Grace, T. 1974. Studies on a nuclear polyhedrosis virus in *Bombyx mori* cells *in vitro. J. Ultrastructure Res.* **47:** 384–399.

Ragona, G., Edwards, S. A., Mercola, D. A., Adamson, E. D., and Calogero, A. 1991. The transcriptional factor Egr-1 is synthesized by baculovirus-infected insect cells in an active, DNA-binding form. *DNA Cell. Biol.* **10:** 61–66.

Ramakrishna, N., Saikumar, P., Potempska, A., Wisniewski, H. M., and Miller, D. L. 1991. Expression of human Alzheimer amyloid precursor protein in insect cells. *Biochem. Biophys. Res. Commun.* **174:** 983–989.

Ramer, S. E., Winkler, D. G., Carrera, A., Roberts, T. M., and Walsh, C. T. 1991. Purification and initial characterization of the lymphoid-cell protein-tyrosine kinase p56lck from a baculovirus expression system. *Proc. Natl. Acad. Sci. U.S.A.* **88:** 6254–6258.

Rankin, C., Ooi, B. G., and Miller, L. K. 1988. Eight base pairs encompassing the transcriptional start point are the major determinant for baculovirus polyhedrin gene expression. *Gene* **70:** 39–50.

Rasmussen, L., Battles, J. K., Ennis, W. H., Nagashima, K., and Gonda, M. A. 1990. Characterization of virus-like particles produced by a recombinant baculovirus containing the GAG gene of the bovine immunodeficiency-like virus. *Virology* **178:** 435–451.

Ray, R., Galinski, M. S., and Compans, R. W. 1989. Expression of the fusion glycoprotein of human parainfluenza type 3 virus in insect cells by a recombinant baculovirus and analysis of its immunogenic property. *Virus Res.* **12:** 169–180.

Redfield, R. R., Birx, D. L., Ketter, N., Tramont, E., Polonis, V., Davis, C., Brundage, J. F., Smith, G., Johnson, S., Fowler, A., Wierzba, T., Shafferman, A., Volvovitz, F., Oster, C., Burke, D. S., and Military Medical Consortium for Applied Retroviral Research. 1991. A phase I evaluation of the safety and immunogenicity of vaccination with recombinant gp160 in patients with early human immunodeficiency virus infection. *N. Engl. J. Med.* **324:** 1677–1684.

Reed, L., and Muench, H. 1938. A simple method for estimating fifty percent endpoints. *Am. J. Hyg.* **27:** 493–497.

Reid-Sanden, F. L., Sumner, J. W., Smith, J. S., Fekadu, M., Shaddock, J. H., and Bellini, W. J. 1990. Rabies diagnostic reagents prepared from a Rabies N-gene recombinant expressed in baculovirus. *J. Clinical Microbiol.* **28:** 858–863.

Reilander, H., Boege, F., Vasudevan, S., Maul, G., Hekman, M., Dees, C., Hampe, W., Helmreich, E. J. M., and Michel, H. 1991. Purification and functional characterization of the human-β2-adrenergic receptor produced in baculovirus-infected insect cells. *FEBS Lett.* **282:** 441–444.

Renauld, J. C., Goethals, A., Houssiau, F., Van, R. E., and Van, S. J. 1990. Cloning and expression of a complementary DNA for the human homolog of mouse T cell and mast cell growth factor p40. *Cytokine* **2:** 9–12.

Reynisdottir, I., O'Reilly, D. R., Miller, L. K., and Prives, C. 1990. Thermally inactivated SV40 tsA58 mutant T antigen cannot initiate viral DNA replication *in vitro*. *J. Virol.* **64:** 6234–6245.

Rice, W. C., and Miller, L. K. 1986. Baculovirus transcription in the presence of inhibitors and in nonpermissive *Drosophila* cells. *Virus. Res.* **6:** 155–172.

Richardson, C., Lalumiere, M., Banville, M., and Vialard, J. 1992. Screening of recombinant baculovirus with vectors which co-express β-galactosidase and foreign genes. In *Baculovirus Expression Protocols*, eds. C. Richardson, and J. Walker. Clifton, NJ: Humana Press. In press.

Richardson, N. E., Brown, N. R., Hussey, R. E., Vaid, A., Matthews, T. J., Bolognesi, D. P., and Reinherz, E. L. 1988. Binding site for human immunodeficiency virus coat protein gp120 is located in the amino-terminal region of T4 CD4 and requires the intact variable-region-like domain. *Proc. Natl. Acad. Sci. U.S.A.* **85:** 6102–6106.

Roberts, T. E., and Faulkner, P. 1989. Fatty acid acylation of the 67K envelope glycoprotein of a baculovirus: *Autographa californica* nuclear polyhedrosis virus. *Virology* **172:** 377–381.

Rodewald, H. R., Langhorne, J., Eichmann, K., and Kupsch, J. 1990. Production of murine interleukin-4 and interleukin-5 by recombinant baculovirus. *J. Immunolog. Meth.* **132:** 221–226.

Rohel, D. Z., Cochran, M. A., and Faulkner, P. 1983. Characterization of 2 abundant messenger RNA species of *Autographa californica* nuclear polyhedrosis virus present late in infection. *Virology* **124:** 357–365.

Rohel, D. Z., and Faulkner, P. 1984. Time course analysis and mapping of *Autographa californica* nuclear polyhedrosis virus transcripts. *J. Virol.* **50:** 739–747.

Rohrmann, G. F. 1986. Polyhedrin structure. *J. Gen. Virol.* **67:** 1499–1514.

Roosien, J., Belsham, G. J., Ryan, M. D., King, A. M. Q., and Vlak, J. M. 1990. Synthesis of foot-and-mouth disease virus capsid proteins in insect cells using baculovirus expression vectors. *J. Gen. Virol.* **71:** 1703–1711.

Rose, R. C., Bonnez, W., Strike, D. G., and Reichman, R. C. 1990. Expression of the full-length products of the human papillomavirus type 6b (HPV-6b) and HPV-11 L2 open reading frames by recombinant baculovirus, and antigenic comparisons with HPV-11 whole virus particles. *J. Gen. Virol.* **71:** 2725–2729.

Rosen, B. S., Cook, K. S., Yaglom, J., Groves, D. L., Volkanis, J. E., Damm, D., White, T., and Spiegelman, B. M. 1989. Adipsin and complement factor D activity: an immune-related defect in obesity. *Science* **244:** 1483–1487.

Ross, T. K., Prahl, J. M., and Deluca, H. F. 1991. Overproduction of rat 1,25-dihydroxyvitamin-D3 receptor in insect cells using the baculovirus expression system. *Proc. Natl. Acad. Sci. U.S.A.* **88:** 6555–6559.

Rota, P. A., Black, R. A., De, B. K., Harmon, M. W., and Kendal, A. P. 1990. Expression of influenza A and B virus nucleoprotein antigens in baculovirus. *J. Gen. Virol.* **71:** 1545–1554.

Rothman, A. L., Kurane, I., Zhang, Y. M., Lai, C. J., and Ennis, F. A. 1989. Dengue virus-specific murine T-lymphocyte proliferation serotype specificity and response to recombinant viral proteins. *J. Virol.* **63:** 2487–2491.

Roy, L. M., Swenson, K. I., Walker, D. H., Gabrielli, B. G., Li, R. S., Piwnica, W. H., and Maller, J. L. 1991a. Activation of p34-cdc-2 kinase by cyclin A. *J. Cell. Biol.* **113:** 507–514.

Roy, P., Adachie, A., Urakawa, T., Booth, T. F., and Thomas, C. P. 1990a. Identification of bluetongue virus VP6 protein as a nucleic acid-binding protein and the localization of VP6 in virus-infected vertebrate cells. *J. Virol.* **64:** 1–8.

Roy, P., Hirasawa, T., Fernandez, M., Blinov, V. M., and Rodrique, J. M. S. 1991b. The complete sequence of the group-specific antigen, VP7, of African horsesickness disease virus serotype-4 reveals a close relationship to bluetongue virus. *J. Gen. Virol.* **72:** 1237–1241.

Roy, P., Urakawa, T., van Dijk, A. A., and Erasmus, B. J. 1990b. Recombinant virus vaccine for bluetongue disease in sheep. *J. Virol.* **64:** 1998–2003.

Royer, M., Cerutti, M., Gay, B., Hong, S. S., Devauchelle, G., and Boulanger, P. 1991. Functional domains of HIV-1 gag-polyprotein expressed in baculovirus-infected cells. *Virology* **184:** 417–422.

Rusche, J. R., Lynn, D. L., Robert-Guroff, M., Langlois, A. J., Lyerly, H. K., Carson, H., Krohn, K., Ranki, A., Gallo, R. C., Bolognesi, D. P., Putney, S. D., and Mathews, T. J. 1987. Humoral immune response to the entire human immunodeficiency virus envelope glycoprotein made in insect cells. *Proc. Natl. Acad. Sci. U.S.A.* **84:** 6924–6928.

Russell, R. L. Q., and Rohrmann, G. F. 1990. A baculovirus polyhedron envelope protein—immunogold localization in infected cells and mature polyhedra. *Virology* **174:** 177–184.

Sambrook, J., Fritsch, E. F., and Maniatis, T. 1989. *Molecular Cloning; A Laboratory Manual.* Cold Spring Harbor, NY: Cold Spring Harbor Laboratory.

Santucci, S., Androphy, E. J., Bonne-Andréa, C., and Clertant, P. 1990. Proteins encoded by the bovine papillomavirus E1 open reading frame: expression in heterologous systems and in virally transformed cells. *J. Virol.* **64:** 6027–6039.

Sánchez-Martínez, D., and Pellett, P. E. 1991. Expression of HSV-1 and HSV-2 glycoprotein-G in insect cells by using a novel baculovirus expression vector. *Virology* **182:** 229–238.

Sárvari, M., Csikós, G., Sass, M., Gál, P., Schumaker, V. N., and Závodszky, P. 1990. Ecdysteroids increase the yield of recombinant protein produced in baculovirus insect cell expression system. *Biochem. Biophys. Res. Commun.* **167:** 1154–1161.

Schaap, D., and Parker, P. J. 1990. Expression, purification, and characterization of protein kinase C-e. *J. Biol. Chem.* **265:** 7301–7307.

Schaefer, E. M., Siddle, K., and Ellis, L. 1990. Deletion analysis of the human insulin receptor ectodomain reveals independently folded soluble subdomains and insulin binding by a monomeric α-subunit. *J. Biol. Chem.* **265:** 13248–13253.

Schetter, C., Oellig, C., and Doerfler, W. 1990. An insertion of insect cell DNA in the 81-map-unit segment of *Autographa californica* nuclear polyhedrosis virus DNA. *J. Virol.* **64:** 1844–1850.

Schmaljohn, C. S., Chu, Y.-K., Schmaljohn, A. L., and Dalrymple, J. M. 1990. Antigenic subunits of Hantaan virus expressed by baculovirus and vaccinia virus recombinants. *J. Virol.* **64:** 3162–3170.

Schmaljohn, C. S., Parker, M. D., Ennis, W. H., Dalrymple, J. M., Collett, M. S., Suzich, J. A., and Schmaljohn, A. L. 1989. Baculovirus expression of the M genome segment of Rift Valley fever virus and examination of antigenic and immunogenic properties of the expressed proteins. *Virology* **170:** 184–192.

Schmaljohn, C. S., Sugiyama, K., Schmaljohn, A. L., and Bishop, D. H. L. 1988. Baculovirus expression of the small genome segment of hantaan virus and potential use of the expressed nucleocapsid protein as a diagnostic antigen. *J. Gen. Virol.* **69:** 777–786.

Schwaller, M., Smith, G. E., Skehel, J. J., and Wiley, D. C. 1989. Studies with cross-linking reagents on the oligomeric structure of the *env* glycoprotein of HIV. *Virology* **172:** 367–369.

Scott, M. R. D., Butler, D. A., Bredesen, D. E., Walchli, M., Hsiao, K. K., and Prusiner, S. B. 1988. Prion protein gene expression in cultured cells. *Protein Eng.* **2:** 69–76.

Sekine, H., Fuse, A., Tada, A., Maeda, S., and Simizu, B. 1988. Expression of human papillomavirus type 6b E2 gene product with DNA-binding activity in insect *Bombyx mori* cells using a baculovirus expression vector. *Gene* **65:** 187–194.

Sekine, H., Fuse, A., Inaba, N., Takamizawa, H., and Simizu, B. 1989. Detection of the human papillomavirus 6b E2 gene product in genital condyloma and laryngeal papilloma tissues. *Virology* **170:** 92–98.

Seppanen, H., Huhtala, M. L., Vaheri, A., Summers, M. D., and Okerblom, C. 1991. Diagnostic potential of baculovirus-expressed rubella virus envelope proteins. *J. Clin. Microbiol.* **29:** 1877–1882.

Shearer, M. H., Lanford, R. L., and Kennedy, R. C. 1990. Monoclonal antiidiotypic antibodies induce humoral immune responses specific for simian virus 40 virus large tumor antigen in mice. *J. Immunol.* **145:** 932–939.

Shiu, S. Y. W., Morikawa, S., Buckley, A., Higgs, S., Karunakarannair, V., Blachere, C., and Gould, E. A. 1991. 17D Yellow fever vaccine virus envelope protein expressed by recombinant baculovirus is antigenically indistinguishable from authentic viral protein. *J. Gen. Virol.* **72:** 1451–1454.

Singh, P. 1977. *Artificial Diets for Insects, Mites, and Spiders.* New York: Plenum Press.

Singh, P., and Moore, R. eds. 1985. *Handbook of insect rearing.* New York: Elsevier.

Sissom, J., and Ellis, L. 1989. Secretion of the extracellular domain of the human insulin receptor from insect cells by use of a baculovirus vector. *Biochem J.* **261:** 119–126.

Sissom, J. F., and Ellis, L. 1991. Biosynthesis of the precursor of a soluble human insulin receptor ectodomain in insect Sf9 cells infected with a recombinant baculovirus. *Biochem. Biophys. Res. Comm.* **177:** 764–770.

Smith, C. ed. 1966. *Insect colonization and mass production.* New York: Academic Press.

Smith, G. E., Fraser, M. J., and Summers, M. D. 1983a. Molecular engineering of the *Autographa californica* nuclear polyhedrosis virus genome: deletion mutations within the polyhedrin gene. *J. Virol.* **46:** 584–593.

Smith, G. E., Summers, M. D., and Fraser, M. J. 1983b. Production of human β-interferon in insect cells infected with a baculovirus expression vector. *Mol. Cell. Biol.* **3:** 2156–2165.

Smith, G. E., Ju, G., Ericson, B. L., Moschera, J., Lahm, H., Chizzonite, R., and Summers, M. D. 1985. Modification and secretion of human interleukin-2 produced in insect cells by a baculovirus vector. *Proc. Natl. Acad. Sci. U.S.A.* **82:** 8404–8408.

Southern, E. 1975. Detection of specific sequences among DNA fragments separated by gel electrophoresis. *J. Mol. Biol.* **98:** 503–517.

Srinivasan, G., and Thompson, E. B. 1990. Overexpression of full-length human glucocorticoid receptor in *Spodoptera frugiperda* cells using the baculovirus expression vector system. *Mol. Endocrinol.* **4:** 209–216.

St. Angelo, C., Smith, G. E., Summers, M. D., and Krug, R. M. 1987. Two of the three influenza viral polymerase proteins expressed by using baculovirus vectors form a complex in insect cells. *J. Virol.* **61:** 361–365.

Steiner, H., Pohl, G., Gunne, H., Hellers, M., Elhammer, A., and Hansson, L. 1988. Human tissue-type plasminogen activator synthesized by using a baculovirus vector in insect cells compared with human plasminogen activator produced in mouse cells. *Gene* **73:** 449–458.

Stewart, L. M. D., Hirst, M., Ferber, M. L., Merryweather, A. T., Cayley, P. J., and Possee, R. D. 1991. Construction of an improved baculovirus insecticide containing an insect-specific toxin gene. *Nature* **352:** 85–88.

Stiles, B., and Wood, H. A. 1983. A study of the glycoproteins of *Autographa californica* nuclear polyhedrosis virus (AcNPV). *Virology* **131:** 230–241.

Stoltz, D., Pavan, C., and Da Cunha, A. 1973. Nuclear polyhedrosis virus: a possible example of *de novo* intranuclear membrane morphogenesis. *J. Gen. Virol.* **19:** 145–150.

Summers, M., and Anderson, D. 1972. Granulosis virus deoxyribonucleic acid: a closed double-standed molecule. *J. Virol.* **9:** 710–713.

Summers, M., Engler, R., Falcon, L., and Vail, P. eds. 1975. *Baculoviruses for insect pest control: safety considerations.* Washington, DC: American Society for Microbiology.

Summers, M., and Kawanishi, C. eds. 1978. *Viral pesticides: present knowledge and potential effects on public and environmental health.* Research Triangle Park, NC: EPA, Publication # 60019-78-026.

Summers, M. D., and Smith, G. E. 1987. A manual of methods for baculovirus vectors and insect cell culture procedures. *Texas Agricultural Experiment Station Bulletin no. 1555.*

Suzuki, K., Shimoi, H., Iwasaki, Y., Kawahara, T., Matsuura, Y., and Nishikawa, Y. 1990. Elucidation of amidating reaction mechanism by frog amidating enzyme, peptidylglycine α-hydroxylating monooxygenase, expressed in insect cell culture. *EMBO J.* **9:** 4259–4265.

Svoboda, M., Przybylski, M., Schreurs, J., Miyajima, A., Hogeland, K., and Deinzer, M. 1991. Mass spectrometric determination of glycosylation sites and oligosaccharide composition of insect-expressed mouse interleukin-3. *J. Chromatog.* **562:** 403–419.

Tacket, C. O., Baqar, S., Munoz, C., and Murphy, J. R. 1990. Lymphoproliferative responses to mitogens and HIV-1 envelope glycoprotein among volunteers vaccinated with recombinant gp160. *Aids Res. Hum. Retroviruses* **6:** 535–542.

Tada, A., Fuse, A., Sekine, H., Simizu, B., Kondo, A., and Maeda, S. 1988. Expression of the E2 open reading frame of papillomaviruses BPV1 and HPV6B in silkworm by a baculovirus vector. *Virus Res.* **9:** 357–368.

Takase-Yoden, S., Kikuchi, T., Siddell, S. G., and Taguchi, F. 1991. Localization of major neutralizing epitopes on the S1 polypeptide of the murine coronavirus peplomer glycoprotein. *Virus Res.* **18:** 99–108.

Takehara, K., Ireland, D., and Bishop, D. H. L. 1988. Co-expression of the hepatitis B surface and core antigens using baculovirus multiple expression vectors. *J. Gen. Virol.* **69:** 2763–2778.

Takehara, K., Morikawa, S., and Bishop, D. H. L. 1990. Characterization of baculovirus-expressed Rift Valley fever virus glycoproteins synthesized in insect cells. *Virus Res.* **17:** 173–190.

Tanada, Y., and Hess, R. 1984. The cytopathology of baculovirus infections in insects. In *Insect Ultrastructure,* Vol. 2, eds. R. King, and H. Akai, pp. 517–556. New York: Plenum.

Tang, W. J., Krupinski, J., and Gilman, A. G. 1991. Expression and characterization of calmodulin-activated type I adenylyl cyclase. *J. Biol. Chem.* **266:** 8595–8603.

Tavare, J. M., Clack, B., and Ellis, L. 1991. Two-dimensional phosphopeptide analysis of the autophosphorylation cascade of a soluble insulin receptor tyrosine kinase. The tyrosines phosphorylated are typical of those observed following phosphorylation of the heterotetrameric insulin receptor in intact cells. *J. Biol. Chem.* **266:** 1390–1395.

Tavernier, J., Devos, R., Van der Heyden, J., Hauquier, G., Bauden, R., Fache, I., Kawashima, E., Vandekerckhove, J., Contreras, R., and Fiers, W. 1989. Expression of human and murine interleukin-5 in eukaryotic systems. *DNA* **8:** 491–502.

Tessier, D. C., Thomas, D. Y., Khouri, H. E., Laliberte, F., and Vernet, T. 1991. Enhanced secretion from insect cells of a foreign protein fused to the honeybee melittin signal peptide. *Gene* **98:** 177–183.

Thiem, S. M., and Miller, L. K. 1989a. Identification, sequence, and transcriptional mapping of the major capsid protein gene of the baculovirus *Autographa californica* nuclear polyhedrosis virus. *J. Virol.* **63:** 2008–2018.

Thiem, S. M., and Miller, L. K. 1989b. A baculovirus gene with a novel transcription pattern encodes a polypeptide with a zinc finger and a leucine zipper. *J. Virol.* **63:** 4489–4497.

Thiem, S. M., and Miller, L. K. 1990. Differential gene expression mediated by late, very late, and hybrid baculovirus promoters. *Gene* **91:** 87–94.

Thomas, C. P., Booth, T. F., and Roy, P. 1990. Synthesis of bluetongue virus-encoded phosphoprotein and formation of inclusion bodies by recombinant baculovirus in insect cells: it binds the single-stranded RNA species. *J. Gen. Virol.* **71:** 2073–2083.

Thomsen, D. R., Post, L. E., and Elhammer, A. P. 1990. Structure of O-glycosidically linked oligosaccharides synthesized by the insect cell line Sf9. *J. Cell. Biochem.* **43:** 67–79.

Tobias, J., Schrader, T., Rocap, G., and Varshavsky, A. 1991. The N-end rule in bacteria. *Science* **254:** 1374–1377.

Tomalski, M. D., Eldridge, R., and Miller, L. K. 1991. A baculovirus homolog of a Cu/Zn superoxide dismutase gene. *Virology* **184:** 149–161.

Tomalski, M. D., and Miller, L. K. 1991. Insect paralysis by baculovirus-mediated expression of a mite neurotoxin gene. *Nature* **352:** 82–85.

Tomalski, M. D., Wu, J., and Miller, L. K. 1988. The location, sequence, transcription and regulation of a baculovirus DNA polymerase gene. *Virology* **167:** 591–600.

Towbin, H., Staehelin, T., and Gordon, J. 1979. Electrophoretic transfer of proteins from polyacrylamide gels to nitrocellulose sheets: procedure and some applications. *Proc. Natl. Acad. Sci. U.S.A.* **76:** 4350–4354.

Tramper, J., Joustra, D., and Vlak, J. 1987. Bioreactor design for growth of shear-sensitive insect cells. In *Plant and Animal Cells,* eds. C. Webb, and F. Mavituna, pp. 125–136. Chichester, England: Ellis Horwood Ltd.

Tramper, J., and Vlak, J. 1986. Some engineering and economic aspects of continuous cultivation of insect cells for the production of baculoviruses. *Annals N.Y. Acad. Sci.* **469:** 279–288.

Tramper, J., Williams, J., Joustra, D., and Vlak, J. 1986. Shear sensitivity of insect cells in suspension. *Enzyme Micro. Technol.* **8:** 33–36.

Tratner, I., Detogni, P., Sassone-Corsi, P., and Verma, I. M. 1990. Characterization and purification of human *fos* protein generated in insect cells with a baculoviral expression vector. *J. Virol.* **64:** 499–508.

Treat, T. L., and Halfhill, J. E. 1973. Rearing alfalfa loopers and celery loopers on an artificial diet. *J. Econ. Ent.* **66:** 569–570.

Tsai, S. Y., Srinivasan, G., Allan, G. F., Thompson, E. B., O'Malley, B. W., and Tsai, M. J. 1990. Recombinant human glucocorticoid receptor induces transcription of hormone response gene *in vivo*. *J. Biol. Chem.* **265:** 17055–17061.

Tsao, T., Hsieh, J. C., Durkin, M. E., Wu, C., Chakravarti, S., Dong, L. J., Lewis, M., and Chung, A. E. 1990. Characterization of the basement membrane glycoprotein entactin synthesized in a baculovirus expression system. *J. Biol. Chem.* **265:** 5188–5191.

Tweeten, K. A., Bulla, L. A., and Consigli, R. A. 1980. Characterization of an extremely basic protein derived from granulosis virus nucleocapsids. *J. Virol.* **33:** 866–876.

Urakawa, T., Ferguson, M., Minor, P. D., Cooper, J., Sullivan, M., Almond, J. W., and Bishop, D. H. L. 1989a. Synthesis of immunogenic, but non-infectious, poliovirus particles in insect cells by a baculovirus expression vector. *J. Gen. Virol.* **70:** 1453–1464.

Urakawa, T., Ritter, D. G., and Roy, P. 1989b. Expression of largest RNA segment and synthesis of VP1 protein of bluetongue virus in insect cells by recombinant baculovirus: association of VP1 protein with RNA polymerase activity. *Nucleic Acids Res.* **17:** 7395–7401.

Urakawa, T., and Roy, P. 1988. Bluetongue virus tubules made in insect cells by recombinant baculoviruses: expression of the NS1 gene of bluetongue virus serotype 10. *J. Virol.* **62:** 3919–3927.

Urakawa, T., Small, D. A., and Bishop, D. H. L. 1988. Expression of snowshoe hare bunyavirus S RNA coding proteins by recombinant baculoviruses. *Virus Res.* **11:** 303–318.

Vail, P., Sutter, G., Jay, D., and Gough, D. 1971. Reciprocal infectivity of nuclear polyhedrosis viruses of the cabbage looper and alfalfa looper. *J. Invertebr. Pathol.* **17:** 383–388.

Vail, P. V., Knell, J. D., Summers, M. D., and Cowan, D. K. 1982. *In vivo* infectivity of baculovirus isolates, variants and natural recombinants in alternate hosts. *Environ. Entomol.* **11:** 1187–1192.

Van Bokhoven, H., Mulders, M., Wellink, J., Vlak, J. M., Goldbach, R., and van Kammen, A. 1991. Evidence for dissimilar properties of comoviral and picornaviral RNA polymerases. *J. Gen. Virol.* **72:** 567–572.

Van Bokhoven, H., Wellink, J., Usmany, M., Vlak, J. M., Goldbach, R., and Van Kammen, A. 1990. Expression of plant virus genes in animal cells: high-level synthesis of cowpea mosaic virus B-RNA-encoded proteins with baculovirus expression vectors. *J. Gen. Virol.* **71:** 2509–2517.

Van Der Wilk, F., Van Lent, J. W. M., and Vlak, J. M. 1987. Immunogold detection of polyhedrin p10 and virion antigens in *Autographa californica* nuclear polyhedrosis virus-infected *Spodoptera frugiperda* cells. *J. Gen. Virol.* **68:** 2615–2624.

Van Drunen Littel-Van den Hurk, S. V., Parker, M. D., Fitzpatrick, D. R., Zamb, T. J., van den Hurk, J. V., Campos, M., Harland, R., and Babiuk, L. A. 1991. Expression of bovine herpesvirus 1 glycoprotein gIV by recombinant baculovirus and analysis of its immunogenic properties. *J. Virol.* **65:** 263–271.

Van Wyke Coelingh, K. L., Murphy, B. R., Collins, P. L., Lebacq Verheyden, A. M., and Battey, J. F. 1987. Expression of biologically active and antigenically authentic parainfluenza type 3 virus hemagglutinin-neuraminidase glycoprotein by a recombinant baculovirus. *Virology* **160:** 465–472.

Vasudevan, S., Reilander, H., Maul, G., and Michel, H. 1991. Expression and cell membrane localization of rat M3 muscarinic acetylcholine receptor produced in SF 9 insect cells using the baculovirus system. *FEBS Lett.* **283:** 52–56.

Vaughn, J. L., and Fan, F. 1989. Use of commercial serum replacements for the culture of insect cells. *In Vitro Cell. Dev. Biol.* **25:** 143–145.

Vaughn, J. L., Goodwin, R. H., Tompkins, G. J., and McCawley, P. 1977. The establishment of two cell lines from the insect *Spodoptera frugiperda* (Lepidoptera: Noctuidae). *In Vitro* **13:** 213–217.

Vernet, T., Tessier, D. C., Richardson, C., Laliberte, F., Khouri, H. E., Bell, A. W., Storer, A. C., and Thomas, D. Y. 1990. Secretion of functional papain precursor from insect cells. Requirement for N-glycosylation of the pro-region. *J. Biol. Chem.* **265:** 16661–16666.

Vialard, J., Lalumière, M., Vernet, T., Briedis, D., Alkhatib, G., Henning, D., Levin, D., and Richardson, C. 1990. Synthesis of the membrane fusion and hemagglutinin proteins of measles virus, using a novel baculovirus vector containing the β-galactosidase gene. *J. Virol.* **64:** 37–50.

Vik, T. A., Sweet, L. J., and Erikson, R. L. 1990. Coinfection of insect cells with recombinant baculovirus expressing $pp60^{v\text{-}src}$ results in the activation of a serine-specific protein kinase $pp90^{rsk}$. *Proc. Natl. Acad. Sci. U.S.A.* **87:** 2685–2689.

Villalba, M., Wente, S. R., Russell, D. S., Ahn, J., Reichelderfer, C. F., and Rosen, O. M. 1989. Another version of the human insulin receptor kinase domain: expression, purification, and characterization. *Proc. Natl. Acad. Sci. U.S.A.* **86:** 7848–7852.

Viscidi, R., Ellerbeck, E., Garrison, L., Midthun, K., Clements, M. L., Clayman, B., Fernie, B., and Smith, G. 1990. Characterization of serum antibody responses to recombinant HIV-1 gp160 vaccine by enzyme immunoassay. *AIDS Res. Hum. Retroviruses* **6:** 1251–1256.

Vissavajjhala, P., and Ross, A. H. 1990. Purification and characterization of the recombinant extracellular domain of human nerve growth factor receptor expressed in a baculovirus system. *J. Biol. Chem.* **265:** 4746–4752.

Vlak, J. M., Klinkenberg, F. A., Zaal, K. J. M., Usmany, M., Klinge-Roode, E. C., Geervliet, J. B. F., Roosien, J., and Van Lent, J. W. M. 1988. Functional studies on the p10 gene of *Autographa californica* nuclear polyhedrosis virus using a recombinant expressing a p10-β-galactosidase fusion gene. *J. Gen. Virol.* **69:** 765–776.

Vlak, J. M., Schouten, A., Usmany, M., Belsham, G. J., Klinge-Roode, E. C., Maule, A. J., van Lent, J. W. M., and Zuidema, D. 1990. Expression of cauliflower mosaic virus gene I using a baculovirus vector based upon the p10 gene and a novel selection method. *Virology* **179:** 312–320.

Vlak, J. M., and Smith, G. E. 1982. Orientation of the genome of *Autographa californica* nuclear polyhedrosis virus: a proposal. *J. Virol.* **41:** 1118–1121.

Volkman, L. E. 1986. The 64-k envelope protein of budded *Autographa californica* nuclear polyhedrosis virus. In *Current Topics in Microbiology and Immunology,* eds. W. Doerfler, and P. Boehm, pp. 103–118. Berlin, Heidleberg, and New York: Springer-Verlag.

Volkman, L. E., and Goldsmith, P. A. 1983. *In vitro* survey of *Autographa californica* nuclear polyhedrosis virus interaction with nontarget vertebrate host cells. *Appl. Environ. Microbiol.* **45:** 1085–1093.

Volkman, L. E., and Goldsmith, P. A. 1985. Mechanism of neutralization of budded *Autographa californica* nuclear polyhedrosis virus by a monoclonal antibody: inhibition of entry by adsorptive endocytosis. *Virology* **143:** 185–195.

Volkman, L. E., Goldsmith, P. A., Hess, R. T., and Faulkner, P. 1984. Neutralization of budded *Autographa californica* nuclear polyhedrosis virus by a monoclonal antibody: identification of the target antigen. *Virology* **133:** 354–362.

Wahl, G., Berger, S., and Kimmel, A. 1987. Molecular hybridization of immobilized nucleic acids: theoretical concepts and practical considerations. In *Guide to Molecular Cloning Techniques,* eds. S. Berger, and A. Kimmel, pp. 399–407. San Diego: Academic Press.

Wang, N. P., Qian, Y. W., Chung, A. E., Lee, W. H., and Lee, E. Y. H. P. 1990. Expression of the human retinoblastoma gene product $pp110^{RB}$ in insect cells using the baculovirus system. *Cell Growth and Differentiation* **1:** 429–437.

Wang, X., Ooi, B. G., and Miller, L. K. 1991. Baculovirus vectors for multiple gene expression and for occluded virus production. *Gene* **100:** 131–137.

Wathen, M. W., Aeed, P. A., and Elhammer, A. P. 1991a. Characterization of oligosaccharide structures on a chimeric respiratory syncytial virus protein expressed in insect cell line Sf9. *Biochemistry* **30:** 2863–2868.

Wathen, M. W., Brideau, R. J., and Thomsen, D. R. 1989a. Immunization of cotton rats with the human respiratory syncytial virus F glycoprotein produced using a baculovirus vector. *J. Infect. Dis.* **159:** 255–264.

Wathen, M. W., Brideau, R. J., Thomsen, D. R., and Murphey, B. R. 1989b. Characterization of a novel human respiratory syncytial virus chimeric FG glycoprotein expressed using a baculovirus vector. *J. Gen. Virol.* **70:** 2625–2635.

Wathen, M. W., Kakuk, T. J., Brideau, R. J., Hausknecht, E. C., Cole, S. L., and Zaya, R. M. 1991b. Vaccination of cotton rats with a chimeric FG glycoprotein of human respiratory syncytial virus induces minimal pulmonary pathology on challenge. *J. Infect. Dis.* **163:** 477–482.

Watson, C. J., and Hay, R. T. 1990. Expression of adenovirus type 2 DNA polymerase in insect cells infected with a recombinant baculovirus. *Nucleic Acids Res.* **18:** 1167–1173.

Webb, A., Bradley, M., Phelan, S., Wu, J., and Gehrke, L. 1991. Use of the polymerase chain reaction for screening and evaluation of recombinant baculovirus clones. *BioTechniques* **11:** 512–519.

Webb, N. R., Madoulet, C., Tossi, P.-F., Broussard, D. R., Sneed, L., Summers, M. D., and Nicolau, C. 1989. Cell-surface expression and purification of human CD4 produced in baculovirus-infected insect cells. *Proc. Natl. Acad. Sci. U.S.A.* **86:** 7731–7735.

Wedegaertner, P. B., and Gill, G. N. 1989. Activation of the purified protein tyrosine kinase domain of the epidermal growth factor receptor. *J. Biol. Chem.* **264:** 11346–11353.

Weiss, S. A., Smith, G. C., Kalter, S. S., Vaughn, J. L., and Dougherty, E. 1981. Improved replication of *Autographa californica* nuclear polyhedrosis virus in roller bottles: characterization of the progeny virus. *Intervirology* **15:** 213–222.

Weiss, S. A., Orr, T., Smith, G. C., Kalter, S. S., Vaughn, J. L., and Dougherty, E. M. 1982. Quantitative measurement of oxygen consumption in insect cell culture infected with polyhedrosis virus. *Biotechnol. Bioeng.* **24:** 1145–1154.

Weiss, S. A., and Vaughn, J. L. 1986. Cell culture methods for large-scale propagation of baculoviruses. In *The Biology of Baculoviruses.* Vol. 2, eds. R. R. Granados, and B. A. Federici, pp. 63–87. Boca Raton, Fla: CRC Press.

Welch, S.-K. W., Crawford, S. E., and Estes, M. E. 1989. Rotavirus SA11 genome segment 11 protein is a nonstructural phosphoprotein. *J. Virol.* **63:** 3974–3982.

Wells, D. E., and Compans, R. W. 1990. Expression and characterization of a functional human immunodeficiency virus envelope glycoprotein in insect cells. *Virology* **176:** 575–586.

Wells, D. E., Vugler, L. G., and Britt, W. J. 1990. Structural and immunological characterization of human cytomegalovirus gp55-116 (GB) expressed in insect cells. *J. Gen. Virol.* **71:** 873–880.

Weyer, U., Knight, S., and Possee, R. D. 1990. Analysis of very late gene expression by *Autographa californica* nuclear polyhedrosis virus and the further development of multiple expression vectors. *J. Gen. Virol.* **71:** 1525–1534.

Weyer, U., and Possee, R. D. 1988. Functional analysis of the p10 gene 5′ leader sequence of the *Autographa californica* nuclear polyhedrosis virus. *Nucleic Acids Res.* **16:** 3635–3654.

Weyer, U., and Possee, R. D. 1989. Analysis of the promoter of the *Autographa californica* nuclear polyhedrosis virus p10 gene. *J. Gen. Virol.* **70:** 203–208.

Weyer, U., and Possee, R. D. 1991. A baculovirus dual expression vector derived from the *Autographa californica* nuclear polyhedrosis virus polyhedrin and p10 promoters: co-expression of two influenza virus genes in insect cells. *J. Gen. Virol.* **72:** 2967–2974.

White, S., Taetle, R., Seligman, P. A., Rutherford, M., and Trowbridge, I. S. 1990. Combinations of anti-transferrin receptor monoclonal antibodies inhibit human tumor cell growth *in vitro* and *in vivo;* evidence for synergistic antiproliferative effects. *Cancer Res.* **50:** 6295–6301.

Whitefleet-Smith, J., Rosen, E., McLinden, J., Ploplis, V. A., Fraser, M. J., Tomlinson, J. E., McLean, J. W., and Castellino, F. J. 1989. Expression of human plasminogen cDNA in a baculovirus vector-infected insect cell system. *Arch. Biochem. Biophys.* **271:** 390–399.

Whitford, M., Stewart, S., Kuzio, J., and Faulkner, P. 1989. Identification and sequence analysis of a gene encoding gp67, an abundant envelope glycoprotein of the baculovirus *Autographa californica* nuclear polyhedrosis virus. *J. Virol.* **63:** 1393–1399.

Whitt, M. A., and Manning, J. S. 1988. A phosphorylated 34-kda protein and a subpopulation of polyhedrin are thiol linked to the carbohydrate layer surrounding a baculovirus occlusion body. *Virology* **163:** 33–42.

Williams, G. V., Rohel, D. Z., Kuzio, J., and Faulkner, P. 1989. A cytopathological investigation of *Autographa californica* nuclear polyhedrosis virus p10 gene function using insertion-deletion mutants. *J. Gen. Virol.* **70:** 187–202.

Wilson, M. E., and Consigli, R. A. 1985. Functions of a protein kinase activity associated with purified capsids of the granulosis virus infecting *Plodia interpunctella. Virology* **143:** 526–535.

Wilson, M. E., Mainprize, T. H., Friesen, P. D., and Miller, L. K. 1987. Location, transcription and sequence of a baculovirus gene encoding a small arginine-rich polypeptide. *J. Virol.* **61:** 661–666.

Wilson, M. E., and Miller, L. K. 1986. Changes in the nucleoprotein complexes of a baculovirus DNA during infection. *Virology* **151:** 315–328.

Wojchowski, D. M., Orkin, S. H., and Sytkowski, A. J. 1987. Active human erythropoietin expressed in insect cells using a baculovirus vector: a role for N-linked oligosaccharide. *Biochim. Biophys. Acta* **910:** 224–232.

Wu, J. G., and Miller, L. K. 1989 . Sequence, transcription and translation of a late gene of the *Autographa californica* nuclear polyhedrosis virus encoding a 34.8K polypeptide. *J. Gen. Virol.* **70:** 2449–2459.

Yamada, K., Nakajima, Y., and Natori, S. 1990. Production of recombinant sarcotoxin IA in *Bombyx mori* cells. *Biochem. J.* **272:** 633–636.

Yang, C. L., Stetler, D. A., and Weaver, R. F. 1991. Structural comparison of the *Autographa californica* nuclear polyhedrosis virus-induced RNA polymerase and the 3 nuclear RNA polymerases from the host, *Spodoptera frugiperda. Virus. Res.* **20:** 251–264.

Yeung, D. E., Brown, G. W., Tam, P., Russnak, R. H., Wilson, G., Clark, L. I., and Astell, C. R. 1991. Monoclonal antibodies to the major nonstructural nuclear protein of minute virus of mice. *Virology* **181:** 35–45.

Yip, M. T., Dynan, W. S., Green, P. L., Black, A. C., Arrigo, S. J., Torbati, A., Heaphy, S., Ruland, C., Rosenblatt, J. D., and Chen, I. S. Y. 1991. Human T-cell leukemia virus HTLV type II rex protein binds specifically to RNA sequences of the HTLV long terminal repeat but poorly to the human immunodeficiency virus type 1 rev-responsive element. *J. Virol.* **65:** 2261–2272.

Yoden, S., Kikuchi, T., Siddell, S. G., and Taguchi, F. 1989. Expression of the peplomer glycoprotein of murine coronavirus JHM using a baculovirus vector. *Virology* **173:** 615–623.

Yoo, D., Parker, M. D., and Babiuk, L. A. 1990. Analysis of the S spike (peplomer) glycoprotein of bovine coronavirus synthesized in insect cells. *Virology* **179:** 121–128.

Yoo, D., Parker, M. D., and Babiuk, L. A. 1991a. The S2 subunit of the spike glycoprotein of bovine coronavirus mediates membrane fusion in insect cells. *Virology* **180:** 395–399.

Yoo, D. W., Parker, M. D., Song, J. Y., Cox, G. J., Deregt, D., and Babiuk, L. A. 1991b. Structural analysis of the conformational domains involved in neutralization of bovine coronavirus using deletion mutants of the spike glycoprotein s1-subunit expressed by recombinant baculoviruses. *Virology* **183:** 91–98.

Yost, H., Petersen, R., and Lindquist, S. 1990. RNA metabolism: strategies for regulation in the heat shock response. *Trends in Genetics* **6:** 223–227.

Yuen, L., Dionne, J., Arif, B., and Richardson, C. 1990. Identification and sequencing of the spheroidin gene of *Choristoneura biennis* entomopoxvirus. *Virology* **175:** 427–433.

Zander, N. F., Lorenzen, J. A., Cool, D. E., Tonks, N. K., Daum, G., Krebs, E. G., and Fischer, E. H. 1991. Purification and characterization of a human recombinant T-cell protein-tyrosine-phosphatase from a baculovirus expression system. *Biochemistry* **30:** 6964–6970.

Zeira, M., Tosi, P. F., Mouneimne, Y., Lazarte, J., Sneed, L., Volsky, D. J., and Nicolau, C. 1991. Full-length CD4 electroinserted in the erythrocyte membrane as a long-lived inhibitor of infection by immunodeficiency virus. *Proc. Natl. Acad. Sci. U.S.A.* **88:** 4409–4413.

Zhang, Y. M., Hayes, E. P., McCarty, T. C., Dubois, D. R., Summers, P. L., Eckels, K. H., Chanock, R. M., and Lai, C. J. 1988. Immunization of mice with Dengue structural proteins and nonstructural protein NS1 expressed by baculovirus recombinant induces resistance to Dengue virus encephalitis. *J. Virol.* **62:** 3027–3031.

Zhao, L. J., Irie, K., Trirawatanapong, T., Nakano, R., Nakashima, A., Morimatsu, M., and Padmanabhan, R. 1991. Synthesis of biologically active adenovirus preterminal protein in insect cells using a baculovirus vector. *Gene* **100:** 147–154.

Zuidema, D., Klingeroode, E. C., van Lent, J. W. M., and Vlak, J. M. 1989. Construction and analysis of an *Autographa californica* nuclear polyhedrosis virus mutant lacking the polyhedral envelope. *Virology* **173:** 98–108.

Zuidema, D., Schouten, A., Usmany, M., Maule, A. J., Belsham, G. J., Roosien, J., Klinge-Roode, E. C., van Lent, J. W. M., and Vlak, J. M. 1990. Expression of cauliflower mosaic virus gene I in insect cells using a novel polyhedrin-based baculovirus expression vector. *J. Gen. Virol.* **71:** 2201–2209.

zu Putlitz, J. Z., Datta, S., Madison, L. D., Macchia, E., and Jameson, J. L. 1991. Human thyroid hormone-β1 receptor produced by recombinant baculovirus-infected insect cells. *Biochem. Biophys. Res. Comm.* **175:** 285–290.

zu Putlitz, J. Z., Kubasek, W. L., Duchêne, M., Marget, M., von Specht, B. U., and Domdey, H. 1990. Antibody production in baculovirus-infected insect cells. *Bio/Technology* **8:** 651–654.

INDEX

References to tables are indicated by t. References to figures are indicated by f.

R

W

X

Y